Computing in Communication Networks

From Theory to Practice

Computing in Communication Networks
From Theory to Practice

Edited by

Frank H.P. Fitzek
Technische Universität Dresden
Dresden, Germany

Fabrizio Granelli
University of Trento
Trento, Italy

Patrick Seeling
Central Michigan University
Mount Pleasant, MI, United States

ACADEMIC PRESS
An imprint of Elsevier
ELSEVIER

Academic Press is an imprint of Elsevier
125 London Wall, London EC2Y 5AS, United Kingdom
525 B Street, Suite 1650, San Diego, CA 92101, United States
50 Hampshire Street, 5th Floor, Cambridge, MA 02139, United States
The Boulevard, Langford Lane, Kidlington, Oxford OX5 1GB, United Kingdom

Library of Congress Cataloging-in-Publication Data
A catalog record for this book is available from the Library of Congress

British Library Cataloguing-in-Publication Data
A catalogue record for this book is available from the British Library

ISBN: 978-0-12-820488-7

For information on all Academic Press publications
visit our website at https://www.elsevier.com/books-and-journals

Publisher: Mara Conner
Acquisitions Editor: Tim Pitts
Editorial Project Manager: Emily Thomson
Production Project Manager: Nirmala Arumugam
Designer: Matthew Limbert

Typeset by VTeX

Working together
to grow libraries in
developing countries

www.elsevier.com • www.bookaid.org

Contents

PART 1 FUTURE COMMUNICATION NETWORKS AND SYSTEMS

PART 6 EXAMPLES

PART 7 EXTENSIONS

List of contributors

Javier Acevedo
Technische Universität Dresden, Dresden, Germany

Riccardo Bassoli
Technische Universität Dresden, Dresden, Germany

Riccardo Bonetto
Technische Universität Dresden, Dresden, Germany

Juan A. Cabrera G.
Technische Universität Dresden, Dresden, Germany

Carl Collmann
Technische Universität Dresden, Dresden, Germany

Tung Doan
Technische Universität Dresden, Dresden, Germany

Frank H.P. Fitzek
Technische Universität Dresden, Dresden, Germany

Fabrizio Granelli
University of Trento, Trento, Italy

Simon Hanisch
Technische Universität Dresden, Dresden, Germany

Thomas Höschele
Technische Universität Dresden, Dresden, Germany

Malte Höweler
Technische Universität Dresden, Dresden, Germany

Sebastian A.W. Itting
Technische Universität Dresden, Dresden, Germany

Bruno Jacobfeuerborn
Deutsche Telekom Group, Bonn, Germany

Alexander Kropp
Technische Universität Dresden, Dresden, Germany

Vincent Latzko
Technische Universität Dresden, Dresden, Germany

Tao Li
Technische Universität Dresden, Dresden, Germany

Truong Giang Nguyen
Technische Universität Dresden, Dresden, Germany

Amr Osman
Technische Universität Dresden, Dresden, Germany

Sreekrishna Pandi
Technische Universität Dresden, Dresden, Germany

Morten V. Pedersen
Technische Universität Dresden, Dresden, Germany

Justus Rischke
Technische Universität Dresden, Dresden, Germany

Hani Salah
Technische Universität Dresden, Dresden, Germany

Roland Schingnitz
Technische Universität Dresden, Dresden, Germany

Robert-Steve Schmoll
Technische Universität Dresden, Dresden, Germany

Patrick Seeling
Central Michigan University, Mount Pleasant, MI, United States

Peter Sossalla
Technische Universität Dresden, Dresden, Germany

Thorsten Strufe
Technische Universität Dresden, Dresden, Germany

Maroua Taghouti
Technische Universität Dresden, Dresden, Germany

Máté Tömösközi
Technische Universität Dresden, Dresden, Germany

Roberto Torre
Technische Universität Dresden, Dresden, Germany

Marian Ulbricht
Technische Universität Dresden, Dresden, Germany

Christian Leonard Vielhaus
Technische Universität Dresden, Dresden, Germany

Huanzhuo Wu
Technische Universität Dresden, Dresden, Germany

Zuo Xiang
Technische Universität Dresden, Dresden, Germany

Dongho You
Technische Universität Dresden, Dresden, Germany

Renbing Zhang
Technische Universität Dresden, Dresden, Germany

Sandra Zimmermann
Technische Universität Dresden, Dresden, Germany

About the editors

Frank H.P. Fitzek is Professor and head of the "Deutsche Telekom Chair of Communication Networks" at TU Dresden coordinating the 5G Lab Germany since 2014. He is the spokesman of the DFG Cluster of Excellence "Center for Tactile Internet" (CeTI). He received his diploma (Dipl.-Ing.) degree in electrical engineering from the University of Technology – Rheinisch-Westfälische Technische Hochschule (RWTH) – Aachen, Germany, in 1997 and his Ph.D. (Dr.-Ing.) in Electrical Engineering from the Technical University Berlin, Germany, in 2002 and became Adjunct Professor at the University of Ferrara, Italy, in the same year. In 2003, he joined Aalborg University as Professor. In 2015, he was awarded the honorary degree "Doctor Honoris Causa" from Budapest University of Technology and Economics (BUTE).

Fabrizio Granelli is Associate Professor at the Department of Information Engineering and Computer Science (DISI) – University of Trento, Italy, IEEE ComSoc Director for Educational Services (2018–2019) and Chair of Joint IEEE VTS/ComSoc Italian Chapter. He is Research Associate Professor at the University of New Mexico, NM, USA. He received the M.Sc. and Ph.D. degrees from University of Genoa, Italy, in 1997 and 2001. He was a visiting professor at the State University of Campinas, Brasil, and at the University of Tokyo, Japan, and IEEE ComSoc Distinguished Lecturer in 2012–2015. He is Associate Editor in Chief of IEEE Communications Surveys and Tutorials.

Patrick Seeling is Professor at the Department of Computer Science at Central Michigan University, USA. He received his diploma (Dipl.-Ing.) degree in Industrial Engineering and Management from the Technical University Berlin, Germany, in 2002 and his Ph.D. in Electrical Engineering from Arizona State University, USA, in 2005. He was an Associated Faculty at ASU until 2008 and an Assistant Professor at the University of Wisconsin-Stevens Point until 2011. In 2011, he joined Central Michigan University as Assistant Professor, where he became a tenured Associate Professor in 2015 and a Full Professor in 2018.

Preface from the editors

This book is the result of a series of tutorials, generally on the topic of *Computing in Communication Networks*, that we offered at several IEEE conferences over the last years. Furthermore, parts of this book are outcomes of our lecture series in Dresden, such as *Communication Networks I, II, and III, Network Coding*, and *Cooperation in Communication Networks*. These particular courses are based on classical lecture elements and aligned with several problem-based learning course elements, where the students work on their individual (mini) projects. Therefore we hope that other educators will find this book helpful in providing easy access to the topic. We similarly hope that students in courses related to computing in communication networks will find this book helpful to provide valuable learning experiences, particularly by means of the examples presented in this book.

This book features several tutorial-style chapters in its beginning to provide (student) readers with a basic understanding of communication systems and technologies, assuming a basic familiarity with the overall content domain. We keep this introductory content to a minimum and focus directly on applied examples for an intuitive approach to the subject matter. Whenever possible, we provide the interested reader with additional background references for further studies. We similarly note that the examples provided in this book are optimized for teaching purposes and are not suited for production. This book is also the outcome of direct requests from our students over the last years to have a complete lecture script rather than a set of, hopefully nice, slides. It is always difficult to aim at a moving target, and the topic of this book is so current that we had to make a choice of the content to include, selecting what will likely have the highest possible impact in the future. We maintain a companion website for this book to improve the provided examples, introduce additional content over time, and to collect your feedback.

https://cn.ifn.et.tu-dresden.de/compcombook

Furthermore, we provide presentation slide decks to aid educators and students of *Computing in Communication Networks* in utilizing this book in their educational endeavors. QR codes next to the page will lead the reader to web pages for further reading or videos.

There are several ways to read this book. Its structure consists of eight major parts and 27 chapters, as illustrated in Fig. 0.1. For students, we propose to read the book sequentially from beginning to the end. Chapter 1 provides a solid introduction to the topic and connects all necessary technologies that will be discussed in greater detail in the subsequent chapters. Chapter 2 lists several standardization activities, completing Part 1. Parts 2 and 3 describe the underlying concepts and enabling technologies for computing in communication networks, whereas Part 4 describes current innovations that are made possible and are likely to be implemented at scale in the near future.

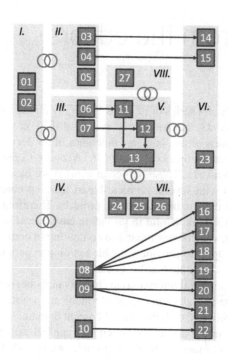

FIGURE 0.1

Structure of this book.

The first four parts are helpful information for the reader to understand the examples and to motivate the need for computing in future communication networks.

The core of this book is the ComNets emulator described in Part 5 together with the underlying software implementations Mininet and Docker. Experts that are familiar with the underlying theoretical topics could directly start with this part. In Part 6, examples are provided to deploy the various technologies described in the prior parts of the book. Part 7 contains extensions to the ComNets emulator. The last part, Part 8, introduces the basic tools used in the aforementioned examples.

Now we wish all students a lot of fun reading this book and trying out the examples. We look forward to new examples that will be generated in our *Problem-Based Learning* courses in the future.

Frank H.P. Fitzek
Fabrizio Granelli
Patrick Seeling
2020

Acknowledgments

First of all, the editors would like to thank the numerous authors that have contributed the different chapters to this book. Most of the authors were Ph.D. students or post-docs at the time of writing with the Deutsche Telekom Chair of Communication Networks. They have contributed significant time in addition to their regular duties to make this book possible. We express deep gratitude to the Deutsche Telekom, especially Tim Höttges, Claudia Nemat, Alex Jin Sung Choi, Antje Williams, and Daniel Brower, for their continuous support over the last years and their insights into the future of communication systems. Several examples presented in this book have been carried out with the 5G Lab Germany, comprised of 23 professors, over 600 researchers, and 20 industrial partners. We would like to explicitly mention Bosch, BMW, DE-CIX, Ericsson, Magna, Nokia, and Volkswagen. The Centre for Tactile Internet with Human-In-The-Loop (CeTI) provided us with an abundance of motivation over the last years, showing the clear need for computation in the network beyond 5G use cases. With the financial support of the center, we are able to conduct research in this exciting field. We thank our design team Jens Krzywinski, Tina Bobbe, Lisa Lueneburg, and their team members for the support in generating designs and graphics for several demonstrators and illustrations presented in the book. Special thanks go to Oleksandr Zhdanenko and Ievgen Kharabet for the wonderful design and illustration of numerous virtual worlds. Their work did not only give the book a nice look, but over the last years helped us to convey the main idea of future communication systems to the public. The work on this book would not have been possible without the endless support of our universities Technische Universität Dresden, University of Trento, and Central Michigan University. We are deeply thankful to Christian Scheunert for his support in managing the LaTeX sources and his patience over the last months. It is his achievement to have all the sources of this book pulled together.

Frank H.P. Fitzek
Fabrizio Granelli
Patrick Seeling
2020-01-01

Acronyms

1G	1st Generation
2G	2nd Generation
3G	3rd Generation
3GPP	3rd Generation Partnership Project
4G	4th Generation
5G	5th Generation
ACK	Acknowledgement
ACM	Association for Computing Machinery
ADC	Analog-to-Digital Converter
AI	Artificial Intelligence
AII	Administrative Instance Identifier
AKA	Authentication and Key Agreement
AMF	Access Management Functions
AMP	Approximate Matching Pursuit
ANN	Artificial Neural Network
ANNs	Artificial Neural Networks
API	Application Programming Interface
AR	Augmented Reality
ARIB	Association of Radio Industries and Businesses
ARP	Address Resolution Protocol
ARPANET	Advanced Research Projects Agency NETwork
ARQ	Automatic Repeat reQuest
AS	Autonomous System
ASIC	Application-Specific Integrated Circuit
ATIS	Alliance for Telecommunications Industry Solutions
BASH	Bourne-Again SHell
BB	BaseBand
BBU	Baseband Unit
BDP	Bandwidth-Delay Product
BGP	Border Gateway Protocol
BOMP	Block Orthogonal Matching Pursuit

BP	Basis Pursuit
BPDN	Basis Pursuit De-Noising
BPF	Berkley Packet Filter
BS	Base Station
BSI	British Standards Institution
BW	BandWidth
CA	Collision Avoidance
CAPEX	CAPital EXpenses
CART	Classification And Regression Tree
CBS	Credit-based Shaper
CCE	Categorical CrossEntropy
CCN	Content-Centric Networking
CCSA	China Communications Standards Association
CDF	Cumulative Distribution Function
CDMA	Code-Division Multiple Access
CDN	Content Delivery Network
CDNs	Content Delivery Networks
CFS	Customer Facing System
Cgroups	Control Groups
CIR	Commited Information Rate
CLI	Command Line Interface
CMP	Constrained Matching Pursuit
CN	Core Network
CNN	Convolutional Neural Network
COIN	Computing in the Network
COINRG	Computing in the Network Proposed Research Group
ComNetsEmu	ComNets Emulator
CoSaMP	Compressive SaMPling
COTS	Commercial Off-The-Shelf
CPRI	Common Public Radio Interface
CPS	Cyber-Physical System
CPU	Central Processing Unit
C-RAN	Cloud Radio Access Network
CRC	Cyclic Redundancy Check
CS	Content Store
CSI	Channel State Information
CSMA	Carrier-Sense Multiple Access

CSMPSP	Compressive Sampling Matching Pursuit with Subspace Pursuit
CSRC	Contributing Source
CT	Core Network and Terminals
CTCP	Compressed TCP
CTS	Clear To Send
CV	Computer Vision
cwnd	congestion window
DAC	Digital-to-Analog Converter
DB	DataBase
DCS	Distributed Compressed Sensing
DCS-SOMP	DCS-Simultaneous Orthogonal Matching Pursuit
DCT	Discrete Cosine Transform
DFS	Depth-First Search
DHCP	Dynamic Host Configuration Protocol
DIN	Deutsches Institut für Normung
DINRG	Decentralized Internet Infrastructure
DNN	Data Network Name
DNS	Domain Name System
DONA	Data-Oriented Network Architecture
DoS	Denial of Service
DPDK	Data Plane Development Kit
DPI	Deep Packet Inspection
DQN	Deep Q-Learning
DSP	Digital Signal Processor
DSS	Distributed Storage System
eBPF	Extended Berkeley Packet Filter
EBS	Excess Burst Size
EC	Edge Computing
ECG	ElectroCardioGram
EGP	Exterior Gateway Protocols
EIR	Excess Information Rate
EM	Element Management
eMBB	Enhanced Mobile BroadBand
EMP	Expander Matching Pursuit
ESO	European Standards Organization
ETSI	European Telecommunications Standards Institute

FEC	Forward Error Correction
FFT	Fast Fourier transform
FIA	Future Internet Architectures
FIB	Forwarding Information Base
FIFO	First In First Out
FISTA	Fast Iterative Shrinkage-Thresholding Algorithm
FNC	Fulcrum Network Coding
FOSS	Free and Open-Source Software
FPGA	Field Programmable Gate Array
FTP	File Transfer Protocol
FW	FireWall
GCL	Gate Control List
GD	Gradient Descent
gNB	Next generation Node B
gNB-CU	Next generation Node B Centralized Unit
gNB-DU	Next generation Node B Distributed Unit
gNB-RU	Next generation Node B Radio Unit
GNU/Linux	GNU is Not Unix with Linux added
GPP	General Purpose Processors
GPS	Global Positioning System
GPU	Graphics Processing Unit
GRE	Generic Routing Encapsulation
HARQ	hybrid automatic repeat request
HDD	Hard Disk Drive
HDL	Hardware Description Languages
HDMI	High Definition Multimedia Interface
HEVC	High Efficiency Video Coding
HFCS	Hierarchical Fair Service Curve
HSPA	High Speed Packet Access
HTB	Hierarchical Token Bucket
HTP	Hard Thresholding Pursuit
HTTP	Hypertext Transfer Protocol
I/O	Input/Output
ICMP	Internet Control Message Protocol
ICN	Information-Centric Networking

ICNRG	Information-Centric Networking Research Group
ICT	Information and Communications Technology
ID	Identifier
IEEE	Institute of Electrical and Electronics Engineers
IETF	Internet Engineering Task Force
IFFT	inverse Fast Fourier transform
IGMP	Internet Group Management Protocol
IGP	Interior Gateway Protocols
IHT	Iterative Hard Thresholding
IMS	IP Multimedia Subsystem
IMT2020	International Mobile Telecommunications-2020
IoT	Internet of Things
IP	Internet Protocol
IPAM	IP Address Management
IPC	Inter-Process Communication
IPFS	InterPlanetary File System
IPV	Internal Priority Value
IPv4	Internet Protocol Version 4
IPv6	Internet Protocol Version 6
IRTF	Internet Research Task Force
ISG	Industry Specification Group
ISI	Inter Symbol Interference
ISM	Industrial, Scientific, and Medical
ISO	International Organization for Standardization
ISP	Internet Service Provider
IST	Iterative Shrinkage-Thresholding
ITU-T	ITU Telecommunication Standardization Sector
JPEG	Joint Photographic Experts Group
JSM	Joint Sparse Model
KCS	Kronecker Compressed Sensing
KPI	Key Performance Indicator
L2	OSI Layer 2
L3	OSI Layer 3
LAN	Local Area Network
LARS	Least Angle Regression
LASSO	Least Absolute Shrinkage and Selection

LCM	Life-cycle management
LDPC	Low-Density Parity-Check
LLC	Logical Link Control
LLR	Log-Likelihood Ratio
LoRa	Long Range
LoRaWAN	Long Range Wide Area Network
LSRWG	Link State Routing Working Group
LTE	Long-Term Evolution
M2M	Machine to Machine
MAC	Medium Access Control
MANET	Mobile Ad-hoc Networks
MANO	MANagement and Orchestration
MCC	Mobile Cloud Computing
MCM	multi-carrier modulation
MDP	Markov Decision Process
MEC	Mobile Edge Cloud
MEM	Mobile Edge Manager
MEO	Mobile Edge Orchestrator
MEP	Mobile Edge Platform
MGD	Mini-batch Gradient Descent
MIMO	Multiple Input Multiple Output
MitM	Man-in-the-Middle
ML	Machine Learning
MME	Mobility Management Entity
mMTC	massive Machine Type Communications
MP	Matching Pursuit
MPTCP	Multipath TCP
MSE	Mean Squared Error
MSEC	MultiService Edge Cloud
MSS	Maximum Segment Size
MTC	Machine-Type-Communication
MTU	Maximum Transmission Unit
MVO	Mobile Virtual Operator
NAT	Network Address Translation
NC	Network Coding
NDN	Named-Data Networking
NEF	Network Exposure Function

NETCONF	Network Configuration Protocol
NetEM	Network Emulator
NetInf	Network of Information
NetServ	Network Service
NF	Network Function
NFV	Network Function Virtualization
NFV-AF	Network Functions Virtualization Architectural Framework
NFVI	Network Function Virtualization Infrastructure
NFVM	Network Function Virtualization Manager
NFV-MANO	Network Functions Virtualization MANagement and Orchestration
NFVO	Network Function Virtualization Orchestrator
NIC	Network Interface Control
NIHT	Normalized Iterative Hard Thresholding
NMLRG	Proposed Network Machine Learning Research Group
NMS	Network Management System
NR	New Radio
NRF	Network Resource Function
NS	Network Slicing
NSF	National Science Foundation
NSI	Network Slice Instance
NSMF	Network Slice Management Function
NSP	Null Space Property
NSSF	Network Slice Selection Function
NSSI	Network Slice Subnet Instance
NSSMF	Network Slice Subnet Management Function
NWCRG	Coding for efficient NetWork Communications Research Group
OFDM	Orthogonal Frequency-Division Multiplexing
OFDMA	Orthogonal Frequency-Division Multiple Access
OMP	Orthogonal Matching Pursuit
ONF	Open Networking Foundation
OOM	Out Of Memory
OPEX	OPerating EXpenses
OS	Operating System
OSGA	One-Step Greedy Algorithm

OSI	Open Systems Interconnection
OSPF	Open Shortest Path First
OSS	Operation Support System
OSS/BSS	Operation/Business Support Scheme
OTT	Over-The-Top
OVS	Open vSwitch
P2P	Peer to Peer
PAT	Port Address Translation
PCAP	Packet Capture
PCF	Policy Control Function
PCRF	Policy and Charging Rules Function
PDCP	Packet Data Convergence Protocol
PDU	Protocol Data Unit
P-GW	Packet data network GateWay
PHY	Physical Layer
PHYchip	Physical Layer Chip
PID	Process ID
PIT	Pending Interest Table
PKI	Public Key Infrastructure
PNF	Physical Network Function
POSIX	Portable Operating System Interface
PSFP	Per-Stream Filtering and Policing
PSIRP	Publish/Subscribe Internet Routing Paradigm
PSTN	Public Switched Telephone Network
PTP	Precision Time Protocol
pub/sub	Publish/Subscribe
QDISC	Queueing Discipline
QoE	Quality of Experience
QoS	Quality of Service
RAID	Redundant Array of Independent Disks
RAN	Radio Access Network
RE	Restricted Eigenvalue
REST	REpresentational State Transfer
RF	Radio Front-end
RFCs	Requests-For-Comments
RIP	Routing Information Protocol

RL	Reinforcement Learning
RLC	Radio Link Control
RLNC	Random Linear Network Coding
RNG	Random Number Generator
RO	Resource Orchestrator
RoHC	Robust Header Compression
RoHCv1	Robust Header Compression version 1
ROI	Return On Investment
RPC	Remote Procedure Call
RRC	Radio Resource Control
RRH	Remote Radio Head
RRI	Request-Routing Infrastructure
RS	Reed-Solomon
RTP	Real Time Protocol
RTS	Request To Send
RTT	Round-Trip Time
SA	Services and Systems Aspects
SBA	Service-Based Architecture
SC	subcommittees
SCR	Software-Controlled Radio
SDAP	Service Data Adaptation Protocol
SDN	Software-Defined Network
SDNO	Software Defined Networking Orchestrator
SDNRG	Software-Defined Networking
SDP	Software-Defined Protocol
SDR	Software-Defined Radio
SDU	Service Data Unit
SF	Service Function
SFC	Service Function Chaining
SGD	Stochastic Gradient Descent
S-GW	Serving-GateWay
SIMD	Single Instruction Multiple Data
SMF	Session Management Function
SMP	Sparse Matching Pursuit
SoA	State-of-the-Art
SOMP	Simultaneous Orthogonal Matching Pursuit
SP	Subspace Pursuit
SRP	Stream Reservation Protocol

SSH	Secure Shell
SSMP	Sequential Sparse Matching Pursuit
STD	Standard Deviation
SVD	Singular Value Decomposition
SVM	Support Vector Machine
SVMs	Support Vector Machines
TAS	Time-Aware Shaper
TBF	Token Bucket Filter
TC	tenant controller
tc	Traffic Control
TCP	Transmission Control Protocol
TCP/IP	Transmission Control Protocol / Internet Protocol
TDMA	Time-Division Multiple Access
TI	Tactile Internet
TP	Trivial Pursuit
TSDSI	Telecommunications Standards Development Society India
TSG	Technical Specification Groups
TSN	Time-Sensitive Networking
TSSDN	Time-Sensitive Software-Defined Network
TTA	Telecommunications Technology Association
TTC	Telecommunication Technology Committee
TTL	Time to Live
TX	Transmission
UCB	Upper Confidence Bound
UDM	Unified Data Management
UDP	User Datagram Protocol
UE	User device
UHD	USRP hardware driver
UPF	User Plane Function
URLLC	Ultra-Reliable Low Latency Communications
USB	Universal Serial Bus
USRP	Universal Software Radio Peripheral
V2X	Vechicle-to-Everything
veth	Virtual Ethernet Device
VIM	Virtualized Infrastructure Manager

VIP	Virtualized Infrastructure Platform
VLAN	Virtual Local Area Network
VLC	VideoLan Client
VM	Virtual Machine
VNF	Virtual Network Function
VNFM	VNF manager
VoIP	Voice over Internet Protocol
VPN	Virtual Private Network
VR	Virtual Reality
VRU	Vulnerable Road User
VXLAN	Virtual eXtensible Local Area Network
WCDMA	Wideband Code-Division Multiple Access
WG	working group
WiFi	Wireless Fidelity
WLAN	Wireless Local Area Network
WNOS	Wireless Network Operating System
WSN	Wireless Sensor Network
WWW	World Wide Web
XML	Extensible Markup Language
YAML	YAML Ain't Markup Language
YANG	Yet Another Next Generation
YOLO	You Only Look Once

Future communication networks and systems

We commence this book by introducing the need for computing in future communication networks via examples of current and upcoming use cases. Furthermore, we provide an overview of standardization activities in the field by different regulatory bodies.

On the need of computing in future communication networks

1

Frank H.P. Fitzek[a], Patrick Seeling[b], Thomas Höschele[a], Bruno Jacobfeuerborn[c]

[a]*Technische Universität Dresden, Dresden, Germany*
[b]*Central Michigan University, Mount Pleasant, MI, United States*
[c]*Deutsche Telekom Group, Bonn, Germany*

> *It is not the strongest or the most intelligent who will survive, but those who can best manage change.*
> **Charles Darwin**

1.1 Evolution of communication networks

To understand the future of communication networks, it is of utmost importance to understand how these networks evolved over time. Communication networks in various forms have a long history. First communication systems were point-to-point oriented, exchanging information using optical (Roman or Greek light fire), acoustic (drumming), or physical media (exchange of letters on stone or paper). Later, the concept of relaying information over different communication hops lead to the first communication networks, successively increasing the attainable communication range. In France, the Chappe telegraph was introduced by Claude Chappe (1763–1805) in 1769, transmitting a 196-combination semaphore code using two wooden arms, referred to as indicators, connected with a crossbar, referred to as regulator. Each indicator had seven predefined possible positions, whereas the regulator had only two possible positions leading to 98 combinations. Two symbols where combined to achieve the 196 words for the code book. The 196 words where used to represent the alphabet with 30 letters and 10 digits. The rest was used for predefined words, predefined sentences, and control sequences. The latter ones have been used for protocol initialization (start/stop), error control (erase symbol), rate control (slower/faster), or flow control (stop-and-wait/go-back-n). The code book had already all the elements of current, modern protocol design. The distance that could be covered by two Chappe telegraphs was limited by sight of the tower and its mounted equipment, which typically required relays for distances beyond about 10 km range. (Assuming binoculars and tower heights of about $h = 10$ m, the visible horizon distance is $d \approx 3.57 \cdot \sqrt{h}$ km. This follows from the simplification of Pythagoras' theorem of $d^2 = (R + h)^2 - R^2$, with R as radius of the Earth and exploiting that $2Rh \gg h^2$, so that h^2 can be dropped.) In 1794 a multihop system from Paris to Lille (225 km) with several relaying stations was installed.

While the Chappe telegraph was under development and in deployment, considerations to employ electricity to convey messages over a distance were also underway. Georges-Louis Le Sage (1724–1803) is attributed to have developed the first working electric telegraph in 1774, connecting two rooms of his house. His version utilized a separate electric wire for each of the individual letters of the alphabet. The following years are characterized by a plethora of experimental approaches to electrically transmitted messages employing different approaches. Francis Ronalds' (1788–1873) 1816 design used static electricity to operate revolving dials to transmit messages and constitutes the first working electrical telegraph.

In the following years, several design and prototype experiments took place, which led to the late 1830s, with independently developed telegraph systems by Cooke and Wheatstone in England and Morse and Vail in the United States. It was not until 1838, when the Cooke and Wheatstone telegraph system was successfully deployed commercially, that the electrical telegraph would begin its commercial success story. A major remaining drawback of the electrical telegraph system was its reliance on mechanical components to convey the messages, which required manually relaying messages when the signal strength on the wire was no longer powerful enough. The rapid expansion of the electrical telegraph systems with multiple implementation detail differences also showcased the need to agree on globalized standards to communicate. This led to the adoption of a modified version of the Morse code by many central European countries in 1851, a major milestone in telecommunications unions' standardization efforts with effects that last to today.

1.1.1 The telephone networks: circuit-switched

The very first commercial, global, and successful communication networks emerged with the well-known telephone services. The initial implementations required direct links between all communication partners; see Fig. 1.1A. In this scenario, each telephone (the communication end point) was directly connected with every other telephone in the network. Obviously, such a solution does not scale, as the number of cabling increases exponentially with the number of installed phones.

Localized central switching was introduced to reuse telephone cables more efficiently; see Fig. 1.1B. The idea of switching at one central point, like the telephone tower in Stockholm, required only one cable per telephone. This cable, however, still had to be long enough to reach the switching center, which limits the solution to the proximity of the switching center for a given region. Another disadvantage inherent to this centralized design is the introduction of a single point of failure, which negatively impacts the overall network resilience. When the telephone tower in Stockholm burned down, it took more than three years to rebuild the structure and reestablish the single local switching center.

Hierarchical switching was introduced into the telephone network with the appearance of circuit switching, as illustrated in Fig. 1.1C. With the introduction of

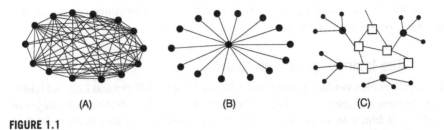

FIGURE 1.1

The evolution of telephone networks architectures. (A) Fully meshed; (B) Star Networks;
(C) Hierarchically switched networks.

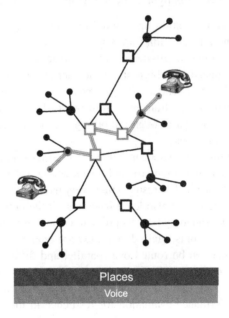

FIGURE 1.2

The evolution of communication architectures: Circuit-switched networks.

hierarchical switching, the problems of cable lengths, coverage, and resilience were
solved. This development ultimately resulted in the emergence of the commercial
Public Switched Telephone Network (PSTN). In circuit-switched networks, there
are always dedicated physical resources available for the exclusive use by a given
communication pair. Though realized by utilizing several *hops* across connecting
equipment, the resulting implementation appears *logically* as a single dedicated virtual cable between the communication partners. PSTNs are referred to as intelligent
networks, as the communication end points (the standard telephones) are kept very
simple, whereas the voice communication services are established within the network. The intelligence is needed to optimize for the routing and establishment of

the circuits. As shown in Fig. 1.2, circuit-switched networks connected *places* and mainly provided *voice services*.

1.1.2 The Internet: packet-switched

The age of packet switching started with the works of Paul Baran and Leonard Kleinrock. In eleven chapters [1–11] of his book, Paul Baran described the advantages of packet switching over circuit switching. The main idea is to make a packet-switched network low cost and simple, but also robust against link failures and outages of single communication nodes. This was achieved through different main design approaches, which we briefly highlight in the following.

Low cost: Every communication link (wired or wireless) can be used by several communication partners – links are not reserved exclusively for one communication pair as in the circuit-switched case. To achieve this, packet communication was introduced. The biggest advancement over circuit switching is that packets from different senders are conveyed over the same medium, that is, they share the underlying transmission resource. The separation of packets in this scenario is achieved through the orthogonal use of resources in terms of time, frequency, or codes.

Simple: Every communication node is connected in a mesh architecture. Nodes follow a simple communication protocol for incoming packets, which is referred to as *store and forward*. The result is that every packet entering an intermediate communication node will also leave it via one of the outgoing paths. The exceptions are represented by packets prone to erroneous transmissions or queue overruns. This concept is referred to as the *node continuity rule*.

Robust: Every packet can be routed over parallel and different paths within the mesh network to create redundancy in case of packet loss or path losses. In his first chapter, Paul Baran describes the advantages of multipath and presents a very first performance evaluation for resilience in meshed communication systems [1].

Leonard Kleinrock and others significantly contributed to the queueing theory of packet-switched communication systems that focus on the delivery of data services (in opposition to the PSTNs that were focusing on voice services). First insights into throughput, resilience, and latency have been introduced [12–16]. The revolutionary idea of Paul Baran and others was to make the network simple and robust, whereas the communication end points became intelligent. The initial message *LO* sent during a night now over 50 years ago started to be the foundation of DARPAnet, the initial implementation of packet-switched networks. Although the first message caused the entire system to crash and added a few extra work hours to everyone experimenting with the system that night, the repeated successful transmission of *LOG IN* ushered in a new era of communications. Packet switching enabled the breaking of long messages into smaller ones and realized the concepts of efficient time-shared resource utilization. However, there still were heterogeneous implementations of packet-switched networks that could not communicate directly. Furthermore, addressing, congestion, error handling, and inter-network routing were not part of the original packet switching designs. A unifying layer was needed on top of these disparate networks to enable interoperability, which was presented by Cerf and Kahn [17] in 1974. In their work, they technically presented two layers as we know them today, namely, i) Transmission Control Protocol (TCP) and ii) Internet Protocol (IP). Jointly, these two layers enabled the process-oriented and reliable communication end-to-end and across different packet-switched networks. This addition to packet switching enabled the seamless interconnection of individual networks to the Internet.

With the evolution of the Internet, the idea of packets and packet switching following the *store-and-forward* policy was fully adopted. Only the idea of multipath communication, also presented by Baran, was not realized in the implementation of the Internet. Baran proposed to use repetition coding (i.e., sending the same packet on different links) for resilience. Given the capacity limits in those days, the performance was too inefficient, and the idea was ultimately withdrawn. However, the idea of multipath transmissions has reemerged recently in different forms, such as multipath TCP [18,19] or coded multipath [20]. As given in Fig. 1.3, the packet-switched networks connected *people*, following the slogan of Nokia, and provided *voice* and *data* services. The figure also shows that the Internet relies mainly on single-path communication, despite the work and effort of Paul Baran.

Baran [6] additionally addressed the realization of communication networks employing wireless communication links. Years later, some of these original ideas were utilized in the design of cellular communication systems. Whereas the Internet's packet-switched approach flourished throughout those years, cellular communication systems originally started as wireless extension networks for PSTNs, focusing on voice services. Later, the cellular networks were connected solely by IP to allow for data transmission, whereas voice services were realized with Voice over Internet Protocol (VoIP) services.

1.1.3 The cellular communication networks

The 1st Generation (1G) of cellular communication networks was based on analog technologies. It was designed to enable voice services, extending the PSTNs. Mobility was hardly supported and had to be announced by the mobile user with dedicated voice calls to an operator. Due to the cost of that technology and the difficult handling, the overall system was not meant for a mass market. The switch to digital technologies was performed with the 2nd Generation (2G) of cellular communication networks. This switch ultimately enabled a mass market adoption through removal of the complicated user-driven handling. The air interface (communication link) was based on Time-Division Multiple Access (TDMA). Though the main service was voice again, the 2G system provided two important new features, namely security and mobility. The following 3rd Generation (3G) had mainly three goals:

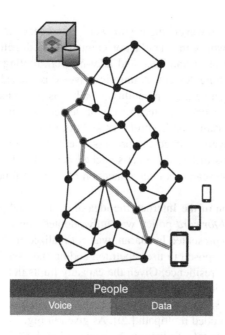

FIGURE 1.3

The evolution of communication architectures: Packet-switched networks.

Air interface: The revised 3G system should switch to a new air interface referred
to as Wideband Code-Division Multiple Access (WCDMA). Qualcomm was
able to convince the standardization bodies that Code-Division Multiple Access (CDMA) would have a better spectral efficiency than TDMA. The claim
was that WCDMA would be better than TDMA by a factor of seven [21]. It
was not. CDMA had some advantages based on the activity factor of human
voice, which later was also exploited by TDMA-based systems.

Global mobility: To have global mobility, rather than regional mobility, 3G system
designers tried to harmonize frequency bands and find a minimum number of
common bands worldwide.

Support of large data: The 3G system should enable data communications to
a large extent in addition to the traditional prior generations' voice services.

The 4th Generation (4G) cellular network system enabled the true and fully IP-
based mobile Internet. Again, the air interface was switched to a new technology,
namely Orthogonal Frequency-Division Multiple Access (OFDMA). This switch was
partially performed to avoid Qualcomm's estimated license fees of three dollars per
device. All four generations of cellular communication systems have in common that:
i) they enabled commercial communication for humans, ii) they are a simple extension to existing networks, such as the circuit or packet-switched networks, iii) there
is one decade between each generation, and iv) they follow the end-to-end paradigm

with intelligent end nodes and dumb and agnostic communication networks in the middle.

With the introduction of 4G, software began to play a dominant role in cellular networks for the very first time. With the widespread adoption of flat rates for Internet connectivity, Over-The-Top (OTT) services began to appear in the fixed Internet. OTT services ran on cloud and device infrastructure only and kept the network operators out of business. A similar trend of flat mobile Internet rates subsequently started to take over the cellular communication market. Though developers around the world already started to program applications for 2G and 3G phones, the fully IP-networked mobile device emerged as a great play-out delivery vehicle for OTT services. Softwarization not only took place at the higher protocol layers. Due to the large number of frequency bands that had to be supported globally, the Software-Defined Radio (SDR) had already been implemented to dynamically support mobile end devices.

Currently, the 5th Generation (5G) of cellular network designs is being implemented. In contrast to its predecessors, it seems to be a real revolution – rather than the evolution from 1G to 4G. First of all, it is not targeting solely services for humans, it also targets services for billions of things, the so-called Internet of Things (IoT). Furthermore, some of the IoT devices require quasi-real-time communication to combine control and communication theory. Secondly, 5G is not just a wireless extension as with prior cellular network generations. The 5G cellular system will be a holistic design comprising the wireless and the wired network worlds. Therefore the 5G communication system is standardized by the 3rd Generation Partnership Project (3GPP) for the wireless part (including the Radio Access Network (RAN)), the so-called New Radio (NR) 5G, whereas the Internet Engineering Task Force (IETF) addressed changes within the Internet for the wired part. An overview of this joint efforts is illustrated in Fig. 1.4.

The IETF approach is solely driven by software. Therefore the packet-switched network core has no notion of a generation, as updates can be executed anytime. In contrast, the cellular domain was characterized by a need for different generations as it was always hardware-driven. Every new generation meant to change a number of 19" racks due to required hardware modifications. With 5G, network softwarization is the dominating factor for both, the wired and the wireless domains. Availability for specialized hardware, such as Radio Front-end (RF)/BaseBand (BB), antennas, or efficient computing (due to hardware acceleration), remains, but the overarching trend is softwarization. The result of this trend is that the wireless part of cellular networks is software-driven, too, and should not require any new generational upgrades in the future. Notwithstanding, some marketing people will make us believe that we need a 6G or 7G wireless network. Nevertheless, the IETF and the 3GPP are advocating to continue to transform communication networks. They propose a transformation from solely conveying information between two places using the *store and forward* paradigm to future communication systems where information is also processed within the communication network following a *compute and forward* paradigm. In the following, we discuss the 5G technology in detail to emphasize the revolutionary change that lies ahead of us. Where we refer to a 5G communication

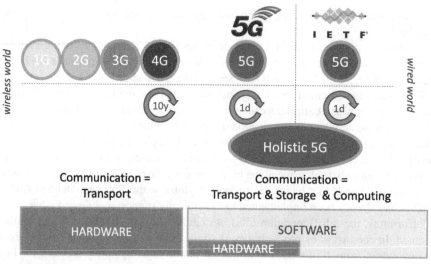

FIGURE 1.4

The evolution of communication architectures: Cellular communication systems.

system in the following, we refer to the wireless and the wired worlds that are joined within 5G. After the circuit-switched and packet-switched networks, the new era of communication networks can be referred to as *computing-centric networks*. As illustrated in Fig. 1.5, the computing-centric networks will focus on connecting *things* and provide *voice*, *data*, and *control* services. The figure addresses computing within the network and multipath communication as a tribute to Paul Baran. At the same time, the need for computing for communication networks was foreseen by Claude Shannon in 1959, Pennsylvania, when he stated *I think that this present century, in a sense, will see a great upsurge and development of this whole information business ... the business of collecting information and the business of transmitting it from one point to another, and perhaps most important of all, the business of processing it* [22]. Today, 60 years later, we bring the *most important business* to our communication networks.

1.2 The 5G communication system

In contrast to the International Mobile Telecommunications-2020 (IMT2020) definition of 5G [23] with its three dimensions of massive IoT, massive multimedia, and massive low latency, we will introduce 5G in the form of an atom. As illustrated in Fig. 1.6, the core of the *5G Atom* is represented by different use cases. These use cases define the first tier of the atom, namely the technical requirements. To achieve each of the technical requirements, novel communication concepts are required, which form the second tier. The third tier is comprised of the softwarized technologies realizing

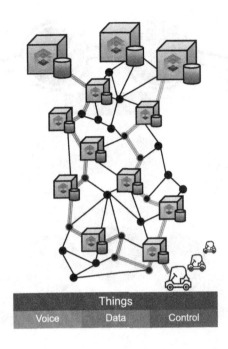

FIGURE 1.5

The evolution of communication architectures: Computing-centric networks.

these new communication concepts. The last tier describes possible innovations and novelties that, due to the concept of softwarization, can be introduced in a straightforward and timely fashion into future communication networks. In the following, we describe the *5G Atom* model in greater detail.

1.2.1 The *5G Atom* core: use cases

In this section, we describe our current view on potential 5G use cases. At this point in time, to make this book readable in the near future, we have to stress that the potential 5G *killer* use case may not be presented here. To our defense, so far no researcher worldwide has ever predicted the next *killer* use case for any generation beforehand. Nevertheless, 5G has not only been defined by network operators and mobile device and equipment manufacturers (as in the generations before), but also by contributions from application-oriented industry partners (which have a significant interest in shaping future communication networks). The initial wave of 5G use cases that move toward implementation will target the classical machine-to-machine IoT application scenarios. Currently well-researched 5G use cases include, among others, connected cars, Industry 4.0, construction, agriculture, education, health care, and energy grids. The following big wave of applications will target human-to-machine applications realizing the Tactile Internet (TI). This next stage of use cases will target

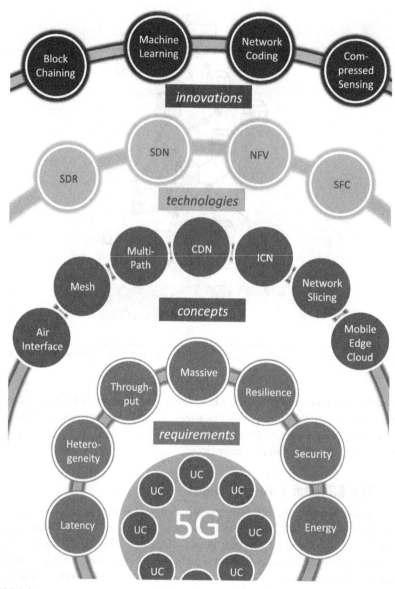

FIGURE 1.6

The *5G Atom*.

a wide range of Virtual Reality (VR)/Augmented Reality (AR) interactions, namely live events, education, health care, human–machine collaboration, and gaming. At the 5G Lab Germany [24] at the Technische Universität in Dresden, we have at least one industrial partner that represents each individual category.

1.2.1.1 *Connected autonomous cars*

In the future, connected autonomous cars will be vital parts of our everyday life. The transportation of people and goods will become safer, it will consume less energy, and it will result in less negative side effects, such as polluting emissions. Current field tests for autonomous driving solely rely on the on-board sensors of individual cars themselves. In turn, the need for communication among vehicles might not be obvious at a first glance. The current isolation of autonomous vehicles, however, is mainly driven by the economic interests of the individual car companies (that would like to aim for their individual world domination and find it hard to cooperate with other brands). There are several reasons that counter pure economic incentives and make compelling arguments for direct or indirect intervehicle communications. Subsequently, it is highly likely that not just autonomous cars, but *connected* autonomous cars will play a significant role in the future of transportation. In the following, we briefly highlight two examples that showcase the need for direct or indirect intervehicle communications. Consider a scenario where several autonomous cars were to interact in close proximity with one another. The individually employed algorithms to perform individual steering and maneuvering would, subsequently, interact with those of other nearby vehicles. This could lead to instability by coupling second-order control loops to higher-order ones, which can be only controlled by lowering the speed or increasing distances between the vehicles. Both of these approaches would decrease the efficiency of employing isolated autonomous vehicles. This is especially true for platooning scenarios, as illustrated in Fig. 1.7.

FIGURE 1.7

The *5G Atom* use cases: Connected autonomous cars platooning example.

Current vehicular sensors are not able to monitor and detect around corners or obstacles, or to monitor distances beyond the sensor range in front of them. The first limitation is obviously important for scenarios of crossing intersections. Without information about other cars entering an intersection, each car would have to slow down at each intersection to initially determine the situation. The second limitation represents a limit on improving the efficiency of traffic, predominantly on highways or motorways. This applies especially for platooning scenarios (e.g., as in multiple vehicles joining for long-haul freight trucking) to jointly control speed and make navigation decisions. The advantage of a sensor fusion using communication networks (compared to a single set of sensors per vehicle) is the prediction of future events that will occur in several-kilometer distance. The use of agnostic sensor sets per vehicle would just react to sensed data from the vehicle in their individual front. Another advantage of connected sensors in different vehicles is the possibility to *look around the corner* in crossing situations, as illustrated in Fig. 1.8 for a Cyber-Physical System (CPS). In Fig. 1.9 a virtual world with more sophisticated crossing scenarios is depicted.

FIGURE 1.8

The *5G Atom* use cases: Connected autonomous cars crossing scenario.

Connected autonomous cars enable new scenarios and offer multiple advantages compared to isolated autonomous cars that are solely controlled employing their individual on-board sensors. Both of the aforementioned scenarios highlight the increased system efficiency, for example, by allowing for higher driving speeds. However, connected autonomous cars can also enhance the safety for vulnerable road users via image recognition and information exchange with other connected cars. The increase in overall mobility will furthermore post demands on cities of the future to coordinate a large number of different road users. The required coordination and control algorithms will have to be performed in proximity, for instance, at crossroads where optimized traffic coordination would result in continuous traffic flow without wasting energy for unnecessary stopping, situation assessment, and resumed

FIGURE 1.9

The *5G Atom* use cases: Connected autonomous cars Barcelona example.

driving. Increasing the dimensions to a city or even nationwide level, communication for autonomous cars remains necessary for navigation purposes. This will reduce street congestion, foe example, by steering via optimized roads within a city or by a timely redirect on motorways. Some of these tasks are latency-critical and have to be computed close to the connected car. Passengers of autonomous cars may want to consume video or virtual reality content while traveling. A reliable and secure communication system will need to distinguish between leisure and car control network traffic, as well as allowing for dynamically inserted public safety network traffic, commonly guaranteeing network quality of service requirements for the latter types of traffic. As we outline in greater detail in Section 1.2.1.4, future energy grids rely heavily on batteries, for example, to stabilize the frequency of an electrical power grid. As every vehicle will have batteries on board, the parking of the vehicles could be optimized based on energy demand and supply within the energy network. Furthermore, every vehicle will have communication infrastructure on board that could potentially act as a base station or access point for other customers. Again, the appropriate placement of cars alone could lead to an improvement (in this case, of the communication network).

1.2.1.2 Industry 4.0

The 5G standard introduces new wireless communication scenarios within the production process, mostly described with the buzzword Industry 4.0. Industry 4.0 is supposed to logistically connect every layer of production processes. Additionally, Industry 4.0 will see a massive rise in wireless communication devices as 5G finally meets the robust communication requirements of industry automation within the pro-

FIGURE 1.10

The *5G Atom* use cases: Example network architecture in Industry 4.0.

duction process. An example of the Industry 4.0 environment is illustrated in Fig. 1.10 with collaboration of humans and machines. The figure additionally includes technical elements, such as wireless communication and computing in the factory. Data collection will be performed by a Remote Radio Head (RRH), and signal processing is performed by an in-house computing element, which is later referred to as Mobile Edge Cloud (MEC). For more information about the MEC, we refer to Chapter 4. The main 5G applications can be divided into three major groups:

Control loop: The control and steering functions of the robot arms are currently hosted locally at the robot. To reduce its costs and create new services, the control functions of the robot will be conveyed to the MEC. Hence the communication link becomes a part of the control loop, which imposes new challenges on the communication link in terms of latency and resilience. Control functions for multiple cooperative robots will also be hosted in the MEC. One candidate for new services is the training or remote control of robots in the virtual room as illustrated in Fig. 1.11.

Massive sensors: A large number of sensors will be deployed within factory or production halls to enhance process control, planning, and production adjustment. These sensors will likely connect to multiple base stations to increase throughput and resilience, and to reduce latency (multiconnectivity). The collected data will be used to simulate or virtualize the production process or factory environment in real time, which enables immediate reactions to production changes or machine/process failures. The collected sensor information will be preprocessed before sending, thus reducing the amount of data required for transmission and storage. Most likely, virtualization will be performed at the MEC. However, virtualization could be performed by dedicated software

FIGURE 1.11

The *5G Atom* use cases: Industry 4.0 training or remote control of robots in a virtual room.

in a data center under certain boundary conditions. Predictive maintenance of machines will require sensor data and control loop data to be collected and transmitted to the machine manufacturer. The machine manufacturer can use this data to provide new services, which could decrease production downtime.

Mobility support: Path planning and collision avoidance are needed for the mobility of platforms or mobile robots. This will be achieved via object detection, recognition, and analysis in cameras. These cameras could send contextual data instead of video streams, to ease stress on the communication link complying with personal privacy requirements of the staff.

Within an Industry 4.0 factory, the communication network needs to support broadly diverse applications. Each of these applications has different requirements regarding latency, throughput, reliability, and so on. Facilitating all requirements at reasonable costs will only be possible within a heterogeneous communication environment that plays into the strengths of the different components.

1.2.1.3 Agriculture

The agricultural sector is facing several challenges today, such as climate change, dramatic transformations in demographics, and immense increases in demands for food. Agriculture 4.0, a name that was created in light of Industry 4.0, aims at addressing these challenges by incorporating cross-industry technologies and applications. Agriculture 4.0 employs information and communications technologies with the ultimate goal of improving crop productions. It defines methods for measuring related data, analyzing the measurements (in near real-time, if needed), and defining and applying actions accordingly. For instance, data about weather and soil conditions can be collected by specialized sensors and subsequently analyzed by machine learning algorithms to determine the proper amounts of water, pesticides, and fertilizers for each individual plant. To realize the outlined example applications, a communication network has to solve several challenges. Starting with unfavorable or no cellular coverage on fields at all, the network must operate independently of a mobile service provider. A MEC (which we discuss later in Section 1.2.3.7) could be deployed on an agricultural machine and thus be physically mobile itself. This has to be supported by network functions and opens new challenges for positioning algorithms. Additionally, applications can also transmit employing different wireless technologies, whereas the main supporting network has to simultaneously coordinate resources for all these applications.

FIGURE 1.12

The *5G Atom* use cases: Agriculture 4.0 machinery platooning with small machines.

1.2.1.4 Energy grid

Over the past decades, the traditional centralized approach to energy generation has given way to more decentralized concepts. This change in the energy supply is due to an increase in renewable energy production methods, such as photo-voltaic, wind energy, or biomass, which are typically based on small, local power plants. In addition to this transformation of energy production means, the introduction of 5G will provide the energy sector with the abilities of building smart energy grids, as in the example illustrated in Fig. 1.13.

Smart energy grids will provide enhanced monitoring capabilities and superior energy distribution methods. As sectors of the grid become enabled to dynamically decouple and reconnect with the main grid, expensive energy transportation will be reduced. New energy storage solutions supported by more efficient battery technologies, maybe based on carbon, will provide the capabilities to store energy in a decentralized fashion. Energy generated locally by renewable sources will be stored locally as well, further decreasing the need for energy transportation. The 5G communication standard with its focus on reduced latency will be an enabler of these new approaches to providing energy in the upcoming years.

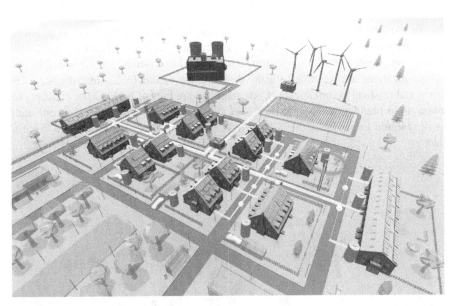

FIGURE 1.13

The *5G Atom* use cases: Virtual power plant model for the smart energy grid.

Several applications based on 5G communication standards are currently being proposed and researched, for example, in the National 5G Energy Hub. New communication technologies will enable wireless connectivity in conventionally hard-to-cover areas, such as basements. This will provide the controllers of local structures (e.g., an individual house) with the means to forecast and monitor local power generation and local load. These structures could be temporally self-supporting and become decoupled from the main grid. As local structures form segments, the management of the local segment within its larger grid can subsequently be performed through power generation and demand forecasts, coordinated across structures and segments. 5G will additionally enable fault detection and fault clearance within these segments due to the low communication latency. Such segments, in turn, can be regarded as self-organizing virtual power plants. Another application scenario is the coordination of smart home devices. Load and generation forecasts can be used to increase the use of decentralized load generation, reducing energy transportation losses.

1.2.1.5 Tactile Internet

The aforementioned used cases are mainly machine-to-machine communication oriented. The Tactile Internet (TI) use case focuses on human–machine communication. One definition of the TI is given by the Institute of Electrical and Electronics Engineers (IEEE) 1918.1 working group: *A network or network of networks for remotely accessing, perceiving, manipulating or controlling real, or virtual objects, or processes in perceived real time by humans or machines.* Whereas the current Internet democratizes access to information for all people independent of location and time, the Tactile Internet aims to democratize access to skills and expertise to promote equity for people of different genders, ages, cultural backgrounds, or physical limitations on a global scale as illustrated in Fig. 1.14. Current developments in 5G communications are beneficial for the development of the Tactile Internet. Low latency and resilient communication, for example, are part of both. Other concepts to attain these foundational requirements are present in both approaches as well, such as the Mobile Edge Cloud (MEC) or Network Slicing (NS), which we will discuss later.

FIGURE 1.14

The *5G Atom* use cases: Tactile Internet.

1.2.2 First tier: the technical requirements

After discussing with a large number of industry partners, we have identified the technical requirements that the highly diverse future application domains will present to their communication systems. The commercial (and noncommercial) applications of 5G will present unprecedented challenges to future communication systems with respect to latency, throughput, resilience, security, heterogeneity, massiveness, and energy. In the remainder of this first tier section, we initially provide brief discussions of the aforementioned technical requirements. We follow this overview with a description of more specific values targeting different use cases we outlined before.

1.2.2.1 Latency and jitter

Data networks have been evaluated and compared for a long time with respect to data rate or data volumes that can be conveyed between two communication end points. This was due to the request for communication services like web browsing, file exchange, or video streaming. Early requests by researchers or gamers to also optimize for latency and jitter, as in [25], have been ignored for a similarly long time. The report in [25] was published in 1996 with the enlightening title *It's the Latency, Stupid* and discussed the need for low latency. Although latency and the derivative jitter did not play any major role in the first four mobile communication systems or the Internet in general, the 5G communication network placed this requirement on the top of the requirements list. The reason why latency has become important now is based on the fact that future communication systems are addressing use cases, such as machine-to-machine or human-to-machine communication, with *integrated control loops*. Latency and jitter need to be addressed in an end-to-end manner and are not solely based on the performance of the communication links. As illustrated in Fig. 1.15, the latency is a sum of several components, such as the sensor/actuator with embedded computing, the wireless (5G) link, the wired link, and the computing within the network. Some use cases allow end-to-end delays of only 1 ms (see Table 1.1).

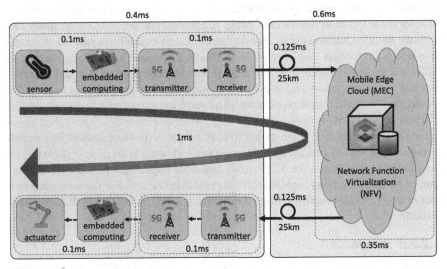

FIGURE 1.15

The *5G Atom* technical requirements: Latency budget for end-to-end communication [26].

In [26], we presented a delay budget for the given setup with 1 ms end-to-end delay. The overall delay budget is composed as follows:

Sensor/actuator: 20% of the millisecond can be used for the sensor and actuator pair with the given embedded hardware. The embedded hardware is used, for example, to compress or decompress the sensor information to avoid additional latency values on the communication links. Current sensors and actuators are often not designed for low latency use cases, and therefore the latency budget might need to be adjusted.

Wireless links: 20% of the millisecond can be used for the wireless link. Although the data rate will often be high enough, challenges remain in the Logical Link

Control (LLC) and the Medium Access Control (MAC). Furthermore, other queueing tasks will add to the overall link delay.

Wired links: 25% of the millisecond are reserved for the transportation of information in the wired domain. As light travels with circa 300 km/ms, the distance between the wireless end points and a computing entity within the network should not exceed 25 km (two way).

Computing within the network: 35% of the millisecond subsequently remains for the computing within the network. Even though this seems enough for most use cases and their related computing tasks, virtualized environments will consume a large part of it, as described in greater detail in [26].

1.2.2.2 Throughput

As with every new generation, the fifth generation of mobile communication systems will provide higher throughput than the predecessors. The request for higher data rates is mainly driven by video streaming services, which are currently responsible for the majority of mobile data traffic. Significant amounts of data are needed to transmit encoded multimedia as mobile device screen capabilities increase to common television formats and beyond. Newer mobile multimedia services, such as in AR or VR application scenarios, will require even higher data rates next to low latencies and computing resources. In contrast to other requirements, such as latency, the requirement of higher throughput is well known to the engineers of next-generation wireless networks. 5G is marketed as having 1000 times more throughput than existing 4G technologies. Even though this sounds dramatic, the increase will be reached by higher base station density, more frequencies, and better utilization of the wireless link. In the current release of 3GPP, peak data rates larger than 10 Gb/s for the Enhanced Mobile BroadBand (eMBB) will be available in a cell. For the individual end user in a cell, 50 Mb/s and 100 Mb/s are available in the uplink and in the downlink, respectively.

1.2.2.3 Resilience

The term resilience is defined in the Oxford dictionary as *the ability to (elastically) recover from difficulties and to spring back into shape*. Sterbenz et al. [27] define network resilience as *the ability of the network to provide and maintain an acceptable level of service in the face of various faults and challenges to normal operation*. As discussed earlier, 5G networks are inherently complex and dynamic, and they are expected to enable revolutionary applications. Therefore the resilience of these networks should be treated as a function of several Key Performance Indicators (KPIs), including: i) throughput, the amount of data that can be transported through the network within a given time; ii) latency, the amount of time taken to transport a data unit through the network, and iii) reliability, the probability of successfully receiving a data unit. According to information theory [28], a network cannot be simultaneously optimized for all the aforementioned KPIs, that is, improving one KPI may negatively affect the others. For example, the use of forward error correction to improve the reliability will decrease the system throughput. Also, queueing the excess data, a widely used technique to handle network congestion, increases the latency. Therefore a trade-off should be made. Furthermore, due to the contrasting nature of the aforementioned KPIs, it is not sufficient to just improve the physical and data-link layers. Instead, the entire network stack, ranging from the radio, networking, transport, and up to the application layer, needs to work cooperatively toward minimizing the latency and maximizing the throughput and reliability of the end-to-end network.

1.2.2.4 Security

The traditional goal of cryptography was transporting messages between two trust zones through an untrustworthy environment, for example, two armies trying to coordinate their attack by exchanging messages across a battlefield. Friend and foe were clearly defined, and therefore the threat to the message could be assessed. In the digital world, this simple threat model no longer applies because of the increase in dependencies and therefore complexity. Nowadays, service providers execute their services *in the cloud*, which is hardware that is run by a third party and shared with an additional fourth party. Furthermore, the cloud provider could even subcontract the management of the servers to an additional fifth party. Since the service providers no longer have direct control over the hardware, they have to trust the cloud provider to separate tenants from each other and to not tamper with the service itself. This complexity prevents the user and the service provider from fully assessing the infrastructure they are working on, to decide if it is trustworthy and secure. With this scenario, the trust model is either naive and simple (like it used to be) or arbitrary complex. In this setting, absolute security can no longer be achieved.

Coming as close as possible to the original trust model might neither be advantageous nor feasible. However, it is necessary to have a handle on security and hence successively decrease the complexity and necessary trust into additional parties. The result is that data, algorithms, and metainformation have to be protected. Since this is a broad topic, formal security goals have been defined to describe the security aspects. The classic security goal triad is abbreviated as *CIA*: i) Confidentiality: data transmitted or stored should only be revealed to the intended audience; ii) Integrity: it should be possible to detect any modification of data; and iii) Availability: services should be available and function correctly at the time of request.

For 5G systems, the confidentiality and integrity of stored or in-transit data need be ensured. Traditional methods, such as encryption and Message Authentication Codes, can deliver protection for these goals. Availability can be more challenging, especially when an ultralow latency is required. Even short unavailability can result in breaking the underlying service model. Furthermore, depending on the application in question, a sudden loss of availability can result in high damages or even the loss of lives. It is therefore essential that the availability of data is protected. Another goal to consider is privacy. For example, applications that use sensors to capture human behavior allow us to analyze and identify humans. To guarantee privacy for users, applications must anonymize data.

1.2.2.5 Massiveness

With the introduction of the Internet of Things (IoT), the number of wireless communication devices will increase dramatically. As the number of mobile phones reached the number of human beings globally in 2014, the number of any kind of wireless communication device is to be expected one or even two orders of magnitude higher. For example, Cisco projects that by 2022, there will be 12.3 billion mobile-connected devices [29]. Ericsson predicts that by 2023, more than 23.3 billion IoT devices will be connected wirelessly, with an additional 10.3 billion other mobile devices [30]. Some other studies go far beyond these numbers. The number of expected IoT devices varies greatly due to the definition of a wireless device and what networks to consider. Some estimations refer to the classical mobile phones, and others also account simple energy harvesting sensors. With Long Range (LoRa)/Long Range Wide Area Network (LoRaWAN) and Narrow-Band IoT in 4G, two examples for dedicated solutions for IoT are introduced, among others [31,32]. For example, Long Range (LoRa) devices can be used by anyone within the Industrial, Scientific, and Medical (ISM) radio bands, which democratizes IoT but makes actual accounting difficult. In turn, these solutions by themselves will not provide alleviation from the massive amounts of IoT devices and increase the complexity and heterogeneity further.

1.2.2.6 Heterogeneity

Due to the large number of expected IoT devices, there will also be a large variety of such devices. This, subsequently, will lead to a massive heterogeneity in the device characteristics. Mobile communication systems to date, on the other hand, can be characterized by their underlying networks being highly standardized. Similarly, the mobile devices utilizing these networks are very homogeneous in terms of connectivity capabilities, battery capacities, display sizes, and so on. In an IoT world, these characteristics will vary dramatically between devices: Whereas a car has a huge battery, ample computational resources, and several communication standards on board, some sensors in the streets might be energy harvesting with absolutely no or very limited computation capabilities on board and rely on a singular, extreme low-power networking standard to communicate. Despite the large heterogeneity of connected device characteristics, the introduction of several networks for each separate communication need is not desirable. First, numerous networks would have to be maintained and supported. Second, separating IoT devices into different networks will cause them to lose knowledge about the pure existence of other communication nodes. Third, numerous IoT solutions will depend on the collective interaction of different types of devices. It is highly likely that cumbersome approaches to re-merging disparate networks will result in more complex structures, which, in turn, would be more prone to errors and failures. A more feasible solution as goal should be the employment of one physical communication network with scalable, agile, and flexible algorithms that perform at different complexity levels.

Table 1.1 3GPP TS 22.261: Performance requirements for low-latency and high-reliability scenarios.

	latency [ms]	jitter [μs]	availability [%]	resilience [%]	Data Rate [Mbps]
Discrete automation	1	1	99.9999	99.9999	1
Process automation – remote control	50	20	99.9999	99.9999	1
Process automation – monitoring	50	20	99.9	99.9	1
Electricity distribution – medium voltage	25	25	99.9	99.9	10
Electricity distribution – high voltage	5	1	99.9999	99.9999	10
Intelligent transport systems	10	20	99.9999	99.9999	10
Tactile interaction	0.5	TBC	99.999	99.999	low

1.2.2.7 Energy consumption

Energy consumption in Information and Communications Technology (ICT) is always a challenge, as network operators have to pay huge energy bills and users are suffering similarly via the operational times of their devices. As mentioned beforehand, we assume a massive increase in the number of devices in the coming years. Even when considering the state-of-the-art communication architecture, connecting each device directly to a base station would lead to a massive, untenable level of energy consumption. Therefore new communication architectures and protocols need to be developed to significantly decrease energy consumption.

1.2.2.8 Technical requirements per use case

To provide some quantitative numbers for each technical requirement, we have consulted different sources and talked to industry partners directly. In [33], for example, numbers are presented for different use cases. We have extracted some of these values in Table 1.1 for Industry 4.0 (first three rows), Energy (rows 4 and 5), Mobility (row 6), and the Tactile Internet (row 7) with respect to latency, availability, resilience, and data rate. The performance requirements for low-latency and high-reliability scenarios are addressed in 3GPP TS 22.261 [33], which is considered to be a living document and subject to ongoing modifications.

Nevertheless, we derive the conclusion from Table 1.1 that the use cases, even within one vertical industry sector, are quite heterogeneous in the required KPIs ranging from very hard latency values even below 1 ms to 50 ms and data rates ranging from 1 Mbps to 10 Mbps. Each vertical industry, such as connected autonomous cars, consists of multiple different use cases. Each use case has multiple different technical requirements that need be supported by the network. We briefly describe the use cases and provide a spider diagram for the technical requirements of each use case. In most vertical industries the use cases will be implemented in parallel on the same physical network infrastructure. Yet, the network has to meet the requirements for each use case individually. This can only be achieved via network slicing.

From our discussions at the 5G Lab Germany with industry partners, we have derived our enriched set of technical parameters. In the following, we present various use cases for five different vertical industries with respect to data rate, latency, resilience, security, energy, massiveness, storage, and computing needs. We will describe the most important known use cases for each industry and provide a spider diagram showing the technical requirements of the use cases. The more on the outer circle the technical requirement is located, the more important or hard is the particular requirement for the use case.

Technical requirements: connected autonomous cars

The most important use cases for connected autonomous cars are i) car control, ii) navigation, iii) in-car entertainment, and iv) predictive maintenance. The use cases and their technical requirements for the connected autonomous cars industry are summarized in the Spider diagram in Fig. 1.16.

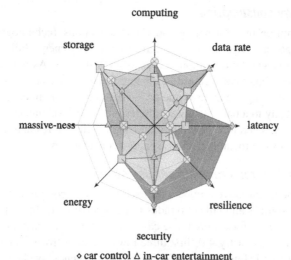

FIGURE 1.16

5G Atom technical parameters and use cases: Connected Cars.

Connected autonomous cars will drive safely through cities without human interaction: they will accelerate, break, and stop on their own. To ensure this scenario of *car control* without accidents, a reliable network connection and Vehicle-to-Everything (V2X) communication will be essential. Controlling these cars requires minimum latency and enough bandwidth to ensure immediate reaction to the changing circumstances of the traffic environment. All the autonomous cars will be *navigated* through the city on the optimal routes considering different parameters, such as energy consumption, time, total traffic, and so on. These routes have to be optimized, including whole city parts, cities, or even regions, to avoid traffic jams or street congestion. Autonomous cars will enable the driver to consume *in-car entertainment* content or perform other tasks while traveling. Latency is not crucial for these applications, but the bandwidth must be large enough to ensure a high-quality user experience. To *predict maintenance* stops and intervals, car manufacturers will retrieve information on the car status. For the transmission of this information, neither latency nor bandwidth is important, as the data will be collected in the car and then sent to the car manufacturer once a day or on demand.

Technical requirements: Industry 4.0

The most important use cases for Industry 4.0 are i) massive sensor networks for virtualization, ii) camera data for object recognition, iii) machine control, and iv) predictive maintenance. The Industry 4.0 vertical use cases are summarized in the Spider diagram in Fig. 1.17 and briefly discussed in the following.

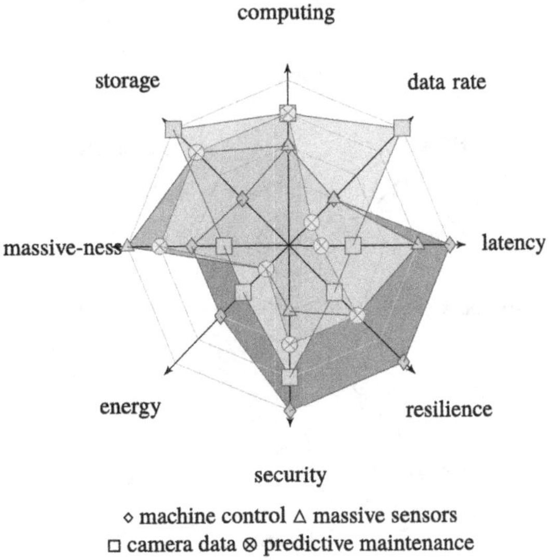

FIGURE 1.17

5G Atom technical parameters and use cases: Industry 4.0.

The virtualization of the manufacturing process to monitor, control, steer, and optimize the production will be based on various kinds of information. This information will be collected, for example, by *massive networks of sensors* or through *object recognition via cameras*. Massive sensors could detect audio data and transmit information for failure detection and process control. Within a factory, process control is a critical task, and therefore real-time interaction is necessary. Each individual sensor requires only small bandwidth, but many sensors are required to collect full information for this analysis. Video data from cameras will be used for several tasks, for example, virtualization of the production process or information for machine control. *Machine control* could include domains such as production tasks performed by a robot or mobility support for autonomous driving vehicles. Robots will produce most goods in Industry 4.0 factories, and autonomous vehicles will transport equipment and goods. Both applications, robot control loop and mobility support, require very low latency as a failure could be expensive. The analysis for movements, trajectories, and control and steering data is computed at the MEC. Sensors integrated on robots and machines within the factory will collect various information on the status of various machines. This data will be used, for example, for predictive maintenance. The data could be collected at the factory and then sent to a third party for analysis and *predictive maintenance* services.

Technical requirements: Agriculture 4.0

In the following, we focus on four exemplary Agriculture 4.0 use cases: i) ground sensing, ii) farm photographing, iii) agricultural machinery automation, and iv) predictive maintenance. As in prior scenarios, we summarize the use cases in a Spider diagram in Fig. 1.18.

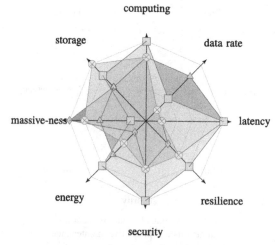

◇ ground sensing △ farm imaging
□ agricultural machine automation ⊗ predictive maintenance

FIGURE 1.18

5G Atom technical parameters and use cases: Agriculture 4.0.

For *ground sensing*, sensors are deployed over a wide area to provide information about current ground conditions, such as humidity. The sensors transmit their data, which are later analyzed to determine optimal soil conditions for growth. Another use case is *farm photographing*, whereby agricultural machinery and the farm are equipped with cameras to take photos and videos for several objects, such as plants and crops. Image recognition and processing can be combined with machine learning approaches to process the captured media, for example, to identify plant diseases or pests. *Agricultural machinery automation* considers, to a high degree, vehicular platooning in the context of harvesting. For instance, platooning can be used to enable an agricultural vehicle to follow a leading tractor, as illustrated in Fig. 1.12. Both vehicles could synchronize in terms of speed, position, and braking, employing methods related to connected autonomous vehicles described in Section 1.2.1.1. The idea of *predictive maintenance* is employing data measured about the agricultural machinery to identify the time of upcoming failures as closely as possible, thus enabling preventive actions before unplanned equipment outages. This approach promises to decrease machinery downtime and thus maintenance costs, when compared with time-based preventive maintenance.

Technical requirements: energy grids

Energy grids are a critical infrastructure where new communication technologies offer the opportunity for a lot of efficiency improvement, to reduce energy demands and waste and increase distribution efficiency. This vertical industry has three major use cases: smart home applications, virtual power plants, and fault detection. The technical requirements for the presented use cases for smart grids are depicted in the Spider diagram in Fig. 1.19.

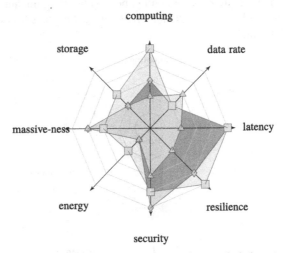

◇ self-organizing virtual power △ smart home □ fault detection

FIGURE 1.19

5G Atom technical parameters and use cases: Smart Energy Grids.

Smart home applications are already adopted widely, yet infrastructure to reduce and optimize energy consumption according to power production forecasts and energy pricing is not widespread. With decentralized self-coordinating cells this will change. Smart home devices will connect to the MEC and will be turned on or off, depending on the market price for energy, current local production, and demands.

Neither latency nor bandwidth is especially critical, as the devices are usually sluggish themselves. Small power generation facilities, such as photovoltaic power plants, wind turbines, and so on, produce energy in a decentralized system. However, currently this energy is injected into the main grid instead of satisfying the local power demand. A *self-organizing virtual power* plant establishes a communication between energy production and demand, balancing supply and demand in a way that power supply from the main grid is minimized. This virtual power plant can dynamically include new users depending on the optimal overall infrastructure, demand/production forecasts, and real-time power metering. For the smart grid backhaul and backbone domain, *fault detection* is a major use case. The communication network must provide secure, reliable, resilient, and latency-critical communication to automatically detect and react to faults in a highly distributed power generation scenario.

Technical requirements: tactile Internet

The Tactile Internet (TI) requires three modals for transmission, namely i) audio, ii) visual, and iii) haptic. Each modal has different technical requirements in terms of the transmission trough the network. The technical requirements are depicted in the Spider diagram in Fig. 1.20. However, the data can be transmitted compressed, which results in smaller data sizes, but increases latency. Uncompressed (raw) data transmission does not require time for coding, but results in significantly larger sizes to transport over networks. The determination of the right balance between these two extremes is a task for the network.

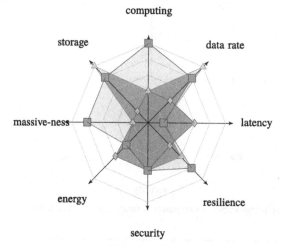

◇ human training △ skill transfer □ human–machine interaction

FIGURE 1.20

5G Atom technical parameters and use cases: Tactile Internet.

The TI has three major use cases: i) human training, ii) skill transfer, and iii) human–machine interaction. All use cases deal with the interaction of humans and machines. Teaching machines certain tasks by programming is a time-consuming and often error-prone activity. In the future, humans will teach these tasks to machines. The human will wear a suit, gloves, and similar devices carrying a multitude of different sensors for multimodal movement tracking. These tasks are learned by *human training*. Human movements are captured via a network of multimodal sensors and translated to machine code, teaching machines their required movements for their designated tasks. Additionally, human movement skills can be stored digitally with this technology. To become an expert in a certain field with a haptic skill, significant amounts of experience and real-world training are necessary. The Tactile Internet training systems will enable the possibility to train these skills in a VR or AR environment with multimodal feedback. The feedback is based on the captured and digitally stored human movement for a particular skill. This enables a *skill transfer* between humans. Besides capturing human expert skills to train others, it is also possible to transfer these skills to machines in order to fulfill human work, for instance, in dangerous environments or to assist humans. For *human–machine interaction*, robots need to understand human movements. This will enable humans and robots to work together and enable robots to assist humans in their daily lives. One possible application is a rehabilitation robot performing repetitive movements with patients for physical therapy.

1.2.3 Second tier: the concepts

Meeting each of the individual technical requirements we described in Section 1.2.1 for the *5G Atom*'s First Tier will be challenging. Unfortunately, meeting a subset of the aforementioned parameters jointly is even more difficult. For a fundamental example of this, we can consider optimizing for latency and throughput simultaneously, which results in the basic trade-off we illustrate in Fig. 1.21. Further detailing this example, we initially evaluate the throughput for different IEEE 802.11 standards that rely on Carrier-Sense Multiple Access (CSMA)/Collision Avoidance (CA) as a function of link-layer frame sizes. As illustrated in Fig. 1.21A for this evaluation, the throughput increases with larger packet sizes, independent of the employed standard. This is an intuitive result, as following the CSMA/CA algorithm provides only a limited time budget for accessing the channel. Assuming that there is always data to be transmitted, an increase of the packet size results in an increase of the system efficiency with respect to throughput (as more time is spent actually transmitting the data). In contrast, a larger packet size will also lead to a higher latency accessing the channel, as illustrated in Fig. 1.21B. As transmissions occupy the channel with larger packets and longer transmission times, the time waiting to access the channel similarly increases as well (as transmissions need to finish before the channel can be accessed). Although IEEE 802.11 technologies are not candidates for cellular access systems, this problem will be encountered for all wireless access systems.

The problem of joint optimization becomes even more evident if we aim for high throughput, low latency, and resilience. Consider a packet-oriented point-to-point communication system with a sender, a receiver, and an error-prone erasure channel.

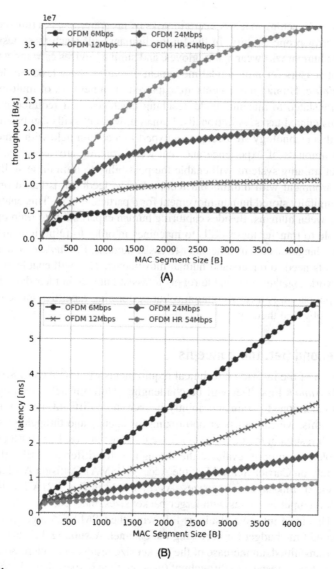

FIGURE 1.21

The *5G Atom* concepts: The trade-off between throughput and latency in an IEEE802.11 Orthogonal Frequency-Division Multiplexing (OFDM)-based system, without Request To Send (RTS)/Clear To Send (CTS). (A) Throughput in IEEE802.11 vs. MAC Protocol Data Unit (PDU); (B) Latency in IEEE802.11 vs. MAC PDU.

If we want to optimize for resilience, erroneous packets have to be either i) repeated by a given Automatic Repeat reQuest (ARQ) or ii) secured by Forward Error Correction (FEC) beforehand by redundancy packets. In the case of ARQ the added channel access and time needed to resend immediately reduces the attainable throughput and latency in case of errors. In case of FEC the ex ante resilience will be improved prior errors but at the cost of the throughput and latency through the added amounts of data. This foundational example immediately exposes the need for new concepts to realize combinations of the technical requirements. In the remainder of this section, we highlight these different concepts (with a more detailed discussion following in dedicated chapters), namely: New Air Interface (Section 1.2.3.1), Mesh (Section 1.2.3.2), Multipath Communication and Multiconnectivity (Section 1.2.3.3), Content Delivery Networks (Section 1.2.3.4), Information Centric Networks (Section 1.2.3.5), Network Slicing (Section 1.2.3.6), and Mobile Edge Cloud (Section 1.2.3.7).

1.2.3.1 New air interface concept

5G will provide a new air interface referred to as New Radio (NR) 5G. Previous generations focused on high data rates to satisfy the requests for bandwidth-demanding video services. However, 5G has also to satisfy other technical requirements, such as those discussed beforehand and listed in Table 1.1. The new air interface will provide improved latency conditions, higher throughput, support of massive numbers of communicating nodes, and improved security. The disruptive feature compared to other generations of the air interface is the low latency feature.

1.2.3.2 Mesh

It would not be possible to connect the expected massive number of communicating devices directly to base stations due to energy constraints. Wireless mesh networks have been extensively researched for several decades by now. Whereas this continuous research commonly focuses on the coverage extension through information relaying, newer research efforts shift the focus to resilient communication and to exploiting multipath communication (which we will addressed in Section 1.2.3.3). In Fig. 1.22 a game is depicted that demonstrates the benefits of the novel wireless mesh communication protocol *Meshmerize* over the well-known protocol *B.A.T.M.A.N.* [34] using multipath and single-path communication, respectively. Furthermore, as we will see in Section 1.2.5.4, there are huge advantages to preprocess sensor data with sensor networks or cooperative clusters to reduce the amount of data that would be necessary to convey it to a back-end cloud. Mesh architecture will be used for automotive use cases in the platooning or crossing scenarios. We would like to underline that mesh should not be seen as a threat to cellular communication, but more as an extension to reduce energy consumption, increase resilience, and provide lower delays in some situations. For example, in scenarios where 5G or any cellular connectivity might not be an option, local wireless mesh connectivity can be an enabler of next-generation services in a noncompetitive fashion, such as in remote Agriculture 4.0 settings.

1.2.3.3 Multipath communication and multiconnectivity

To increase the throughput and the resilience of the communication system, multipath and multiconnectivity are viable solutions. As described beforehand, Paul Baran

(A)

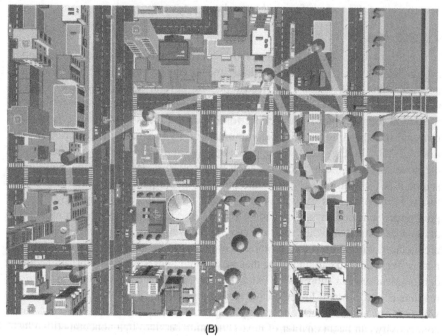

(B)

FIGURE 1.22

The *5G Atom* concepts: Meshmerize (multipath) vs. B.A.T.M.A.N. (single-path)
demonstrator. (A) 3D view; (B) Overview.

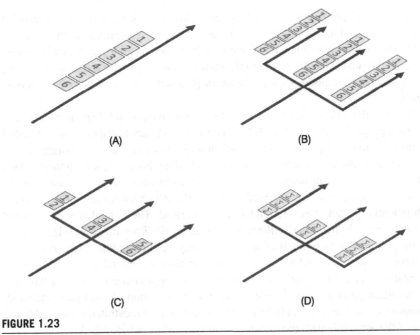

FIGURE 1.23

The *5G Atom* concepts: Multipath examples conveying six packets. (A) Single path; (B) Multipath repetition; (C) Multipath selection; (D) Multipath coding.

already exploited diversity in the communication path to increase resilience at the cost of throughput. Even though the idea never made it to the initial Internet, recently multipath communication is being discussed by the IETF for multipath TCP [35], and multiconnectivity is discussed by the 3GPP. The difference between multipath communication and multiconnectivity is the protocol layer of deployment. Multipath communication refers to the network layer and above, whereas multiconnectivity is more closely related to radio aspects.

In Fig. 1.23, different forms of multipath communication are illustrated. In Fig. 1.23A the state-of-the-art single-path communication is depicted to convey six communication packets. The system resources are used in a reasonable manner. In terms of security, a potential attacker simply needs to have access to the given path to overhear the communication. Furthermore, resilience is directly impacted by the single channel characteristics. As Paul Baran suggested, diversity can be exploited to increase the resilience by conveying the six packets repetitively over three channels at the same time, as shown in Fig. 1.23B. Compared to the single-path communication, the loss probability for a given packet decreases from p to p^3. This simple calculation holds if the channels are uncorrelated, which is more likely for the IP level than for radio channels. The drawback, as mentioned several times beforehand, is the loss of system capacity. Following our example, three times the original system resources are used to achieve the increase in resilience. Furthermore, the security level is also reduced, as a potential attacker needs to hack only into the weakest communication path to overhear it all. In Fig. 1.23C, channel diversity is exploited at the same capacity level as for the single path. Security is increased as all three channels need to be attacked to understand the whole information. But in this case the resilience is the lowest of all, as the poorest communication path will dominate the overall performance.

To break the well-known trade-off between throughput and resilience, coding is a viable option. As given in Fig. 1.23D, coded packets are transported over parallel channels. To maintain privacy, no channel should convey the full information. In that way, potential attackers need to hack into several (all in this example) channels, which requires more effort. Furthermore, to convey the information of the original six packets, only six packets need be received to decode the information successfully. To deal with potential packet losses, redundancy is generated. The redundancy per channel does not have to be the same as illustrated in Fig. 1.23D. Therefore the coded solution has the potential to be optimal in terms of latency, resilience, and efficient usage of resources. As described in [36], the coded multipath communication will have several advantages over existing multipath communication approaches in terms of throughput, resilience, and security. In Fig. 1.24B the combination of multipath and mesh communication systems in an Industry 4.0 environment is illustrated compared to the state-of-the-art single-path (centralized) communication in Fig. 1.24A.

1.2.3.4 Content delivery networks

A Content Delivery Network (CDN) refers to a set of globally distributed servers that helps to increase content delivery quality by storing copies of popular content near users likely interested in it. This approach is beneficial for network service providers, original content providers (e.g., YouTube and Netflix), and end users. Specifically, handling the requests by CDN servers increases content availability, reduces the volumes of traffic (both legitimate and malicious) that are forwarded to the origin content providers, and also enables the end users to receive content faster. A wide variety of CDN designs have been proposed over the last two decades. They mainly differ in their sizes and architectures, server placement algorithms, content selection algorithms, content-to-server assignments, and request routing protocols. Chapter 5 of this book provides more details on CDNs.

1.2.3.5 Information-centric networks

Information-Centric Networking (ICN) (Chapter 5 for more details) is a new networking paradigm that aims to improve both the efficiency of content distribution and network security. To this end, it proposes to start a fresh Internet design based on three concepts, namely: i) networking named contents, ii) content-based security, and iii) in-network caching. These concepts are the basis of remarkable ICN architectures, such as Named-Data Networking (NDN), Data-Oriented Network Architecture (DONA), Publish/Subscribe Internet Routing Paradigm (PSIRP), and Network of Information (NetInf). Among these architectures, NDN is vastly treated as a potential

(A)

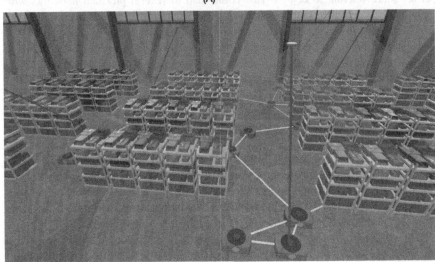

(B)

FIGURE 1.24

The *5G Atom* concepts: Combination of multipath and mesh communication system in an Industry 4.0 environment. (A) State of the art centralized networks; (B) Hybrid communication with centralized and mesh networks.

surrogate for the current Internet. NDN adjusts the classic Transmission Control Protocol / Internet Protocol (TCP/IP) model by three major changes. First, it places named contents in the middle (i.e., the thin-waste) of the model and moves the IP protocol below. Second, it adds a new layer dedicated to security functions just below the named contents. Third, it introduces another layer (termed the strategy layer) between the security layer and the underlying networking technologies (IP, User Datagram Protocol (UDP), etc.) to become responsible for forwarding and transporting. With these changes, content becomes independent of location, application, and means of transportation.

1.2.3.6 Network slicing

The network slicing concept (Chapter 3 for more details) was introduced to support a broad variety of networks with different characteristics in a spectrum-efficient manner. As described in the beginning of this section, realizing the whole *5G Atom* would require one network with low delay, high throughput, high resilience, and many more requirements. Achieving this goal would require a large frequency spectrum to support the maximum requirements simultaneously. Luckily, most services need no low latency, high resilience, and high throughput *at the same time*, as explained in Section 1.2.2.8. For example, most control applications need low latency, but do not require high throughput. On the other hand, video services demand high throughput but are more tolerant to delays. This could lead to several physical networks with individually determined Quality of Service (QoS) parameter sets. Unfortunately, the large heterogeneity in service requests renders such approach nonscalable. The vast range of potential service parameters does not allow for an a priori assignment of frequencies to a large number of physical networks. The solution to this problem is the network slicing concept, which will slice one physical network into several logical (virtual) networks, each with different QoS parameters. As resulting requirements, the whole network infrastructure needs to i) provide provisioning, ii) manage the association to slices, iii) offer interoperability, and iv) support performance and isolation.

Software-Defined Network (SDN) and Network Function Virtualization (NFV) play critical roles in realizing the concept of network slicing in 5G systems. SDN contributes a control plane, which has the complete view and control of network resources (such as network functions and computation infrastructure to quickly set up a configurable data plane on-demand to adapt to various requirements from applications). NFV provides tools to manage and orchestrate computation and storage resources needed to instantiate network functions. In Fig. 1.25 the dynamic assignment of slices is illustrated for two time instances. In Fig. 1.25A, two slices are created to support the connected driving for cars with latency requirements of 1 ms and 70 Mbit/s and multimedia services for people with latency requirements of 30 ms and 220 Mbit/s. In Fig. 1.25B the original slice resources are reassigned as an emergency requires significant resources to be guaranteed for public safety. As the overall available resources do not change, the resources of the two prior services need to be reconfigured with changes to their resulting characteristics. It is worth noting, however, that reallocation of resources needs to take the use case into account. In the presented example the impact on channel utilization and delay is more significant for the multimedia services than for the autonomous driving slice (which follows their intuitive importance). Furthermore, the new public safety slice is generated with

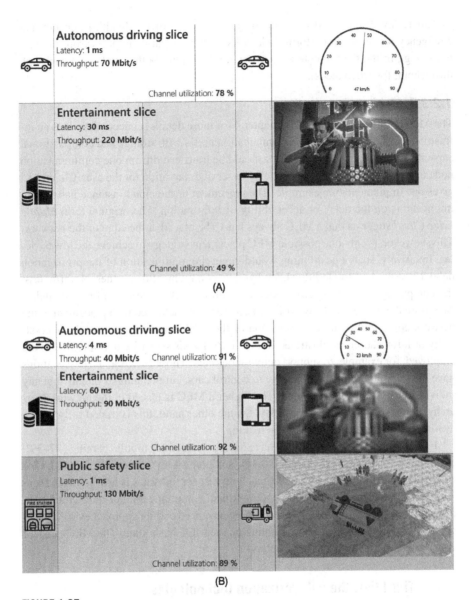

Autonomous driving slice
Latency: **1 ms**
Throughput: **70 Mbit/s**

Channel utilization: **78 %**

Entertainment slice
Latency: **30 ms**
Throughput: **220 Mbit/s**

Channel utilization: **49 %**

(A)

Autonomous driving slice
Latency: **4 ms**
Throughput: **40 Mbit/s**   Channel utilization: **91 %**

Entertainment slice
Latency: **60 ms**
Throughput: **90 Mbit/s**

Channel utilization: **92 %**

Public safety slice
Latency: **1 ms**
Throughput: **130 Mbit/s**

Channel utilization: **89 %**

(B)

FIGURE 1.25

The *5G Atom* concepts: Network slicing. (A) Two slices for 5G connected cars and 5G multimedia services. The slice for the cars offers 1 ms latency leading to a velocity of nearly 50 km/h. The multimedia slice offers 220 Mbit/s, leading to high-quality streaming video; (B) Third slice added in emergency case at the cost of the other two slices. The latency in the car slice drops to 4 ms, and the data rate of the multimedia slice drops to 90 Mbit/s, which in turn lead to half the velocity for the cars and low video streaming quality, respectively.

extremely low latency and generous overprovisioning of bandwidth to ensure that emergency services can perform their tasks without negative impacts from the network (e.g., in-the-field remote telesurgery might require both, low latency and high throughput, for video feeds).

1.2.3.7 Mobile edge cloud

The Mobile Edge Cloud (MEC) (Chapter 4 for more details) concept describes an infrastructure of one or multiple communication nodes with storage and computational capacities. Applications and services should be transferred from one communication node to another without noticeable delay or service starvation for the user. Chapter 15 gives one implementation example. The placement of the cloud instance has several implications on the delay or accessibility of information. The original term *Mobile Edge Cloud* suggests that a MEC always has to be placed at the edge of the network. This, however, is only one possible MEC placement strategy to achieve the lowest delay. Intuitively, such a positioning would target the minimization of the propagation delay between the Mobile Edge Cloud and its communication partner. Nevertheless, the computing delay is equally important with respect to the overall service delay. Therefore the minimum delay will not necessarily be achieved by a placement at the network edge. Furthermore the location of the MEC has an impact on the accessibility of information, as illustrated in Fig. 1.26. In this scenario, multiple MECs are employed, for example, to support autonomous cars. Those MECs at the base stations have the minimum propagation delay to control cars, but the number of cars that any individual MEC can *see* is limited. The higher a MEC is placed in the network hierarchy, the larger the propagation delay. On the other hand, this is traded off with the availability of more data for decision making.

First implementations of MECs have been presented for mobile gaming [37–39], control in industrial environments [40,41], and control for connected cars [42]. One of the most important research questions with respect to MECs is how to make them resilient against failures. In contrast to multipath communication or multicloud storage solutions, repetition coding or even more advanced coding approaches will not be applicable anymore if the service running on the MEC has a state. The MEC concept is described in greater detail in Chapter 4.

1.2.4 Third tier: the softwarization technologies

In the previous Section 1.2.3, we described some of the concepts needed for the fulfillment of the first tier of *5G Atom*. We now shift to the third tier of the *5G Atom*, which is the most important tier considering the theme of this book. As part of the third tier realizations, computing and storage will become integral parts of every communication network, also referred to as softwarization. This layer will have the largest impact on the well-established relationships between network operators, mobile manufacturers, and service providers.

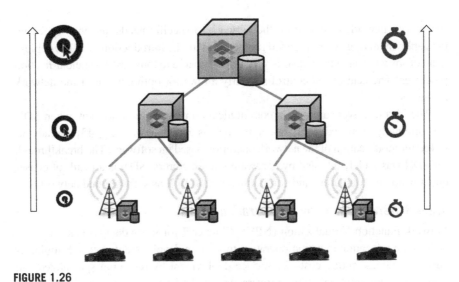

FIGURE 1.26

The *5G Atom* concepts: Hierarchical Mobile Edge Cloud (MEC) concept for 5G-connected cars with optimal placement for latency and cognition.

1.2.4.1 Software-defined radio

The Software-Defined Radio (SDR) was the first communication system entity for which hardware became enriched with software. In the beginning of SDR, two main concepts were considered, reuse of hardware components and flexible control of components. One example for the first concept is the reuse of Viterbi codecs for voice coding and channel coding purposes. An example for the second concept is baseband processing, where the characteristics of the sender and receiver can be changed on-the-fly. This concept was implemented in software rather than in hardware and nowadays allows for a high degree of flexibility in hardware components. A real first need for SDR appeared in the area of 3G, where globally different frequency bands were allocated. To limit the number of potential baseband configurations, SDR was used.

1.2.4.2 Software-defined networks

A Software-Defined Network (SDN) (Chapter 6 for more details) decouples the control plane of switches and routers from their data plane, enabling the control and orchestration of those devices from a central entity. A central (not necessarily one physical) SDN controller is in charge of one single network formed by several SDN switches on which softwarization takes place. It is interesting that the centralization approach impacts the communication network architecture, which subsequently has more similarities with the circuit-switched than with the packet-switched architecture. The SDN approach additionally advocates for centrally controlled network protocols replacing the current state-of-the-art distributed protocols. The centralized approach plus the softwarization of the network protocols make it easier to experiment with new ideas and adopt the network to specific needs and thus speed up the deployment of new or upgraded protocols. With the introduction of SDN, deployment of software and novel ideas is very fast compared to long-lasting standardization processes. The centralized control similarly allows for optimization of the network resources.

The idea of a software-driven communication network was not new when SDN was originally proposed. Early attempts, such as active networking [43], already allowed for modification of the network operation based on software. The breakthrough for SDN was widely adopted by hardware manufacturers. SDN is a result of consequent fusion of computing and communication, as well as software and networking.

1.2.4.3 Network function virtualization

Network Function Virtualization (NFV) (Chapter 7 for more details) is a direct request from telecommunication operators to shorten development cycles. Simultaneously, it enables cutting costs for service deployment by replacing specialized and static hardware solutions with software on standard hardware using virtualization concepts. The softwarization fosters quick deployments of new services, whereas the virtualization allows relocation, live migration, and upgrades and downgrades of services wherever and whenever they are needed. Furthermore, the softwarization will cut the cost of exchanging and maintaining new services, reducing the CAPital EXpenses (CAPEX) and OPerating EXpenses (OPEX) of network operators.

1.2.4.4 Service function chaining

Service Function Chaining (SFC) enables flexible and efficient deployment of network functions for different applications. With NFV, the elements of the chain can be provisioned in virtual environments on any commercial off-the-shelf hardware. SFC facilitates practical use cases that normally require a complete Network Service (NetServ) consisting of several Service Functions (SFs) in a specific order, for example, packets initially traversing a FireWall (FW) followed by a Deep Packet Inspection (DPI). NFV has to be capable of forcing packets to traverse through the different SFs in the predefined order. The traffic in an SF chain traverses between running SF-Interfaces, which are probably distributed over different physical compute nodes.

1.2.5 Fourth tier: innovation and novelties

The fourth tier of the *5G Atom* describes potential novelties and innovations that can be applied in the now softwarized networks. Commonly, solutions in softwarized networks are open-source projects. The added value provided typically either stems from testing and supporting software or original solutions that are applied into these novel networks. Here we introduce four potential innovation drivers, three of which we will employ in this book as examples. Any other innovation or novelty could be listed here as well, and we do not claim completeness in the list of novelties.

1.2.5.1 Block chaining

Block chaining is an interesting approach to enable security concepts in a distributed manner. Several aforementioned ideas, such as multipath communication, distributed storage, distributed SDN controllers, and others, will require new security concepts. Block chaining allows us to have contracts between different entities even in a distributed manner. Exploiting the computing capabilities in the network will allow for energy-efficient and fast block chaining solutions. In this book, we will not contribute any example of block chaining, but wanted to list it for completeness.

1.2.5.2 Machine learning

Due to the increased level of flexibility in the communication network, its optimization can neither be static nor be performed manually. Machine Learning (ML) (Chapter 8 for more details) is one interesting possibility to learn from previous events to optimize the communication network operation. This increased level of flexibility is a result of several changes in the communication network – we briefly highlighted a small subset of them. For example, with the introduction of network slicing, the number of logical networks will vary based on user demands, that is, become highly variable. Similarly, new services will be a superposition of several well-known services. As one example for these amalgamated services, the Tactile Internet will require video, audio, and haptic feedback information as discussed in Section 1.2.1.5.

1.2.5.3 Network coding

Network Coding (NC) (Chapter 9 for more details) is a new concept that breaks with the end-to-end communication paradigm and allows for distributed coding within the network. Network Coding requires computing within the network, as coding does not only happen at the end nodes. Every intermediate node is able to *recode* information, which requires computation capabilities within the network. As explained later in its dedicated chapter, NC will achieve the *min-cut max-flow* capacity of any given communication network. Network Coding is one example for the shift from *store-and-forward* concept to the *compute-and-forward* concept.

1.2.5.4 Compressed sensing

Compressed Sensing (Chapter 10 for more details) is a novel technique that exploits sparsity in the information to reduce the required bandwidth, compared to the Nyquist–Shannon sampling theorem. The approach is highly efficient if the compression is not only done at the end points, but also within the network (which again requires computing). Compressed sensing and network coding can be combined to reduce the amount of data transmitted in a wireless meshed sensor network [44, 45].

1.3 Softwarization: the game changer for network operators

With the introduction of softwarization into communication networks, network operators have the chance to regain the strength levels they had in the past. As explained beforehand, the OTT service providers were able to outplay the network operators for a long time, maneuvering them into a very difficult situation. The enabler for the OTT success has been the flexibility of the software on the mobile device and the cloud. The communication network, connecting those two, became the *dumb pipe* in the middle as given in Fig. 1.27.

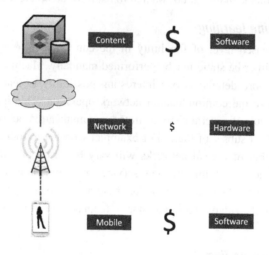

FIGURE 1.27

The *5G Atom* game changer: Softwarization in the cloud and in mobile devices.

The market potential was therefore in the hands of the OTT players such as Uber, AirBnB, WhatsApp, Twillio, and so on, creating the digital layer on top of the data transport services. Due to the pure existence of Application Programming Interfaces (APIs), it was easy for developers to create new services with little effort or entrance barriers in the cloud and as apps on the mobile device. The network operators tried to answer the lack of APIs by introducing IP Multimedia Subsystem (IMS) and several other market campaigns to introduce the *smart pipe*. Simply speaking, IMS was the attempt to give developers APIs to the communication network. But this attempt did not lead to any success, as those APIs were not open to every developer. An additional hurdle was that compared to device or cloud APIs, the network-centric APIs were very hard to use. The other winners were the hardware manufacturers of the mobile devices, such as Amazon, Apple, or Samsung. Network equipment manufacturers, on the other side, had a tough time with less market potential.

The architecture given in Fig. 1.27 will evolve into the more advanced architecture illustrated in Fig. 1.28. First, the back-end cloud will be realized in a more distributed fashion to respond to regional law requirements and to increase resilience in data storage and computing. Furthermore, computing will now be integrated into the communication network with access for 3rd party developers to imagine use cases that have not even been named within this book chapter. The placement of computing at a given place will have impact on the security level, the resilience, and the latency of a given application. Even at the end points, there will be not just a single-user device, but possibly also a cooperative communication cluster (like a platoon of vehicles on the highway or massive amounts of IoT devices) that will be directly served from the communication network. As presented in Fig. 1.28, the computing could be also extended to the user end device, coming back to the example of controlling a vehicle platoon. With this new architecture, two disruptive features will become integrated into future communication systems, namely i) the introduction of *computing within the network* and ii) the extension of point-to-point communication to *more advanced communication architectures*.

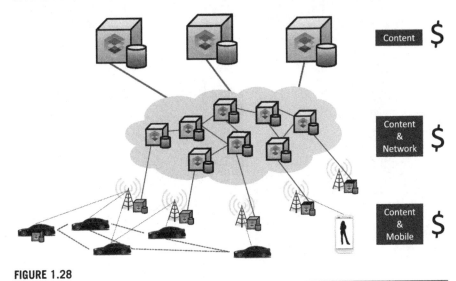

FIGURE 1.28

The *5G Atom* game changer: Softwarization future.

Coming back to the quote in the beginning of this chapter, the challenges and opportunities for service providers, network operators, and network and device manufacturers are the upcoming changes in the overall communication business. Network operators have enormous chances to improve their position due to the massive number of sites they have for their communication equipment, especially for low-latency applications. Nevertheless, other players have a deep understanding of how to make software products that will dominate these new communication services. Other dynamics, such as campus or small cell solutions, allow service providers to become network operators in small regions. Whoever will manage this change, the best will be the winner of the 5G wave – as Charles Darwin predicted.

Standardization activities for future communication networks

2

Fabrizio Granelli[a], Patrick Seeling[b], Frank H.P. Fitzek[c], Riccardo Bassoli[c]

[a]*University of Trento, Trento, Italy*
[b]*Central Michigan University, Mount Pleasant, MI, United States*
[c]*Technische Universität Dresden, Dresden, Germany*

> *Just as modern mass production requires the standardization of commodities, so the social process requires standardization of man, and this standardization is called equality....*
> **Erich Fromm**

2.1 Introduction

To discuss standardization for future generation networks, it is important to understand the intrinsic meaning of standardization, why standardization is important, and the societal/economic/technological impacts that standardization activities have on society. The Oxford English Dictionary defines *standard* (initially appeared in 1154) as a *flag, sculptured figure, or other conspicuous object, raised on a pole to indicate the rallying point of an army (or fleet), or of one of its component portions; the distinctive ensign of a king, great noble, or commander, or of a nation or city*. Moreover, its etymology comes from the old French word *estandard*, which meant stable or fixed, because in the Middle Ages, it was fixed in the ground. Next, a definition of standard, which dates back to 1429, states: *the authorized exemplar of a unit of measure or weight; e.g. a measuring rod of unit length; a vessel of unit capacity, or a mass of metal of unit weight, preserved in the custody of public officers as a permanent evidence of the legally prescribed magnitude of the unit*.

Side by side, the derived verb *to standardize* (initially appeared in 1873) means *to bring to a standard or uniform size, strength, form of construction, proportion of ingredients, or the like* (according to Oxford English Dictionary). After that, nowadays, the most common definition of the word standardization is *the process of implementing and developing technical standards based on the consensus of different parties that include firms, users, interest groups, standards organizations and governments* (Wikipedia). According to this definition, standardization helps to achieve compatibility, interoperability, safety, repeatability, quality, and commoditization of

Computing in Communication Networks. https://doi.org/10.1016/B978-0-12-820488-7.00012-8

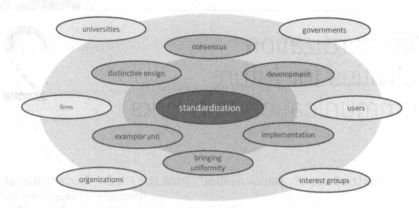

FIGURE 2.1

Main aspects behind standardization (internal ring) and main entities currently involved in standardization activities (external ring).

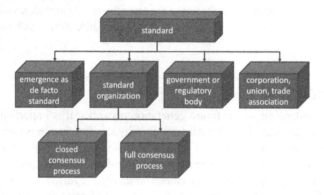

FIGURE 2.2

Different ways to realize a standard.

processes. In fact, in social sciences and economics, standardization represents an optimal solution to the coordination problem. Fig. 2.1 summarizes all the concepts, expressed by the above definitions. The standard represents a fundamental reference (distinctive ensign) and an exemplar unit, and it brings uniformity for customized massive implementation and development of something. Moreover, standards come out from consensus of groups of people. Without loss of generality, the typical protagonists of standardization activities can belong to academia, industry, and interest groups, or they can also be governmental bodies, standards organizations, and end users.

There are different ways for a standard realization, as illustrated in Fig. 2.2. First, a thing can become very popular or part of traditions so that the community decides that it represents a standard. In particular, there are things that pass from *de facto* to *de jure* standards, whereas others remain just *de facto*, thus not becoming part of the legal corpus. Some examples of the former are the QWERTY/QWERTZ keyboard, the MP3 digital audio format, or the HTML and the PDF file formats. On the other hand, examples of the latter are the .DOC and .TEX formats for documents, the phone connector (3.5-mm jack), and the proprietary audio/video interface High Definition Multimedia Interface (HDMI). Second, a thing can become a standard after being discussed and approved within standards organizations. The produced standard can come out after a process involving either a limited number of members (closed consensus) or all the members of the organization (full consensus). In general, a standards organization is composed of working groups (WGs) of experts, who prepare working drafts. Moreover, subcommittees may have several working groups, which can consist of subgroups (SGs). Next, international standards are also developed by tenant controllers (TCs) and subcommittees (SCs) normally following a customized process. Proposal and preparatory stages are the starting point of discussion. Next, committee and inquiry stages develop the draft of the document stating the standard, which subsequently passes through approval and publication stages. Finally, after official publication, the document is periodically reviewed, and once the standard becomes obsolete, it is withdrawn. Some examples of this kind of standard are IEEE 802.11 (physical layer protocols for wireless local area network), IEEE 802.16 (wireless broadband communications), and Long-Term Evolution (LTE) (wireless broadband communication for mobile devices and data terminals). Next, a standard can also be written by a government or governmental regulatory body. This is the case for the first international environmental management standard (now ISO 14001), which was released by British Standards Institution (BSI) in 1992 in the document BS 7750. Other examples are the standards produced by Deutsches Institut für Normung (DIN) referred to international paper sizes (now ISO 216), originally called DIN 476, and referred to regulating typeface used by German railways and on traffic signs (DIN 1451).

Finally, a standard can also be defined by a private entity, such as a corporation, a union, or a trade association. That is what happened, for example, with Bluetooth, originally developed in 1989 at Ericsson Mobile in Sweden. The process of standardization is very helpful, but it also has some significant drawbacks that are meaningful to highlight. Three main perspectives could be chosen for this discussion, such as the firms', the consumers', and the technology/innovation point of view. In case of companies, once a thing is standardized, the competition among the parties shifts from the overall system to its internal individual components. If this shift occurs before standardization, a company's design and production approaches have more freedom (but products suffer from incompatibility), whereas after standardization, processes at each company focus on providing an individual component that is compatible with those from competitors. Next, standardization makes design and production of products shifting to a modular approach, with the objective of supplying other companies with subsystems or components. Shifting to the consumers' perspective, standards have the advantage of increasing compatibility and interoperability between products, allowing information to be shared within a larger network, and attracting more consumers to use the new technology. Moreover, standardization reduces uncertainty, because consumers have more warranties on products. Finally, end users can also benefit from being able to mix and match components of a standardized system to align it with their needs. On the other hand, the process of standardization forces a lack of variety. There is no guarantee that the chosen standard is capable to satisfy all consumers' needs or even that the defined standard is the best solution. In turn, consumers must adapt to the conditions made available by the products. Next, if a standard is published before products are in the market, then consumers are deprived of the penetration pricing, which often results from rivals' competition. Finally, a consumer can still choose a product that is based on a standard that fails to become dominant, which results in spending resources on a product that becomes less useful or, worse, out of the market.

Standardization also has significant impact on innovation and technology. It can be a useful platform to transfer knowledge and to translate it into policy measures. Moreover, the adoption of a new standardized technology can avoid the competition of rival and incompatible solutions in the marketplace, which can slow or even stop the growth of a technology (so-called market fragmentation). On the other hand, the publication of standards restricts technological innovation. Furthermore, it shift competition from design of new technological features to just variation of price because characteristics of products are defined by the standard. Finally, standardization also rules out alternative technologies while enforcing the adoption of those following a specific standard.

2.2 Standardization in telecommunications

Major international standardization bodies in telecommunications and networking are described in the following:

International Organization for Standardization (ISO) The International Organization for Standardization was founded in 1947, and it has its headquarter in Geneva, Switzerland. It is an international standardization body composed of representatives from various national standard organizations. The members of the ISO are in 164 countries. This standards organization was responsible to publish the Open Systems Interconnection (OSI) model, which is a conceptual model that characterizes and standardizes the communication functions of a telecommunication or computing system without regard to its underlying internal structure and technology. The OSI model was defined in the document ISO/IEC 7498.

Institute of Electrical and Electronics Engineers (IEEE) The Institute of Electrical and Electronics Engineers was formed in 1963 from the unification between the American Institute of Electrical Engineers and the Institute of Radio Engineers. It is a professional association for electronic engineering and electrical engineering (and associated disciplines) with its corporate office in New

York City and its operations center in Piscataway, New Jersey. As of 2018, it is the world's largest association of technical professionals with more than 423 thousand members in over 160 countries around the world. The IEEE is composed of societies related to different research areas such as the IEEE Communications Society, the IEEE Computer Society, the IEEE Aerospace and Electronic Systems Society, and so on. This organization was responsible for publishing famous standards such as IEEE 802.11 for Wireless Local Area Network (WLAN), IEEE 802.3 (defining the physical layer and medium access characteristics of wired Ethernet), and IEEE 802.16 (for Wireless Wide Area Networks, so-called WiMAX).

3rd Generation Partnership Project (3GPP)   The 3GPP unites seven telecommunication standard development organizations, namely: Association of Radio Industries and Businesses (ARIB), Alliance for Telecommunications Industry Solutions (ATIS), China Communications Standards Association (CCSA), European Telecommunications Standards Institute (ETSI), Telecommunications Standards Development Society India (TSDSI), Telecommunications Technology Association (TTA), and Telecommunication Technology Committee (TTC). Jointly, they are referred to as *Organizational Partners*. 3GPP paves the ground to produce reports and specifications that define 3GPP technologies. 3GPP covers cellular and mobile telecommunication technologies, including radio access, core network, and service capabilities. The three Technical Specification Groups (TSG) in 3GPP are i) RAN, ii) Services and Systems Aspects (SA), and iii) Core Network and Terminals (CT). 3GPP is the main driver for the wireless 5G standardization process with the current Release 15/16/17.

European Telecommunications Standards Institute (ETSI)   ETSI is a European Standards Organization (ESO). It represents the recognized regional standard bodies dealing with telecommunications, broadcasting, and other electronic communications networks and services. ETSI partners with 3GPP to develop 4G and 5G mobile communication systems.

ITU Telecommunication Standardization Sector (ITU-T)   The mission of the ITU-T is to ensure the efficient and timely production of standards covering all fields of telecommunications and ICT on a worldwide basis and defining tariff and accounting principles for international telecommunication services.

Internet Engineering Task Force (IETF)   The mission of the IETF is to make the Internet work better by producing high-quality, relevant technical documents that influence the way people design, use, and manage the Internet. The IETF currently is the main driver for computing elements in communication networks through their standardization activities on SDNs and NFV. Outcomes of the IETF activities are often incorporated by 3GPP, and there are bilateral meetings between those two bodies.

Internet Research Task Force (IRTF)   The Internet Research Task Force (IRTF) focuses on longer-term research issues related to the Internet, whereas its parallel organization IETF focuses on the shorter-term issues of engineering and standards development.

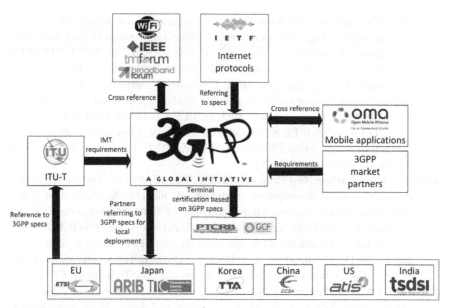

FIGURE 2.3

Taxonomy of 5G standardization activities.

2.3 Standardization of future generation networks

The role of standards organizations in the standardization activities of future generation networks was not decided by chance or during a meeting. However, it resulted from the outcome of the commercial battle between standards organizations (especially, 3GPP and IEEE) to impose their respective standards as 4G networks. So, before discussing ongoing standardization activities for future generation networks, it is important to introduce the premises that depicted existing standardization roles. Between 2006 and 2012, the candidate technologies for 4G networks were LTE (from 3GPP) and WiMAX (from IEEE). Even if WiMAX was originally proposed for fixed wireless connections, between 2007 and 2009 mobile WiMAX was released offering comparable specifications to 3GPP High Speed Packet Access (HSPA) and LTE. Nevertheless, the growing commercial adoption of 3GPP standards in 4G equipment (both from end users and network provider side) made WiMAX more and more obsolete and reduced the role of the IEEE in standardization efforts for future generations of wireless cellular networks. Fig. 2.3 depicts the current organization of standardization activities for future generation networks. From this figure it is possible to derive the current dominant role of 3GPP in comparison to the IEEE in ongoing standardization efforts.

The intrinsic characteristics of future generation networks are very different from current and previous generations. In fact, future generation networks not only include cellular networks, but are comprised of all wired and wireless technologies. This clearly emerges from the illustration in Fig. 2.3, taken from [46], where the standardization process is a coral work of many standardization bodies worldwide, with the 3GPP coordinating and managing the individual activities. Standardization bodies operating in areas related to future networks will be forced to collaborate for the first time to build a new softwarized communication ecosystem. In such ecosystem the central position of network virtualization makes computing a fundamental aspect to be considered in standardization as well. Deployment of computing functionalities in communication networks requires the availability of standardized interfaces to enable proper control of the virtualized and softwarized infrastructure. The fact that different entities from different owners should interface with softwarization and virtualization functionalities clearly indicates the need for standards. Nevertheless, today, network virtualization and softwarization are yet under research and development, and no single standard exists that covers all the functionalities and interfaces required to build an effective programmable and softwarized network infrastructure. The introduction of disruptive architectures for deploying computing within communication networks is expected to open new opportunities. Achieving this goal will be a big step forward in the evolution of the telecommunication infrastructure. Several actors will be involved in different domains, and their interactions will shape the architectures and protocols employed throughout the networks of the future. Softwarization and virtualization represent key technologies to integrate into mobile networks and wired networks. The following subsections describe the ongoing standardization efforts in the field of all elements of the *5G Atom* presented in Section 1.2, which represent proper references for the management and development of novel solutions in the field of programmability and softwarization of network infrastructures.

2.3.1 3GPP standardization

3GPP is deeply involved in the standardization activities related to computing in communication systems. Indeed, the fifth generation of mobile networks is expected to provide an architecture to enable virtualization and slicing to support external tenants to operate and deploy services on the underlying network infrastructure. The 5G system architecture specified by 3GPP and described in [47] is designed to address a wide set of use cases, which can be typically clustered into three groups: i) eMBB, ii) Machine-Type-Communication (MTC), and iii) Ultra-Reliable Low Latency Communications (URLLC). As discussed in the first chapter of this book, supporting all use cases with a common architecture would require significant changes in design philosophies, both for the RAN and the Core Network (CN). In the 5G system specification, there are two options available for the architecture: one with the traditional reference point and interface approach (which represents an evolution of 4G LTE standard IP architecture) and the other where the core network functions interact with each other using a Service-Based Architecture (SBA). Indeed, the SBA represents a big step forward in the virtualization and softwarization of the architecture.

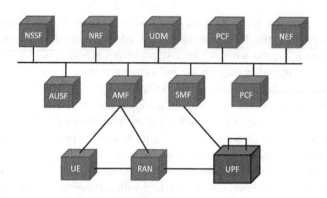

FIGURE 2.4

The 3GPP 5G Service-Based Architecture (SBA).

Details on the SBA option of the 5G system architecture are provided in a white paper by ETSI in [48]. From a general point of view, the SBA framework is built around functions that consume services and/or produce services. Any network function can offer one or more services. The SBA framework provides the necessary functionality to authenticate the consumer and to authorize its service requests, as well as flexible procedures to efficiently expose and consume services. For simple service or information requests, a request-response model can be used. For any long-lived processes, the framework also supports a subscribe–notify model. The 5G SBA with its individual elements is illustrated in Fig. 2.4.

The Network Functions (NFs) and their services are registered in a Network Resource Function (NRF). In MultiService Edge Clouds (MSECs), on the other hand, the services produced by the MSEC applications are registered in the service registry of the MSEC platform. The 5G Network Exposure Function (NEF) acts as a centralized point for service exposure and has a key role in authorizing all access requests originating from outside of the system, too. One of the key concepts in 5G is Network Slicing, which allows the allocation of the required resources from the available network functions to different services or to tenants that are using the services. The Network Slice Selection Function (NSSF) is the function that assists in the selection of suitable network slice instances for users and things and in the allocation of the necessary Access Management Functions (AMF). An application hosted in the distributed cloud of a MSEC system can belong to one or more network slices that have been configured in the 5G core network. The Unified Data Management (UDM) function is responsible for generating the 3GPP Authentication and Key Agreement (AKA) authentication credentials, handling user identification related information, managing access authorization (e.g., roaming restrictions), registering the user serving NFs through serving AMF and Session Management Function (SMF), supporting service continuity by keeping record of SMF/Data Network Name (DNN) assignments, and performing subscription management procedures.

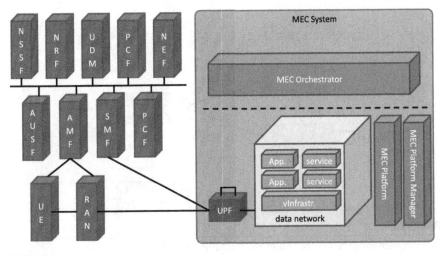

FIGURE 2.5

MSEC deployment in the 5G architecture as proposed by ETSI.

The User Plane Function (UPF) has a key role in service deployments in a 5G network. A UPF can be seen as a distributed and configurable data plane from the service perspective. The Policy Control Function (PCF) provides policy rules for control plane functions, including network slicing, roaming, and mobility management. It is similar to the Policy and Charging Rules Function (PCRF) in 4G networks. The control of that data plane, that is, the traffic rules configuration, now follows the NEF-PCF-SMF route. Consequently, in some specific deployments, the local UPF may even be part of the service implementation.

Basically, the architecture illustrated in Fig. 2.4 enables the deployment of services in different locations between a Base Station (BS) and a remote data center. Nevertheless, all deployments have in common that the UPF is used to steer the traffic toward the targeted applications and networks. It should be noted that the service management system, which orchestrates the operation of service hosts and applications, may decide dynamically where to deploy its applications and services. As an example, Fig. 2.5 illustrates how a MSEC application could be deployed by exploiting the 5G SBA.

3GPP TR 28.801 [49] describes an information model where a network slice contains one or more network slice subnets. Each of the subnets in turn contains one or more network functions and can also contain other network slice subnets. These network functions can be managed as VNFs and/or PNFs. An NFV Network Slicing (NS) can thus be regarded as a resource-centric view of a network slice, for the cases where a Network Slice Instance (NSI) would contain at least one virtualized network function. Fig. 2.6 illustrates this relationship.

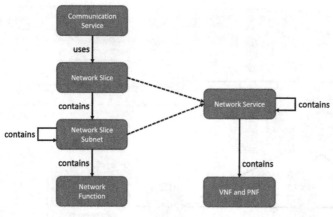

FIGURE 2.6

Network slice instance information model by 3GPP (TR28.801).

2.3.2 ETSI standardization

As mentioned before, ETSI collaborates with 3GPP on the standardization efforts for future generation mobile networks. In this framework, ETSI started an Industry Specification Group (ISG) focused on NFV. The ETSI ISG on NFV report *Network Functions Virtualisation (NFV) Release 3; Evolution and Ecosystem; Report on Network Slicing Support with ETSI NFV Architecture Framework* [50] analyzes use cases related to network slicing, and how these use cases could be mapped to NFV concepts and supported by the ETSI Network Functions Virtualization Architectural Framework (NFV-AF) architectural framework [51] and by Network Functions Virtualization MANagement and Orchestration (NFV-MANO) [52]. The overall architecture is illustrated in Fig. 2.7, which includes bindings to 3GPP functionalities as well.

To properly interface with the NFV-MANO, the Network Slice Management Function (NSMF) and/or Network Slice Subnet Management Function (NSSMF) need to determine the type of NS or set of NSs, Virtual Network Function (VNF), and Physical Network Function (PNF), which can support the resource requirements for an NSI or a Network Slice Subnet Instance (NSSI). Additionally, NSMFs and/or NSSMFs need to determine whether new instances of these NSs, VNFs, and the connectivity to the PNFs need to be created or whether existing instances can be reused.

2.3.3 ITU-T standardization

ITU-T Study Group 13 (SG13) created a focus group with a mandate to research the areas that needed standardization for the nonradio aspects of 5G. The focus group addressed the operation through software control, referred to as *softwarization* of all of the components of the 5G network, which is now being more formally considered by SG13. Many of the areas involved in the softwarization process are not

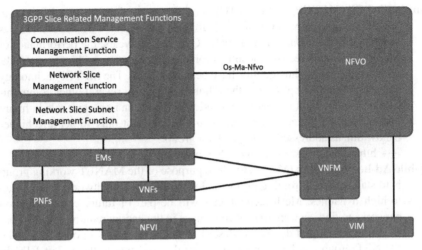

FIGURE 2.7

Network slice management in the ETSI NFV model.

uniquely wireless components, but are also involved in the service providers' other end-to-end-businesses. For example, each slice's interconnecting cloud and transport networks will require new forms of control to ensure that the packet interconnections and computing resources will be adequately dimensioned to meet the Quality of Service (QoS) demands. ITU-T SG13 supports the entire ecosystem of 5G slicing technology to foster not only standardization of the slice management system, but also standardization of the corresponding end-to-end applications and services. Given this broad task, the group continuously produces rapports on their current progress and meeting outcomes.

2.3.4 IETF/IRTF standardization

The IETF represents the main standardization body-generating protocols and architectures to be used on the Internet, and it is the de facto body driving the evolution of the Internet. Its main focus is on protocol layers higher than the physical layer. The philosophy applied by these bodies is to *build stuff and test it out*. Several working groups are currently operating to define the future of the Internet. Some of those working group, listed in the following, are aligned to the points mentioned beforehand in Chapter 1.

Software-Defined Networking (SDNRG)   The SDNRG defines several RFCs [53, 54] to build efficent solutions for SDN networks. Furthermore, the benchmark methodology for SDN controller performance is addressed in [55]. The working group is quite active, and it can be regarded as the main driver in the field of SDNs.

⟶ https://datatracker.ietf.org/rg/sdnrg/about/

Multipath TCP (MPTCP)   The MPTCP working group develops mechanisms to exploit multiple paths simultaneously to form one regular TCP session [19], as suggested by Paul Baran in the 1960s. Over the last years, the performance of MPTCP has increased significantly. Problems, such as the combination of links with heterogeneous link quality, have been addressed. The MPTCP technology has been initially employed in the iPhone and later in several other communication systems. Multipath communication will have an important role in future communication systems, not only with TCP and not only to achieve higher bandwidth, as addressed in the working group.

→ https://datatracker.ietf.org/wg/mptcp/about/

Mobile Ad-hoc Networks (MANET)   The purpose of the MANET working group is to standardize IP routing protocols for wireless mesh networks from no up to high dynamics. Mesh technologies will be part of future communication networks to find answers to new challenges in the energy required due to massive numbers of end devices as described beforehand in Section 1.2.3.2. The MANET group has been active for several years now with its first RFC in 1999 [56], and it is still very active.

→ https://datatracker.ietf.org/wg/manet/about/

Whereas the IETF considers actual problems in the Internet, the IRTF looks into disruptive technologies that might be interesting in the future. Here, we list those that have a clear relationship to this book as a whole or for individual chapters.

Computing in the Network Proposed Research Group (COINRG)   COINRG carries out research on programmable communication networks to implement network functions for improved Internet performance. This research group is rather young but draws its motivation from the same reasoning as this book. First proposals have been submitted to the working group.

→ https://datatracker.ietf.org/rg/coinrg/about/

Decentralized Internet Infrastructure (DINRG)   DINRG reflects the paradigm shift in the Internet from distributed and decentralized systems (packet switched) to centralized and hierarchical systems (computing centric) as introduced in Chapter 1. One of the research fields is the application of Inter-Planetary File System (IPFS) to respond to the paradigm shift.

→ https://datatracker.ietf.org/rg/dinrg/about/

Information-Centric Networking Research Group (ICNRG)   ICNRG proposes communication networks with access to data by name, regardless of origin server location as introduced in Chapter 5. This should overcome the famous 404 web error message and also allow for more efficient and resource-saving content delivery.

→ https://datatracker.ietf.org/rg/icnrg/about/

Coding for efficient NetWork Communications Research Group (NWCRG)
NWCRG applies network coding principles and methods as introduced in Chapter 9, which can benefit Internet communication in general. The research group has already completed several RFCs for efficient transport and low-

latency transport. Network coding is one of the technologies that needs to be deployed within the network rather than in the end points and will be one of the technologies benefiting from the openness of the communication network due to computing capabilities at each node.

$\longrightarrow$ https://datatracker.ietf.org/rg/nwcrg/about/

Currently, new research groups are proposed, such as the Proposed Network Machine Learning Research Group (NMLRG), which we describe in Chapter 8. Other aspects, such as network slicing, are currently discussed in several groups, for example, the Link State Routing Working Group (LSRWG).

Concepts

Outline

We continue from the motivation in Part 1 to a discussion of the important and foundational concepts for future communication systems in greater detail, namely network slicing, mobile edge cloud, and content distribution together with information-centric networks. We dedicate a separate chapter to each of these overarching concepts, as they are typically lesser known than other concepts, such as air interface, mesh, or multipath, which we briefly highlighted beforehand.

Network slicing

3

Fabrizio Granelli
University of Trento, Trento, Italy

In the earliest days, this was a project I worked on with great passion because I wanted to solve the Defense Department's problem: it did not want proprietary networking and it didn't want to be confined to a single network technology....

Vint Cerf

3.1 Introduction

The majority of today's online services converge toward provisioning through packet-switched networks based on the IP and connect, in most cases, through the public Internet. Indeed, the Internet represents the de facto standard platform for provisioning all types of services worldwide across a plethora of devices and connections. However, due to its nature and design features, the Internet basically offers a *best-effort service*, without any guarantees on timeliness or delivery of data. Subsequently, it is commonly not capable of supporting QoS, especially on the complete end-to-end path between a service producer and its consumer. Several solutions are available in the literature and in actual protocol and architecture specifications (e.g., IntServ and DiffServ architectures; see, e.g., [57] for a brief introduction). However, these mechanisms are typically only applied in isolated regions of interconnected networks, such as within individual Autonomous Systems (ASs). The overall consideration of end-to-end QoS is still limited by the need to introduce relevant modifications to the entire Internet architecture and its protocols. For this reason, deployment of those technologies is severely limited. In addition, the services expected to grow and gain support by IP-based networks are increasingly requiring a different balance among their requirements. For example, some services might require high throughput, some might require high reliability, and other services might require low latency. It is well known that a single architecture or configuration is not capable of supporting all such requirements at the same time, since it would need to balance among diverging solutions. On the other hand, applications and services are continuously evolving, and it is extremely difficult to foresee future service requirements.

The need to support different services and to enable rapid deployment of the related network configurations led to the paradigm of network slicing. A network slice

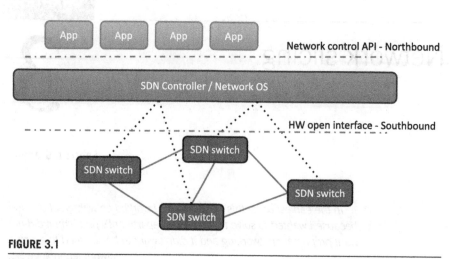

FIGURE 3.1

Illustration of the SDN paradigm.

is an isolated end-to-end network tailored to fulfill diverse requirements requested by a particular application. Such a network is typically built by reorganizing a subset of resources (e.g., communication, switching, computation, or storage) from the underlying network infrastructure. Historically, the concept of a network slice was introduced in the late 1980s with the concept of Overlay Networks. Overlay Networks provided the first example of network slicing, as they combined heterogeneous network resources to create virtual networks over a common infrastructure. Overlay Networks evolved to the definition of Virtual Local Area Networks (VLANs) [58] or more modern Virtual eXtensible Local Area Networks (VXLANs) [59]. Nevertheless, Overlay Networks lacked the feature of programmability. A step forward in the development of the concept of network slicing was introduced by Planet-Lab [60], which introduced a virtualization approach capable of allowing its users to create network functions to program their slices. Today, with the introduction of network virtualization and resource abstraction, the scenario seems to have matured enough for the introduction of real-world programmable network slices. Indeed, Fig. 3.1 illustrates how the introduction of the SDN paradigm enables virtualization and abstraction of resources: the SDN controller (also more generally defined as Network Operating System) provides the interface for applications to use the network resources. In particular, the SDN controller can abstract the physical topology and resources of the network and provide virtualized and abstract *visions* of the network through the Southbound interface to applications. In this chapter, we i) analyze the basic concepts related to network slicing and its architecture, ii) provide some information about the different scopes of network slicing, and iii) introduce reference architectures to generate, maintain, and reallocate network slices.

3.2 Network slice: concept and life cycle

Network slicing enables the definition of multiple virtual networks on a single physical networking infrastructure. A network slice represents an independent virtualized
instance defined by allocation of a subset of the available network resources. Typically, network slices are tailored to meet specific requirements of a set of applications
and services. Network slicing consists of defining an isolated subset of the available
virtual resources (computation, networking, storage) and a set of rules for identifying
the traffic that will run on those. A network slice consists of a set of virtual resources
and the traffic flows associated with it. A network slice can be defined by slicing the
available resources in the forms of:

Bandwidth: Each slice should have its own fraction of bandwidth on a link.
Topology: Each slice should have its own view of network nodes (switches, routers)
     and the connectivity between them.
Device CPU: Each slice should be assigned proper computational resources.
Storage: Each slice might have varying levels of storage capacity.
Forwarding tables and other control plane resources: The forwarding tables and
     other control plane functionalities should be sliced as well.
Traffic: A specific portion of the traffic to one (or more) virtual networks should
     be associated with a slice in order to be cleanly isolated from the remaining
     underlying network.

A network slice is viewed as a logical end-to-end network that can be dynamically
created. A given end user or Internet host may get access to multiple slices over the
same shared infrastructure depending on the needs of his/her services. The end-to-end
architecture enabling network slicing is built upon several key domains. On top of a
shared infrastructure, network slicing for logical networks is defined by means of
allocations of core network applications and by means of the RAN and transport
network partitioning. It is possible to define *partial* network slices, which are not
end-to-end. Those represent the same functionalities as end-to-end network slices,
but they are limited to a reduced scope. For example, it would be possible to define
a slice of the Radio Access Network – and not of the whole cellular network. An
effective procedure to build and manage network slices is to leverage the principles
of NFV and SDN, described in the corresponding chapters of this book. In brief, the
combination of the abstraction possible through SDN with the freedom in deployment
of functionalities deriving from NFV allow the proper level of control on the network,
computation, and storage resources to build and manage slices.

Fig. 3.2 illustrates an example of the above concepts in the framework of
an SDN/NFV scenario. The figure outlines the three core layers in the network slicing
paradigm:

Service Instance Layer: The Service Instance Layer hosts the services or applica-
     tions provided to the end user.
Network Slice Instance Layer: A Network Slice Instance represents a collection
     of resources from the layer below to form a network slice.

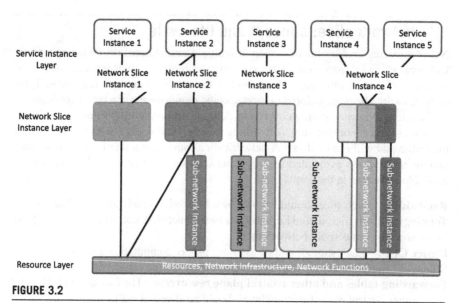

FIGURE 3.2

Illustration of the concept of network slicing.

Resource Layer: The Resource Layer hosts different subnetwork instances. Each subnetwork instance represents a (network, computation, storage) resource, typically, as one or a group of Virtual Network Functions (VNFs) or Physical Network Functions (PNFs).

A Network Slice Instance (NSI) may be composed by none, one, or more NSSIs, which may be shared by another NSI. Similarly, the NSSI is formed of a set of Network Functions, which can be either VNFs or PNFs. A communication service typically uses one NSI. The network slice controller is defined as a network orchestrator, which interfaces with various functionalities performed by each layer to coherently manage each slice request. It enables an efficient and flexible slice creation that can be reconfigured during its life cycle. As we discuss later in this chapter, the complexity of the required tasks of the network slice controller might generate a potential performance bottleneck in the case of a single entity. For this reason, the network slice controller can be composed of multiple orchestrators, each managing a subset of functionalities for each layer. To guarantee service requirements, different orchestration entities will coordinate with each other by exchanging high-level information about the operations involved in slice creation and deployment.

Slice isolation is typically an important requirement in the case of simultaneous coexistence of multiple slices sharing the same infrastructure. Slice isolation commonly consists of imposing that each slice performance must not have any impact on the other slices' performance. This property enhances the network slice architecture in terms of *slice security* (i.e., cyberattacks or fault occurrences affect only the target slice) and *slice privacy* (i.e., private information related to each slice, its state, and

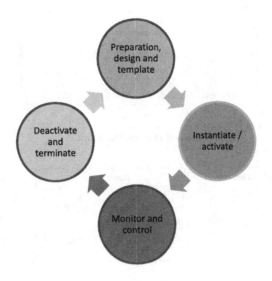

FIGURE 3.3

Conceptual life cycle of a network slice.

its traffic are not shared among slices). The process of requesting a slice, instantiating the slice, and managing it follows a classical operational flow as illustrated in Fig. 3.3. The process starts with design/preparation of a slice template, which is followed by an instantiation request to create, configure, and activate the slice. Once in operation, the slice is monitored and controlled to meet QoS requirements and finally deactivated when it is not required anymore. Associated resources can then be released. This process is handled by the network slice controller or slice manager with a number of interactions (i.e., through request/response or subscribe/notify notifications) between itself and different tenants and any potential underlying management systems.

Fig. 3.4 illustrates the detailed flow diagram defining the life cycle of a Network Slice Instance. The actual life cycle of an NSI is preceded by a preparation phase, where the instance is designed and preprovisioned by preparing the network environment. Then each NSI follows this process flow until it is decommissioned and terminated.

3.3 Network slicing architectures

The basic idea of network slicing is to *slice* the original network architecture in multiple logical and independent networks that are configured to effectively meet various services requirements. To quantitatively realize such concept, several techniques are employed:

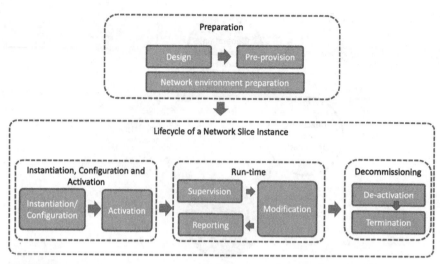

FIGURE 3.4

Illustration of the network slicing components within the life cycle.

NF: NFs express elementary network functionalities that are used as *building blocks* to create every network slice.

Virtualization: Virtualization provides an abstract representation of the physical resources under a unified and homogeneous scheme. In addition, it enables a scalable slice deployment relying on NFV that allows the decoupling of each network function instance from the network hardware it runs on.

Orchestration: Orchestration is the process that coordinates all the different network components that are involved in the life cycle of each network slice. In this context, SDN is employed to enable a dynamic and flexible slice configuration.

These concepts enable the design of the main components of the Network Slicing Architecture, which is as an instantiation of these concepts. The corresponding architecture for the management of network slices is illustrated in Fig. 3.5 with the following main components:

Virtualized Infrastructure Platform (VIP): This platform provides virtual resources (e.g., vComputing, vStorage, vNetwork) to assign to one or more slices. Virtualization is performed through a Virtual Infrastructure Manager.

Network Slice Instance (NSI): A collection of resources from the Virtualized Infrastructure Platform (VIP) organized to form a network slice. The different functionalities are allocated by the NFV Manager based on the directives by the NFV orchestrator.

MANagement and Orchestration (MANO): This main component contains submodules dedicated to the Virtualized Infrastructure Manager (VIM) and the

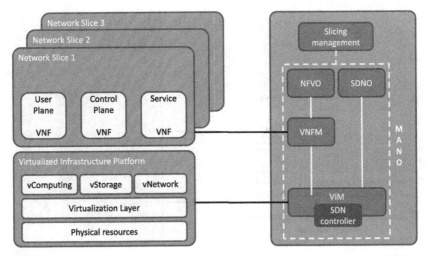

FIGURE 3.5

The generic architecture for network slicing.

NFV Orchestrator and Manager, Network Function Virtualization Orchestrator (NFVO), and Network Function Virtualization Manager (NFVM), respectively, the SDN Orchestration, Software Defined Networking Orchestrator (SDNO), and the Slicing Managemement function. Each Virtualized Infrastructure Manager (VIM) includes one or more SDN controllers to enable virtualization.

The following subsections describe some alternatives to implement this overall architecture. Each of them provides advantages and disadvantages, which are briefly reviewed further.

3.3.1 Single owner, single controller

SDNs typically enable direct network slicing functionalities, as in most cases the SDN environment provides an abstraction of the network resources through the Northbound interface of the SDN controller. The management and orchestration functionalities are subsequently implemented on top of the SDN controller by exploiting the Northbound interface. In this case, the SDN controller operates as the SDN Orchestrator. The corresponding architecture is depicted in Fig. 3.6.

This solution is appropriate for limited regions of the network, especially in the case of a single owner infrastructure, since the SDN controller completely controls all the different slices. It may, however, represent a bottleneck in terms of performance and reliability. The presence of a single controller limits the programmability of the networking infrastructure in case multiple tenants desire deploying network services.

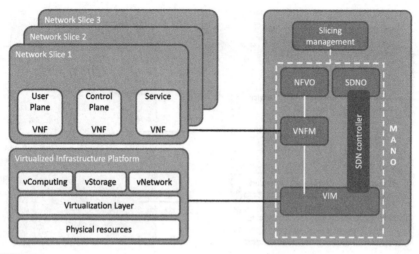

FIGURE 3.6

A Network slicing architecture with a single controller/orchestrator.

3.3.2 Single owner, multiple tenants – SDN proxy

Supporting multiple virtual networks is now becoming common in many settings, from data centers to service provider networks. In this framework an alternative technology to implement network slicing is the usage of an SDN proxy, typically, controlled by the owner of the physical infrastructure. In this case, the SDN proxy provides an abstraction of the network forwarding path that allows the SDN proxy to slice the network. The proxy employs the SDN protocol to define a hardware abstraction layer that logically sits between control and forwarding paths on a network device to enforce the rules and agreements defining the network slices and to maintain isolation. The resulting architecture is presented in Fig. 3.7.

The advantage of this scenario is that it enables multiple virtual tenants to deploy their own controllers/SDN orchestrators on the shared infrastructure while maintaining the isolation between different slice instances. An example of this solution is FlowVisor [61], illustrated in Fig. 3.8. Indeed, the FlowVisor is implemented as an OpenFlow proxy that intercepts messages between OpenFlow-enabled switches and OpenFlow controllers.

The FlowVisor defines a slice as a set of flows running on a topology of switches. It sits between each OpenFlow controller and the switches to make sure that a guest controller can only observe and control the switches it is supposed to. FlowVisor partitions the link bandwidth by assigning a minimum data rate to the set of flows that make up a slice and the flow-table in each switch by keeping track of which flow-entries belong to each guest controller. See Fig. 3.9 for an example of OpenFlow message exchange in an architecture using FlowVisor.

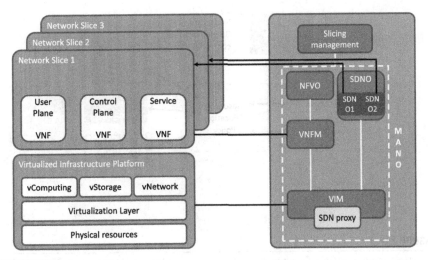

FIGURE 3.7

A network slicing architecture with SDN proxy and multiple orchestrators. The SDN proxy provides topology and capacity slicing to the orchestrators of the different slices (SDNO1 and SDNO2).

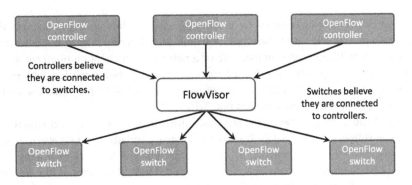

FIGURE 3.8

Conceptual FlowVisor architecture.

3.3.3 Multiple owners, tenants

Virtualization is essential to allow multiple tenants to specify how they desire their resources to be connected, independently from the service or infrastructure provider. The previously reviewed architectures are focused on generating slices of the underlying network infrastructure by dividing or isolating link resources and allowing one or more controllers to operate on the resulting network slices. Nevertheless, the additional requirement for enabling complete freedom for tenants to define their desired network topology implies the need to explicitly introduce an advanced virtualization layer capable of providing mapping between actual infrastructure resources and vir-

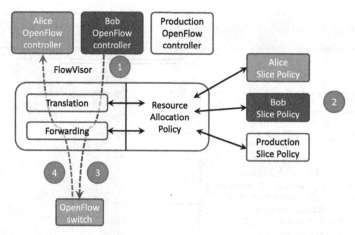

FIGURE 3.9

FlowVisor intercepts OpenFlow messages from guest controllers (1) and, using the user's slicing policy (2), transparently rewrites (3) the message to control only a slice of the network. Messages from switches (4) are forwarded only to guests if they match the corresponding slice policy.

tual overlay topologies required by tenants. In essence, this concept enhances the idea of programmable virtual networks and aims at making them even more flexible. OpenVirteX [62] represents an example of a network virtualization platform which allows tenants to specify their desired topology and addressing while enabling the infrastructure owner(s) to retain control of its own virtual SDN network. The corresponding architecture is presented in Fig. 3.10.

OpenVirteX sits in between the physical infrastructure resources and virtual network controllers (see Fig. 3.11). It allows to i) create isolated virtual networks with the topology specified by tenants, ii) use any controller or Network Operating System, iii) use the whole address space, iv) change the virtual network at runtime, and v) automatically recover from physical failures.

An alternative architecture proposed by the IETF is to perform Network Slicing using Segment Routing [63]. The proposed mechanism uses a unified Administrative Instance Identifier (AII) to distinguish between different virtual network resources for both intra- and interdomain network slicing scenarios. Combined with the segment routing technology, the mechanism could be used for both best-effort and traffic engineered services for tenants.

3.4 Network slicing examples

We continue our introduction to network slicing with a presentation of the utility in a practical example of deployment within the 5G mobile network [47]. The read-

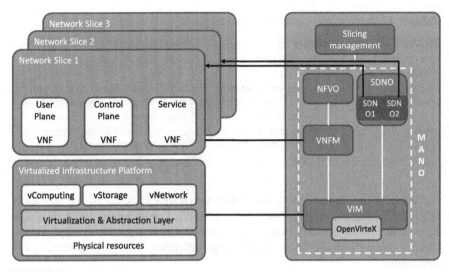

FIGURE 3.10

A network slicing architecture with an enhanced virtualization and abstraction layer. The SDN orchestrators have more degrees of freedom in defining topology and configuration on their slices.

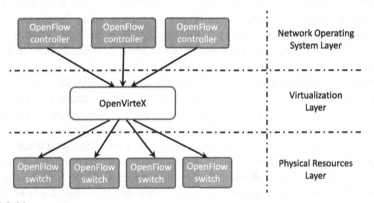

FIGURE 3.11

OpenVirteX architecture: OpenVirteX defines a flexible virtualization environment to provide to tenants' controllers or Network Operating Systems.

ers should already understand the contrasting needs of the different requirements of 5G, which lead to the design of the 5G Service-Based Architecture, incorporating the concept of end-to-end slicing (including RAN and core network segments). The requirements of the main use cases of 5G mobile network are described in Table 3.1. The table illustrates the major 5G use cases and provides some examples and the related main requirements.

Table 3.1 Main 5G use cases and requirements.

5G Use Case	Example	Requirements
Mobile Broadband (eMBB)	4K/8K UHD, hologram, AR/VR	High capacity, video cache
Massive IoT (mMTC)	Sensor network (metering, agriculture, building, logistics, city, home, etc.)	Massive connection (200,000/km^2), mostly immobile devices, high energy efficiency
Mission-critical IoT (URLLC)	Motion control, autonomous driving, automated factory, smart-grid	Low latency (ITS 5ms, motion control 1 ms), high reliability

Clearly, such requirements are associated with different use cases, and they apply only to the corresponding data flows. As an example, low latency and high reliability are associated with the data flows generated by autonomous driving applications, whereas massive connectivity and high energy efficiency are typical of massive IoT applications.

As a consequence, it is possible to define different configurations of network slices to support the different 5G use cases. Indeed, 5G envisages the design and implementation of at least the three types of network slices illustrated in the following figure, one for each use case [64]. In turn, different services will be associated with different slices depending on their requirements and KPIs. Each slice will be configured in terms of topology, capacity, and services o satisfy the requirements of one or more specific service classes (eMBB, massive Machine Type Communications (mMTC), URLLC). The conceptual design of the network slices associated with the three identified use cases [65] is illustrated in Fig. 3.12.

The programmability and isolation properties of the network slices will enable the deployment of different network configurations and service functions, as illustrated in Fig. 3.13. Isolation among network slices provided by SDN is exploited to allocate different functionalities (or Network Functions) and reconfigure the slices to satisfy the requirements of the services they are supposed to host. The figure illustrates that VNFs can be positioned in the core or in the edge cloud, depending on the requirements of each slice, and subsequently be interconnected through SDN. Fig. 3.13 proposes the example of four network slices, dedicated to mobile Ultra-High Definition streaming, mobile phone data services, massive IoT, and mission-critical IoT. The Ultra-High Definition slice allows a Mobile Virtual Operator (MVO) to deploy its streaming service by deploying dedicated functionalities (VNFs: core functionalities, cache nodes, Next generation Node B Distributed Unit (gNB-DU) at the edge of the network to facilitate resource management, reduce latency, and avoid network congestion. Indeed, the 5G base station Next generation Node B (gNB) is designed for flexibility and softwarization, and it can be split into three main functional modules: the Next generation Node B Centralized Unit (gNB-CU), the gNB-DU, and the Next generation Node B Radio Unit (gNB-RU). The gNB-CU can be placed in the cloud infrastructure, and it consists of Radio Resource Control (RRC), Service Data Adaptation Protocol (SDAP), and Packet Data Convergence Protocol (PDCP)

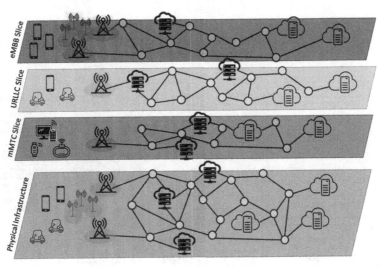

FIGURE 3.12

A pictorial design of the different network slices associated with the 5G use cases.

modules. The gNB-DU is a logical entity that consists of Radio Link Control (RLC), Medium Access Control (MAC), and Physical Layer (PHY) layers. The gNB-DU can support one or multiple cells. In the case of the phone slice, most functionalities are virtualized in the core cloud (mobility and IMS server), whereas the massive IoT slice is simpler, employing only light duty services in the Core Cloud. For the mission-critical IoT slice, in contrast, most services are moved to the Edge Cloud for minimizing transmission delay and improving reliability.

Fig. 3.14 presents a more detailed vision on how to implement network slicing in 5G. A hypervisor located in the Core cloud manages a virtual switching facility (labeled in the figure as vSwitch/vRouter) and performs provisioning of the virtualized servers and network resources. SDN tunnels (i.e., Generic Routing Encapsulation (GRE) and/or VXLAN) are built between each VM in the Core cloud (e.g., 5G IoT Core) and the data center gateway router. The gateway router then performs mapping between these tunnels and the corresponding VPNs (e.g., IoT VPN). The concept is the same in the Edge Cloud (possibly, with different VNFs) and allows us to build the end-to-end slices required for different services.

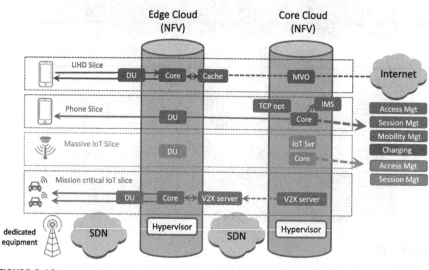

FIGURE 3.13

Conceptual example of the deployment of network slices in 5G.

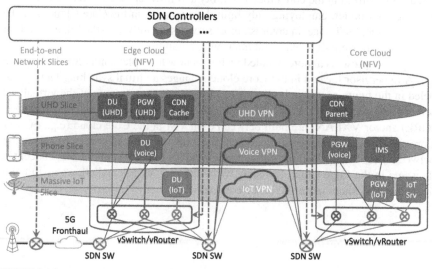

FIGURE 3.14

Detailed example of the deployment of network slices in 5G.

Mobile edge cloud

4

Roberto Torre, Tung Doan, Hani Salah
Technische Universität Dresden, Dresden, Germany

To err is human, but to really foul things up you need a computer.
Paul R. Ehrlich

4.1 Introduction

Increasing requirements from new application types and demands for services have triggered the unfold of the 5th Generation (5G) of mobile communications. In recent years the most influential telecommunication companies have established broad sets of specifications for 5G that aim at incorporating these new demands and requirements from their different customers. The most notable requirements from 5G applications include ultrahigh bandwidth, ultrahigh storage capacity, ultrahigh reliability, low energy consumption, and ultralow latency. Moreover, 5G will have to deal with a massive increase of the number of devices connected to the network, often known as IoT, which can be considered as an ultrahigh density requirement.

Initial generations of mobile communications until 4G placed network emphasis on the communication between the User device (UE) and the network operator. In particular, each UE in these architectures is connected to the core network through a RAN. Along with the developments of virtualization technologies and cloud computing, the concept of a Cloud Radio Access Network (C-RAN) was proposed by a few operators. The characteristics of a C-RAN include centralized processing, collaborative radio, real-time computing, and energy efficiency [66]. The integration of cloud computing into mobile networks was labeled as Mobile Cloud Computing (MCC) [67], which enriches the capabilities of UEs by empowering them with additional storage, energy, and computation resources. However, MCC implements a centralized service management, which reduces flexibility and introduces a significant execution delay. This, in turn, makes MCC hardly applicable for services that require high availability, high mobility, multiconnectivity, and accessibility to multiple devices. Moreover, MCC cannot fulfill the requirements of popular real-time applications, such as AR or VR.

Edge Computing (EC) is a new networking architecture that addresses the aforementioned limitations of MCC by deploying network resources (storage, computing, etc.) at a network's edge. This enables the network to supply high computing and storage capabilities at the edge, as well as high reliability, low latency, and low en-

Computing in Communication Networks. https://doi.org/10.1016/B978-0-12-820488-7.00015-3

ergy consumption, especially with consideration of connected mobile user terminals. EC was first introduced in a white paper by the ETSI and the ISG groups in 2014 [68]. Since the original inception, EC has been used as an acronym for several names such as Mobile Edge Cloud, Mobile Edge Computing, and Multi-Access Edge Computing. For example, Multi-Access Edge Computing extends the scope of the technology so that the benefits of EC reach beyond LTE and 5G and include Wireless Fidelity (WiFi) and other fixed access technologies. As overall, the different names refer to the same underlying concept as it slowly moves from conception to implementation, and we employ the term Mobile Edge Cloud (MEC) throughout this book, as it resembles the general application most closely, without restricting it to a specific implementation.

The use cases introduced by ETSI in 2018 [69] are classified into three different categories, namely i) consumer-oriented services (e.g., immersive media), ii) operator and third-party services (e.g., device tracking, big data, external services), and iii) network performance and QoS improvements (e.g., content and DNS caching). The interested reader is referred to Chapter 1 for an overview of popular use cases enabled by EC, whereas standardization efforts are introduced in Chapter 2. The remainder of this chapter is structured as follows. In Section 4.2, we introduce Mobile Edge Cloud (MEC) concepts, characteristics, challenges, and architecture. Next, in Section 4.3, we focus on the MANagement and Orchestration (MANO) of the MEC, introducing the ETSI framework and its most popular implementations. After that, we give an overview of demonstrators showcasing MEC concepts and challenges.

4.2 Mobile edge cloud

The term MEC was standardized by two of the most important telecommunication standardization groups, ETSI and ISG. Their white paper [70], which was published in 2015, describes MEC as follows: *Mobile edge computing provides an IT service environment and cloud computing capabilities at the edge of the mobile network, within the RAN and in close proximity to mobile subscribers.* This enables a user's device to connect to a nearby server and to offload traffic from the core network to the edge.

4.2.1 Similar concepts

In the following, we describe four terms referring to popular technologies that are similar, but are not identical, to the MEC:

MCC: MCC combines cloud computing and mobile computing. The principal idea is establishing an isolated virtualized environment in the cloud with different resources (e.g., computing, storage, and communication) that end users can access remotely. The main difference to the MEC is that the resources in MCC are placed inside the cloud, whereas in the MEC, they are placed at the edge of the network.

Local cloud: A local cloud is a cloud service administrated in the local network and normally connected to a remote server or to a remote cloud. It provides more privacy and security than a remote cloud since the running software is placed locally. The local cloud is normally a copy of the remote cloud and can be synchronized with it. The main difference to the MEC concept is that the application of the local cloud runs locally in a Local Area Network (LAN), whereas the MEC runs at the RAN's edge.

Cloudlet: Satyanarayanan [71] coined the term *cloudlet* in his works. It refers to a small middleware deployed close to the UEs with the purpose of bringing cloud capabilities closer to an end user. It focuses on latency-critical and real-time applications. However, cloudlets do not focus on interactions with the cloud and thus can also act as standalone clouds. This is one of the differences between cloudlets and the MEC. Another difference is that cloudlets use virtualization based on virtual machines only, whereas the MEC may also use other lightweight approaches, such as containers.

Fog computing: Cisco Systems introduced the term *fog computing* to describe bringing cloud resources closer to end users. The processing is mostly performed in the LAN, either at the gateway or at nodes. Although fog computing and MEC are similar and the two terms are widely used interchangeably, there are small differences between them. Most importantly, the intelligence and management in fog computing are located in a LAN, which results in a better utilization for IoT and Machine to Machine (M2M) communications. In contrast, in the MEC the intelligence and management are located in the RAN, which is more suitable for server-to-client applications.

4.2.2 Characteristics

The main characteristics of the MEC can be summarized as follows [69]:

NFV alignment: Whereas the MEC offers cloud services, NFV provides a framework to virtualize network functions. The same infrastructure can be used for both network functions and MEC applications.

Mobility support: The UEs communicate with MEC servers through wireless channels. Subsequently, users moving out of the coverage areas of the currently utilized edge server would normally cause service interruption. Various life service migration techniques (e.g., [72–74]) have been proposed to solve this problem.

Optimal application placement: Application placement plays an important role for the MEC. However, making a decision on where to deploy the MEC application is not trivial, as various requirements need to be considered. For instance, some MEC applications may require certain computing resources, whereas others may have strict latency requirements. These challenges have been tackled in the literature by orchestration and management frameworks (e.g., see [75–77] and the references therein).

Application mobility: Relocation of application instances to/from an external cloud environment is an important feature of the MEC. However, this can be challenging for several reasons: i) the mobility can face a compatibility issue if the current and target environments, that is, Virtual Machines (VMs) or containers, are not identical, ii) the mobility might be limited by resource constraints, for example, lack of network bandwidth can prolong or interrupt a relocation, or iii) the relocation of latency-sensitive applications should be done with a minimal downtime.

All these characteristics support the shift in networking from MCC to the MEC. This is prone to be included in the new 5G networking paradigms.

4.2.3 Key enablers

To enable the aforementioned characteristics, computing resources must be moved from the cloud to the edge. This shift can be enabled by two technologies, SDN (discussed in Chapter 6) and NFV (discussed in Chapter 7). An SDN network controller is able to monitor the traffic and make decisions, such as anticipation of possible congestion in the network and moving the flows accordingly. With the SDN controller, questions such as *Where to place the server?*, *Where to move the application?*, or *When to perform the handover?* inside the MEC network can be answered. NFV enables different network functions to execute regardless of the underlying hardware, which enhances mobility. Since the application service can be placed and moved to the optimal physical location, the result is a better utilization of the network resources. In a mobile communication environments, such as 5G, placing the service in the optimal physical device close to the user might not resemble the optimal location in the near future. Therefore migration techniques enabled by NFV allow the service to constantly move and be relocated to the optimal physical location. Apart from SDN and NFV, the recent hardware advancements have also enabled smaller devices to run resource intensive applications.

4.2.4 General architecture

Fig. 4.1 illustrates a three-layer network architecture with the MEC layer between the cloud layer and end devices. The mobile devices are connected to the core network via the edge network (i.e., RAN and MEC). Meanwhile, cloud services are provided by a private cloud, which is also connected to the core network. Considering the evolution of the RAN on the LTE system, the deployment of MEC becomes more flexible when it is located close to the mobile subscribers.

As shown in Fig. 4.2, the edge platform represents an edge cloud, which provides applications and services specific to the target use case and mobile environment. The MEC is featured with geodistributed (virtual) servers, which can be deployed at any location close to the end users. The MEC can also leverage cellular network elements, such as the base station or WiFi access points, to provide cloud services.

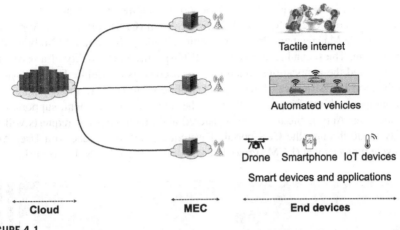

FIGURE 4.1

Three-layer architecture: Cloud, MEC, and mobile end devices (from left to right).

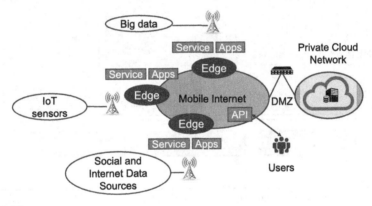

FIGURE 4.2

MEC architecture (adapted from [78]).

4.3 MANO frameworks

Researchers from both the industry and academia proposed MANO frameworks for the MEC that take into account the requirements of cloud services, such as low provisioning time, high availability, and high scalability. The key idea of the proposed frameworks is capturing a global view of the resources and using this view to place the services and to manage the resources efficiently. In the following, we first describe a reference MANO framework architecture for MEC, provided by ETSI. After that, we give an overview of notable MANO frameworks. This reference architecture is initiated by ETSI and is the most popular one. As illustrated in Fig. 4.3, the architecture has two levels: i) the Mobile Edge System Level and ii) the Mobile Edge Host

Level. The first level provides a global view of the entire MEC system. It includes user interface components (Customer Facing System (CFS) portal, User App, User App Life cycle Management Proxy), Operations Support System, and Mobile Edge Orchestration. The second level manages MEC-specific functionality of a particular MEC host and the applications running on it. It consists of Mobile Edge Manager, Virtualization Infrastructure Manager, and Mobile Edge Host. In the beginning the users interact with the framework through the interface environment, supported by CFS and User App. If the users are registered and identified, their requests will be directly made through the CFS portal. The requests are forwarded to a User App Life-cycle management (LCM) Proxy before being handled by the framework.

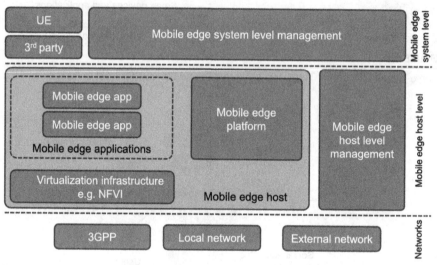

FIGURE 4.3

Reference MANO framework architecture for MEC (adapted from [79]).

To handle the incoming requests, first, the Operation Support System (OSS) translates the requests to a language understandable by the back end. In addition, the OSS determines resource requirements that can be used to serve the requests. Afterwards, the requests are forwarded to the Mobile Edge Orchestrator (MEO), which is responsible for orchestrating the applications. After the MEO specifies the infrastructure used to deploy the MEC application, the requests are then sent to the Mobile Edge Manager (MEM). Since the MEM is in charge of the life cycle management of the MEC applications, it then creates the applications on Mobile Edge Host through VIM. The MEC framework, from bottom to top, consists of the following three function elements:

Mobile edge host: It consists of a Mobile Edge Platform (MEP) and a virtualization infrastructure. The MEP is a collection of essential functionalities for mobile edge applications such as service registry, traffic rules control, and

DNS. The virtualization infrastructure offers compute, storage, and network resources, which are used to deploy the applications. The MEP is mainly responsible for offering an environment that mobile edge applications can leverage, consume, or discover. In addition, it is in charge of receiving traffic rules from other entities such as MEM, applications or services, and then dispatching them to the data plane. The MEP also obtains DNS records from the MEM and configures DNS servers accordingly. As described in the mobile edge host, mobile edge applications are considered as virtual machines or containers running on top of the virtualization infrastructure offered by the mobile edge host. Through the interaction with the MEP, it can consume, discover, and advertise resources. The mobile edge applications can be assigned with a certain number of rules and requirements such as required resources and maximum processing delay.

Mobile edge system level management: It consists of the MEO, the OSS, and the user application life cycle management proxy. The MEO is responsible for providing and maintaining a global view of the mobile edge system, which includes the deployed edge hosts, available resources, and network topologies. Owing to this capability, MEO is also in charge of on-boarding application packages. Specifically, it first validates the integrity and authenticity of the packages, analyzes the application rules and requirements, and possibly modifies them to adapt to the operator polices and the availability of resources. Afterwards, it determines a specific VIM and dispatches a request to a MEM to on-board the applications. The OSS is under the control of an operator, which manages the mobile edge system. After receiving the requests from the CFS portal and from UE Applications, it is the first entity of the mobile edge system that makes decisions on allocating resources to these requests. Afterwards, the OSS forwards the requests to the MEO for further processing. The user application life cycle management proxy allows UE applications to make requests for on-boarding, instantiation, and termination. It is also responsible for deciding on the relocation of the application into or out of the mobile edge system. To handle UE requests, the user application life cycle management proxy first verifies them and then interacts with other components, such as the OSS and the MEO for further processing.

Mobile edge system level management: It includes the MEM and VIM. The MEM is mainly responsible for performing the life cycle management of applications (i.e., instantiation, termination, and update). In addition, it provides element management functions to MEPs. Specifically, it dispatches the application rules and requirements such as traffic rules and DNS configuration. VIMs account for allocating and managing compute, storage, and networking resources of the virtualization infrastructure. To ease the management, each resource type is managed by a specific component in a VIM. For instance, in OpenStack, compute services, storage services, and networking services are managed by Nova, Swift, and Neutron, respectively.

Table 4.1 Reference points for MEC.

Reference Point	Related Components	Operational Objectives
Mp1	MEP and applications	Provides service registration, service discovery, and communication support.
Mp2	MEP and data plane	Instructs data plane for traffic routing between applications.
Mp3	MEPs	Controls the communication between applications.
Mm1	MEO and OSS	Triggers the instantiation and the termination of the application in the mobile edge system.
Mm2	OSS and MEM	Configures the MEP and performs fault and performance management.
Mm3	MEO and MEM	Manages the application life cycle, application rules, and requirements.
Mm4	MEO and VIM	Manages the virtualized resources of the mobile edge hosts.
Mm5	MEM and MEP	Accounts for various configurations, such as platform configuration, application rule configuration, application support procedures, and management of application relocation.
Mm6	MEM and VIM	Dispatches the requests of life cycle management at a high level.
Mm7	VIM and virtualized infrastructure	Manages the virtualized infrastructure at a low level.
Mm8	LCM proxy and OSS	Handles UE requests for running applications.
Mm9	LCM proxy and MEO	Manages the applications requested by UEs.
Mx1	OSS and CFS portal	Used by third parties to deploy applications.
Mx2	LCM proxy and UE applications	Used by UE applications to request the mobile edge system to deploy an application.

To operate a mobile edge system, reference points are needed to connect mobile edge components with each other; see Fig. 4.3. The reference points are detailed in Table 4.1. Motivated by the importance of MEC for their businesses, different companies put huge efforts on designing their own MANO frameworks. The most popular MANO frameworks are described further and gathered in Table 4.2.

Akcraino: Akraino Edge Stack is an open-source software stack that offers high-availability cloud services in terms of edge computing systems and applications. It delivers a deployable and fully functional edge stack for edge use cases

Table 4.2 Popular MANO frameworks for MEC.

	Foundation	Operational Objectives	Technology Features
Akraino	Linux	Supports high-availability cloud services optimized for edge computing systems and applications.	OpenStack, K8s
StarlingX	OpenStack	Focuses on easy deployment, low-touch manageability, rapid response to events, and fast recovery.	OpenStack, OpenDaylight
Airship	OpenStack	Provides automated cloud provisioning and life cycle management in a completely declarative and predictable way.	OpenStack, K8s, Calico
EdgeX	Linux	Simplifies and standardizes the foundation for edge computing architectures in the industrial IoT market.	SDK, MQTT, SNMP, ModBus
OpenEdge	N/A	Provides temporary offline, low-latency computing services, and includes device connect, message routing, remote synchronization, function computing, AI inference.	Docker, MQTT
KubeEdge	N/A	Extends native containerized application orchestration capabilities to hosts at the edge.	K8s, Mosquitto, Docker
MobiledgeX	N/A	Global, privacy-first, trusted workload orchestration that is aware of users and locations.	SDK, DME
vCO	Linux	Produces an OpenDaylight-based reference architecture that, when combined with other functional elements (such as NFV), can support the delivery of residential, business and mobile services.	OpenStack, OpenDaylight

such as IoT, Telco 5G Core & vRAN, uCPE, SDWAN, edge media processing, and carrier edge media processing.

StarlingX: The first release of StarlingX was launched in October 2018. From 2019 onward, its releases are aligned with OpenStack releases. StarlingX offers a virtualization platform that allows easy deployment, low-touch manageability, rapid response to events, and fast recovery. Its use cases include ultralow latency communications and industrial IoT applications, high-bandwidth and large-volume applications, and multiaccess edge computing.

Airship: The Airship community announced its v1.0 release in 2019. Airship provides a collection of loosely coupled but interoperable open-source tools that automate cloud provisioning. Starting from raw bare metal infrastructure, Airship manages the full life cycle of data center infrastructure to deliver a production-grade Kubernetes cluster with Helm deployed artifacts, including OpenStack-Helm. Airship enables operators to manage their infrastructure deployments and life cycle through declarative YAML documents that describe Airship environments.

EdgeX: EdgeX offers an open-source software platform at the edge of the network that interacts with various IoT objects, such as sensors and actuators. EdgeX makes it easier to monitor physical world items, send instructions to them, collect data from them, move the data across the fog up to the cloud where it can be stored, aggregated, analyzed, and turned into information, actuated, and acted upon.

OpenEdge: OpenEdge is an open edge computing framework that extends cloud computing, data, and services seamlessly to edge devices. It can provide temporary offline, low-latency computing services and includes device connect, message routing, remote synchronization, function computing, video access preprocessing, and AI inference.

KubeEdge: KubeEdge is an open-source system for extending native containerized application orchestration capabilities to hosts at the edge. It is built upon Kubernetes and provides fundamental infrastructure support for the network, application deployment, and metadata synchronization between the cloud and the edge.

MobiledgeX: MobiledgeX provides global, privacy-first, trusted workload orchestration that is aware of the users and locations. MobiledgeX aggregates mobile operator infrastructures on a global scale, harmonizes usage, and exposes exciting new edge functionality.

vCO: The goal of the virtual Central Office (vCO) project is producing an OpenDaylight-based reference architecture that, when combined with other functional elements (such as NFV and Orchestration software stacks), can support the delivery of residential, business, and mobile Services.

4.4 MEC example implementations

In this section, we provide an overview of three notable demonstrators developed to showcase the concept of MEC.

4.4.1 Tron demonstrator

The Tron demonstrator shows the impact of the latency on the performance by means of the popular Tron game. It was first presented by Pandi et al. [38] in the IEEE Consumer Communications and Networking Conference (CCNC) in 2017. A 3D version of the demonstrator was later presented by Schmoll et al. [39] in the same conference in 2018. Fig. 4.4 shows the displays of the 3D version (right) and the 2D version (left).

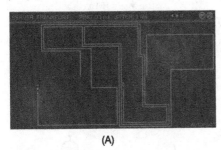
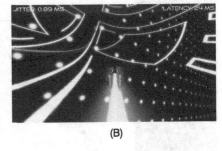

(A)   (B)

FIGURE 4.4

Tron demonstrator presented at CCNC 2017 and CCNC 2018 in Las Vegas, USA. (A) 2D version (2017); (B) 3D version (2018).

The demonstrator emulates five cloud providers located in five different places (Edge Cloud, Germany, Japan, Canada, and Brazil). The audience controls a motorbike using a controller, attached to a device that acts as client. They experience the impact of the inherent latency that appears when the application (i.e., the server side of the Tron game) is placed far away from the client.

4.4.2 Ball sorting machine

The ball sorting machine in Fig. 4.5 (left) demonstrates the impact of latency by comparing the performance of two different technologies, namely the legacy cloud and the MEC. The demonstrator was presented by Kropp et al. [41] at the IEEE Consumer Communications and Networking Conference (CCNC) in 2019. The upper part of the machine contains a ball dispenser filled with white and orange balls. After a ball is dispensed, it runs through a camera (i.e., sensor). The camera sends the information of a ball to a server, which is tasked to determine the color of the ball. The decision is sent to four servo motors (the actuators) that accordingly move a servo flipper to sort the balls by their colors. The demonstrator emulates two scenarios (for the server location), namely the legacy cloud and the MEC. The audience selects one of them via a mobile application. Fig. 4.5 (right) shows a diagram of the process from the moment where the ball is detected to the action from the servo flipper.

4.4.3 Ambulance demonstrator

This demonstrator showcases 5G connected cars as a use case for the MEC. It foresees a futuristic city where smart cars talk to the MEC servers while an ambulance drives toward an accident. The demonstrator was presented by Zhdanenko et al. [42] in the IEEE Consumer Communications and Networking Conference (CCNC) in 2019. As can be seen in Fig. 4.6, the demonstrator has virtual and physical setups.

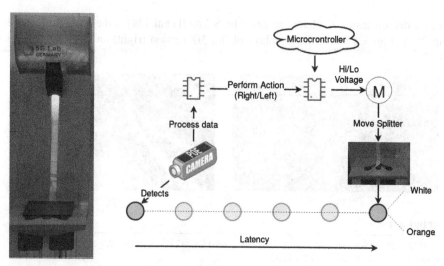

FIGURE 4.5

The ball sorting machine: The box with camera (left) and the servo flipper (right).

The virtual setup consists of a large screen showing the city of Munich, where cars move according to traffic rules. Two smaller screens show the roads from different points of view. The ambulance connects to one active MEC server (i.e., base station) at a time, identified by a blue cloud on top of it. As for the physical setup, it consists of the city map (printed on a board) and Raspberry Pi units, each acting as a MEC sever. There are also light toy fans attached to the Raspberry Pi units used to identify the running MEC server on the board. The audience can steer the ambulance using a controller.

The demonstrator supports four MEC server selection modes: legacy cloud, closest MEC server, least-loaded MEC server, and hybrid MEC. Each mode runs a different server selection algorithm. In the last three modes (i.e., MEC modes) the selected server may change dynamically while the ambulance is moving in response to changes in the network and servers conditions. When this happens, the MEC server migrates from the previous base station to the recently selected one, as shown in Fig. 4.7.

4.4.4 Seamless migration for autonomous cars

Fig. 4.8 demonstrates the impact of the MEC on the seamless migration for the autonomous driving. The demonstrator consists of two displays and a single-rack testbed, which was placed behind the display. The testbed includes seven mini-PCs that are used as follows: two are used as the clients for the MEC and central cloud scenarios, four act as the MEC nodes, and one acts as the central cloud node. The display on the left-hand side shows central cloud scenario, whereas the one on the

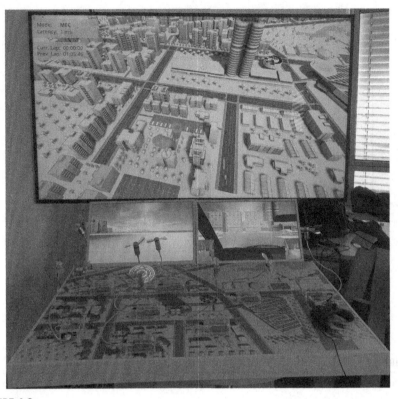

FIGURE 4.6

MEC ambulance demonstrator (CCNC 2019, Las Vegas, USA).

right-hand side shows the MEC scenario. The demonstrator will be presented by Tung et al. [80] at the IEEE Consumer Communications and Networking Conference (CCNC) in 2020. For the MEC scenario, it is worth noting that the car is driven smoothly and kept in the center of the track even when the migration of the steering control application between the MEC nodes is taking place. For the central cloud scenario, due to the high latency of the control loop, the car is sometimes accidentally out of the track. In addition, it takes a longer time to reach the destination compared to the MEC scenario.

Fig. 4.9 shows the demonstrator design, which consists of MEC components and client components. The MEC components are responsible for remotely controlling the steering angle and speed of the autonomous car. Meanwhile, the client components account for simulating the behavior of the car on the track. The cooperation of the MEC and client components results in a seamless migration. A MEC node consists of a key-value store, a controller, and a steering control application. Considering the high mobility of the autonomous car and that the container is much more light-weight than the virtual machine, the authors first adopted Docker [81] (a con-

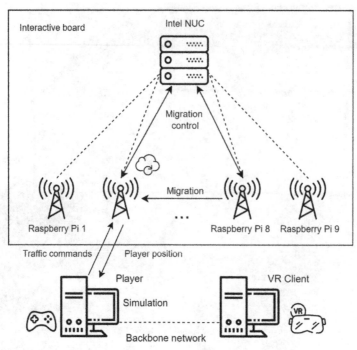

FIGURE 4.7

Layout of the ambulance demonstrator.

tainer platform) to build the steering control application. A proactive strategy was then used to deploy the application and periodically synchronized its states over a set of MEC nodes. Such an approach guarantees the consistency regarding application states, thus allowing for the seamless migration when the car performs handover process to the next serving area. Toward this end, the authors adopted ETCD [82], which is a low-latency, distributed, and reliable key-value (KV) store system. More importantly, since the controller is deployed on each MEC node, the KV store is also effectively used to guarantee the consistency to the life cycle events between MEC nodes.

The client's design also plays an important role in the seamless migration of the autonomous driving application. Client components include a client controller, a web proxy, a network configuration daemon, and an autonomous driving simulator. Since the first three components are totally independent of the simulator, the proposed design is transparent to the clients and potentially applicable for various use cases in MEC. For the simulator, the authors first adopted the Udacity self-driving car [83], which is an open-source project supported by many leading automobile companies such as Mercedes-Benz and BMW. The simulator was then extended to simulate the involvement of MEC in the demonstration. To allow the client to flexibly switch between MEC nodes without any interruption, the client used Nginx [84]

FIGURE 4.8

Demonstration setup of the seamless migration for the autonomous cars.

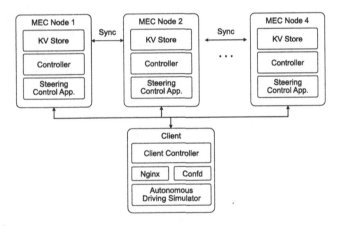

FIGURE 4.9

Demonstration layout of the seamless migration for autonomous cars demonstrator.

and Confd [85], which are a web proxy and a network configuration daemon, respectively. Afterwards, a light-weight Python-based client controller was used to work in co-operation with the controller in the MEC node. The client's controller injects the distances between the cars and base stations from the simulator and adapt them to simulate the latency between the client and the MEC nodes using the Linux kernel.

Content distribution

5

Hani Salah, Sandra Zimmermann, Juan A. Cabrera G.
Technische Universität Dresden, Dresden, Germany

*Change is painful, but nothing is as painful as staying stuck somewhere you
don't belong.*
Mandy Hale

5.1 Introduction

The Internet was primarily developed to connect a small number of hosts to exchange
data or to share special devices (e.g., mainframe computers, supercomputers, or card
readers). The goal at that time was to efficiently transfer information from one host to
another. Toward that end, networking researchers proposed the *packet switching* approach in which separate data packets are sent and forwarded using the address of the
destination host. The TCP/IP protocol suite, which was a remarkable move in the development of the Internet architecture, specifies how packets are formed and handled
until they reach their final destinations. As can be seen in Fig. 5.1, the IP protocol
is located in the middle of the protocol stack (i.e., its thin waist). The simplicity of
the IP protocol and its location in the protocol stack allow us to support a variety of
physical and access technologies and, at the same time, to deploy different types of
services. This has coined the phrase *Everything over IP and IP over everything.*

In the early 1990s, the emergence of the World Wide Web (WWW) started to
change the Internet usage from a network for connecting hosts to a network for
distributing and retrieving content. This trend has increased over the years until it
became the Internet's dominant usage. Watching and uploading videos on YouTube,
publishing information on Facebook and Twitter, sharing photos and videos on Flickr
and Instagram, and online gaming are all representative and content-centric examples
of what the Internet is used for today. These services produce massive and ever-increasing volumes of data and constitute the largest part of the global IP traffic,
which is expected to reach 396 exabytes per month in 2022 [87].

Typically, a trade-off exists between the amount of data required to be delivered and content or media quality (translated into network bandwidth). The delivery
of higher-quality media generally requires more network bandwidth (e.g., consider
a low-resolution video stream versus a 4K video stream). In addition, the variations of bandwidth and delay can have a significant impact on how users experience

Computing in Communication Networks. https://doi.org/10.1016/B978-0-12-820488-7.00016-5

93

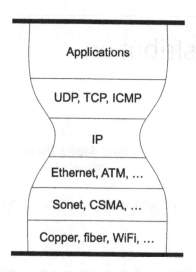

FIGURE 5.1

Protocol stack of TCP/IP (adapted from [86]) illustrating IP as the thin waist of today's Internet.

their service consumptions (see, e.g., [88] for a high-level video streaming example). These network-centric service characteristics are commonly regarded as QoS, which is required to adequately derive a parameter space needed for a successful service offering. For example, reductions in the available data client-side can readily lead to the starvation of the video playout and result in a negative Quality of Experience (QoE). As the QoE can be seen as a subjective metric that could be used to gauge a customer's willingness to pay for a networked service, it has become popular in the determination of service quality [89]. As a psycho-physiological and subject-oriented metric, the QoE determination typically requires human subject experimentation, which is infeasible at scale. In turn, efforts were made to approximate the QoE from the QoS through a generalized quantitative relationship; see, for example, [90,91]. It is, however, not uncommon for actual (i.e., nonacademic) service providers to follow the simple rule of trying to simply maximize the throughput attainable at a given time, such as through HTTP-DASH [92], for video streaming.

In spite of the aforementioned rapid changes and continuous traffic growth, the Internet design (including its TCP/IP protocol stack) stayed without important changes since its invention. This inconsistency between the Internet (host-centric) design and its current (content-centric) usage translates into additional costs, performance and security problems, and degradation in the user's QoE [93–95]. This situation has attracted the attention of researchers and networking experts from academia and industry over the last two decades. The proposed solutions can be classified into two main classes: one suggests batches for the existing Internet architecture, whereas the other calls for clean-slate Internet architecture. In Section 5.2, we describe Content

Delivery Networks (CDNs), the most notable example of the first class. After that, in Section 5.3, we give an overview of Information-Centric Networking (ICN), the common term of the proposed clean-slate architectures.

5.2 Content delivery networks

The basic approach that Content Delivery Networks (CDNs) take to improve the Internet service quality, which is typically measured by the content delivery delay, is to place the content near the user interested in it. In greater detail, the original content provider (e.g., YouTube) needs to sign up with a dedicated CDN provider (e.g., Akamai [96], MaxCDN [97], or CoralCDN [98]) to host a well-chosen set of its contents. The CDN provider, in turn, replicates the chosen contents (or part of them) to a group of servers spreading across different countries[1] and possibly across different Internet Service Provider (ISP) networks, as illustrated in the example in Fig. 5.2. Subsequently, the requests for these contents will be forwarded to the most appropriate (e.g., the closest) replica servers, rather than to the original content provider. These servers thus keep the content items in their local storage, that is, they provide a dedicated cache.

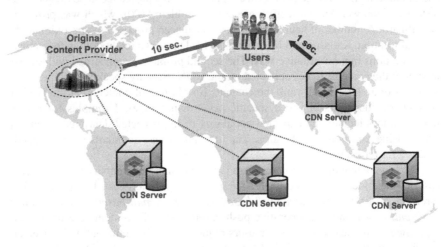

FIGURE 5.2

The principal idea of CDNs: Content is placed closer to the user by employing the servers of the CDN provider than possible using the server of the original content provider.

With the approach described above, CDNs can lower the load on the original content providers, decrease the bandwidth consumption, and improve the user's QoE.

---

[1] The CDN providers usually implement a hierarchical architecture to jointly improve the content delivery performance and robustness of the network.

In addition, malicious traffic, for example, resulting from Denial of Service (DoS) attacks, can be partially handled by the CDN servers, which provide a degree of protection for the original content providers [99]. Motivated by the aforementioned gains, CDNs have become widely used in the last years, and their popularity is constantly increasing. Today they are employed when, for example, watching videos, shopping and gaming online, or browsing and publishing contents on online social networks. In numbers, according to Cisco [87], CDNs will carry 72 percent of Internet traffic by 2022 (compared to 56 percent in 2017), including traffic from major content providers like Facebook, YouTube, and Netflix. In the following, we discuss two key design aspects of CDNs: *content distribution* (Subsection 5.2.1) and *request routing* (Subsection 5.2.2). For an overview of other design aspects, we refer the interested reader to [94] and the references therein.

5.2.1 Content distribution

Content distribution deals with three critical problems [94]: *placement of replica servers*, *content selection*, and *content-to-server assignments*. In general, the placement of replica servers is guided by two goals, namely minimizing the bandwidth consumption for delivering the contents from replica servers to the requesting entity and minimizing the average content access latency. This problem can be treated as an instance of the well-known facility placement problem [100], which was proved to be NP-hard. A number of heuristic solutions have been proposed to approximate the optimal solution. Karlsson and Karamanolis [101] present a methodology to compare these solutions in a quantitative way. The second problem is concerned with the selection of the contents to be replicated on the CDN. Broadly speaking, the proposed solutions (e.g., [102–104]) rank the contents (or clusters of contents) at the original content provider according to a specific performance index. Examples for performance indices include the expected popularity or the expected delay reduction that can be achieved through replication. The top-ranked set of contents is subsequently selected for replication.

The last content distribution problem is determining the number of content replicas and their locations in the CDN. The proposed solutions can be classified into three classes [94]: *cooperative push-based*, *cooperative pull-based*, and *noncooperative pull-based*. In the cooperative push-based approach, contents are proactively replicated on replica servers before users request them. Here the optimal placement of given contents on a given set of replica servers is an NP-complete problem, and therefore heuristic solutions have been proposed (see [94] for an overview). Most of these solutions assume knowledge about users' locations and request rates, which translates into high storage and management overhead for the CDN providers. This explains why this approach is not adopted by commercial CDNs. In cooperative pull-based solutions (such as in CoralCDN [105,98]), nearby replica servers, upon a cache miss, cooperate to locate other servers that likely store the requested content. If a server is found, then it transfers the content to the server that encountered the cache miss. It is important to note that the overall traffic overhead of this approach is relatively low because cooperation takes place only among nearby servers.

Last but not least, in the noncooperative pull-based approach, content is fetched from the original content provider by a replica server and subsequently cached only when the server cannot serve the request (i.e., in the case of a cache miss). As its name states, each replica server works independently in this approach, that is, there is no cooperation among replica servers. Due to the simplicity of this approach and because it is compatible with popular request routing approaches (see Subsection 5.2.2), it is employed by most commercial CDNs [106].

5.2.2 Request routing

Request routing deals with identifying the most appropriate replica server to fulfill a content request and routing the request to the identified server. Notable request routing techniques include Domain Name System (DNS)-based indirection [107], anycasting [108,109], HTTP redirection [110], and Global Server Load Balancing [111]. We restrict our discussion here to DNS-based indirection, the most popular content routing technique. In this technique the following sequence of events takes place [110]:

1. The user (through a browser or another application) sends a DNS query to the local DNS server.
2. The local DNS server forwards the query to the CDN Request-Routing Infrastructure (RRI).
3. The RRI asks the replica servers to examine their routes to the local DNS server and also to perform some measurements allowing selection of the most appropriate replica server.
4. Each replica server performs the measurements and then sends the measurement results to the RRI.
5. The RRI compares the received measurement results, selects the most appropriate replica server according to some policy, and then sends a DNS response to the user's local DNS server.
6. The user's local DNS server forwards the response to the user.
7. The user requests the content from the identified server.

The policy that RRI uses to select the replica server has a significant influence on the overall CDN performance. The policy can be as simple as round-robin, or can be more complicated considering one or more factors, such as latency and number of hops to the requesting entity, current or predicted server's load, or server's capabilities.

5.3 Information-centric networking

In stark contrast to CDNs, which work as an overlay atop the existing sender-driven and host-centric Internet, ICN suggests modifying the Internet core itself. This mod-

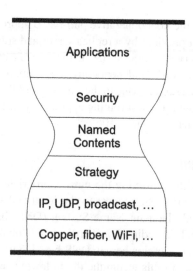

FIGURE 5.3

Protocol stack of NDN (adapted from [86]) where content replaced IP as new thin waist of the future Internet.

ification targets a redesign of the Internet from scratch as a receiver-driven and content-centric network. Through this design change, ICN aims to address several issues facing the current Internet, such as network congestions and bottlenecks, inefficient content distribution, and different security problems. Several ICN architectures have been proposed in the last 15 years, including NDN [93,112], DONA [113], PSIRP [114], and NetInf [115]. Whereas the design details of these architectures are different, all of them are essentially based on two ideas, *in-network caching* and networking based on *named contents*.

Among the aforementioned architectures, NDN has received the largest attention from the research community. Furthermore, it is widely treated as a possible replacement for the current Internet architecture. This is likely because its design is very detailed and addresses several critical issues in the host-centric networking paradigm. Motivated by its popularity and promises, we will focus only on NDN in the remainder of this section. NDN is a project funded by the United States' National Science Foundation (NSF) for Future Internet Architectures (FIA) [116]. It extends on Xerox PARC Content-Centric Networking (CCN) architecture [93] by detailing protocols and algorithms and by providing a completely functional prototype. This explains why the terms NDN and CCN are widely used interchangeably in the literature.

As can be seen in Fig. 5.3, NDN modifies the traditional (i.e., TCP/IP) protocol stack by placing named contents in the thin waist and also by adding two new layers, the *security layer* and the *strategy layer*. The security layer is placed above the named contents to apply security functions on the content itself, instead of securing the communication channel. This enables addressing several traditional security issues, such as reflection attacks and prefix hijacking, by design (see Subsection 5.3.5). The strategy layer is in charge of forwarding and transporting functions. Being placed above the underlying networking technologies (IP, UDP, etc.), the strategy layer can adapt these functions according to access networks and applications.

5.3.1 Operation primitives and packet types

The communication model in NDN, similar to those in other ICN architectures, is derived from the Publish/Subscribe (pub/sub) model [117]. In pub/sub, data sources (called publishers) do not send the data directly to specified receivers (called subscribers). Instead, they just classify the data and publish them as messages without knowledge of subscribers, if there are any. The subscribers, in turn, show interest in one or more contents (or content types) and only receive messages that are of interest, without knowing the publishers. This decoupling between content requests and responses with respect to time and location enables for asynchronous and location-independent content distribution.

Inspired by the operation primitives described before, NDN applies a consumer-driven content-centric communication model with two packet types, *interest* packets and *data* packets, as illustrated in Fig. 5.4. An interest packet is initially employed to request content by the unique name of content. Then the requested content is encapsulated inside a data packet and transmitted using the same path over which the interest packet was forwarded, but in the opposite direction. Since each interest packet can result in only one data packet (or none), *flow balance* is achieved in NDN by design. The interest packet's *nonce* field includes a randomly generated string. By combining this field and the *content name* field each interest packet can be uniquely identified. This is used to detect looping interest packets. The data packet's *digital signature* and *signed information* fields are used for security purposes, as we describe in Subsection 5.3.5.

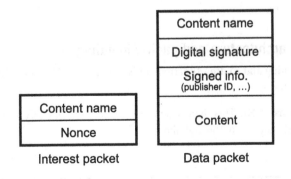

FIGURE 5.4

NDN's interest packet and data packet (adapted from [86]).

5.3.2 Content naming

There are two ways to name contents in ICN, *hierarchical* naming and *flat* naming. NDN adopts the hierarchical way, which often uses URL-like, human-readable names. `/de.tu-resden/communication-book/icn.pdf` is an example of a hierarchical name, where `/de.tu-dresden` is a globally routable prefix whereas `/communication-book/icn.pdf` is an organizational name including the file name. The hierarchical names enable name aggregation, which reduces the number of entries in the routing table and thus improves scalability. However, these names correlate content names with underlying network topology, which makes content multihoming[2] difficult to implement. This problem can be solved if flat names are used. However, the name aggregation becomes infeasible in that case. In Subsection 5.3.5, we discuss the security implications of the two naming types.

5.3.3 In-network caching

In-network caching is an essential feature in all ICN architectures. In NDN, in-network caching works as follows: when content is delivered, a copy of it is cached at each node located along the delivery path between the original content provider and the consumer. Caching-related decisions, for example, content replacement, are performed by each caching node independently (i.e., without coordination with other nodes), according to the node's local knowledge of cache state and content requests. This approach is called on-path (or en-route) caching. The caching approach described before, although it is straightforward and does not incur coordination overhead, has three problems [118]. First, it is expected to result in high (unnecessary) caching redundancy, which translates into bad utilization of the available (already limited [119]) storage. Second, the resulting caching-related decisions, being taken independently based on a local view, likely are low-performing. Third, the cached contents can be used only if they are located on the default routes of the interest packets. In response to these problems, coordinated (often off-path) ICN caching schemes have been proposed since the original inception of caching in ICN (see [120] and the references therein).

5.3.4 Node architecture and packet handling

To perform caching and routing functions in NDN, each NDN node is equipped with three data structures:

1. **Content Store (CS):** This data structure is used to temporarily store (i.e., cache) the data packets passing through the node.

---

[2] Content multihoming enables to select the best path between the consumer and different copies of the requested content, which can significantly improve the efficiency of content delivery.

2. **Pending Interest Table (PIT):** In addition to caching the data packets in the CS, the nodes also cache the interest packets in the PIT. In greater detail, the node creates a PIT entry for each requested content, mapping its name to the interface(s) through which it was requested. The PIT enables *interests aggregation*, that is, if multiple interest packets having the same content name are received, then only the first one is forwarded toward the original content provider. This can significantly minimize the traffic forwarded to upstream nodes. The node deletes a PIT entry either when the corresponding data packet is received or when its timeout is caught.

3. **Forwarding Information Base (FIB):** This is a routing table where each FIB entry maps a content name (or a name prefix) to a list of one or more potential outgoing interfaces.

With this node architecture, NDN nodes handle interest packets and data packets as follows:

- **Interest packets:** When the node receives an interest packet, it initially searches for the name of the requested content in its CS. If the name is found, then the node sends the respective data packet through the interface from which it was requested. If no matching data packet exists in the CS, then the node searches for a matching content name in the PIT. If it exists, then the node checks whether the identifier of the interface from which the interest packet was received is listed in the entry or not. If so, then no further actions are required. Otherwise, the interface is attached to the same PIT entry. If no matching content name is found in the PIT, then the node creates one and forwards the packet through one or more outgoing interface(s) according to the information provided in the FIB.

- **Data packets:** When a data packet arrives, the node first extracts the content name and uses it to search for a matching entry in the PIT. If the same content name is found, then the node stores a copy of the data packet in the CS, forwards the packet through the interfaces that are listed in the PIT entry, and finally removes the PIT entry. If there is no matching PIT entry, then the node simply rejects the data packet.

The example illustrated in Fig. 5.5 illustrates both the node architecture and how interest packets and data packets are handled by the nodes. There are two NDN nodes (N_1 and N_2), three clients (C_1, C_2, and C_3), and two servers (i.e., content providers) responsible for two globally routable prefixes: the upper server is responsible for /com/p1/, whereas the lower one is responsible for /org/p2/. The names of the cached contents and the PIT entries, after each timestamp,[3] are shown. We assume that the caches and PITs were empty before t_1. The FIB entries of N_1 and N_2 are also illustrated.

---

[3] t_{i+} denotes the time period starting directly after a timestamp t_i and ending just before the start of the next timestamp t_{i+1}.

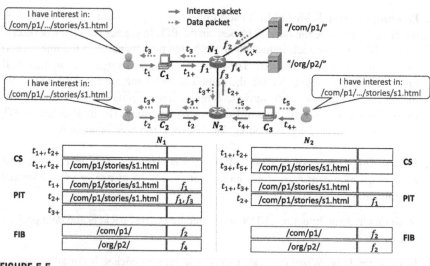

FIGURE 5.5

Handling interest and data packets in NDN (adapted from [86]).

This particular scenario can be summarized as follows:

1. At t_1, the user connected to C_1 requests the content /com/p1/stories/s1.html. Since no matching content name will be found in CS of N_1 nor in its PIT, a PIT entry will be created, and the interest packet will be forwarded (according to the information provided in the FIB) to the content provider (the upper server).

2. At t_2, before satisfying the aforementioned request, the user connected to C_2 requests the same content. The new request will be first forwarded to N_2, which will result in a PIT entry there. The packet will be then forwarded to N_1 in which a matching PIT entry will be found. Therefore the packet will not be forwarded further. The interface f_3 of N_1 will be appended to the matching PIT entry.

3. At t_3, the content provider encapsulates the requested content in a data packet and after that will send it to N_1. There, according to the matching PIT entry, two copies of the packet will be forwarded. A copy of the content will be cached in the CS, and subsequently the PIT entry will be deleted. One of the two ongoing copies will be forwarded through f_1 toward C_1. The other one will be forwarded through f_3 initially to N_2. There a copy of the packet will be initially cached, a matching PIT entry will be found, and the packet will be forwarded accordingly through f_1 toward C_2. After that, the PIT entry will be removed.

4. At t_4, the user connected to C_3 requests the same content. A matching content will be found in CS of N_2 , which will be sent back via f_3 to the client. No further actions will be taken, except updating the content access information in the CS.

5.3.5 **Content-based security**

NDN implements security on the content itself instead of securing communication channels. Specifically, the digital signature field in the data packet (Fig. 5.4) is used to authenticate the origin and to verify the integrity of the content.[4] The digital signature is computed by the original content provider over the content name and the content fields, thus binding them to each other. The provider's public key can be retrieved using the information contained in the packet (Fig. 5.4). This approach allows verifying the packet's authenticity and integrity, regardless where the packet is retrieved from. However, the approach described before requires a Public Key Infrastructure (PKI) to bind content names to the keys, which is considered a major drawback of hierarchical naming. This problem is solved with flat naming, since flat names are self-certifying. Specifically, the key is bound to the content name itself. Self-certification can be realized in this case by embedding the hash of content into its name.

---

[4] Interest packets do not include a digital signature filed. Hence, it is not possible to verify their origin nor their integrity.

Enabling technologies

Outline

In this part of the book, we introduce two technologies, software-defined networks and network function virtualization, as the main building blocks for the concepts we discussed in Part 2. We dedicate a full chapter to each topic, as we will reuse them in the examples given in the book.

Software-defined networks

6

Justus Rischke, Hani Salah

Technische Universität Dresden, Dresden, Germany

There are some things that can't be controlled.
Leonard Kleinrock

6.1 Networking in today's Internet

The current trend in mobile communications is a tenfold increase in bandwidth every five years [121]. 5G as the next generation of mobile communication implementations has high demands on future networks in terms of bandwidth, latency, and reliability, to name a few. Jointly with the increases in demands and capabilities on the mobile side, the performance of wired networks has increased from connections using 56K dial-up modems in 2000 to fiber connections with 1 Gbit/s today. This overall increase in network performances enables new services ranging from simple web pages to video streaming today while providing the foundations for future connected autonomous cars and the *Tactile Internet* with high demands on latency and reliability.

A remaining problem that stems from the original conception of the Internet is the missing evolution of Internet Protocol (IP)-based networks, which do not easily support new technologies and approaches, for example, multipath communications, network slicing, or MECs. The current principles of the Internet remain based on the ideas of the Advanced Research Projects Agency NETwork (ARPANET), which was formed around the notions of decentralized and self-organizing nodes.

This self-organization can be thought of as the source for considering the Internet as the *network of networks*, based on the IP protocol as unifier. We can readily think of a home network or an organizational network as being one of these individual networks constituting the overall Internet. Different individual networks typically belong to a specific owner, be it a person or organization, that manages the network and provides its resources. Additionally, the nodes inside such an individual network would be able to communicate with one another without relying on the rest of the Internet, forming an Autonomous System (AS).

To determine how to deliver an IP packet from source A to destination B, routers need to decide the mapping between incoming and outgoing network interfaces, that is, decide on the path of the packet through the network. For increased speed, routers rely on lookup or forwarding tables that provide this mapping. Overall, routing challenges can be easily solved with the *Dijkstra* or *Bellman–Ford* algorithms, especially if considering static routes. In reality, however, even on the inside of a large AS, the network topology is typically neither known nor static and requires updates of the routers' forwarding tables. This problem becomes even more pronounced when considering more than just an individual AS.

In accordance with the principles of the ARPANET, that is, decentralized and self-organizing nodes, routing protocols were developed. These protocols are utilized to provide information to the routers to determine the mapping and continuously update their forwarding tables. Commonly, we can differentiate between routing protocols that are geared toward optimizing routing *within* an individual AS and those that are designed to enable the interconnection of multiple ASs via routing. We can also consider the implications for the owners of the individual networks: although the inside routing of an AS network should follow the owner's policy and business needs, the AS as a whole can communicate its capabilities to external entities, That is, other ASs to interconnect. However, not every AS needs to know about the inside of another AS, just its capabilities and associated costs, to make a decision of whether to route packets. In turn, routing nodes connecting different ASs, the gateway nodes, do not forward the routing protocol information from the inside of the network. Individual ASs should typically not contain more than 50 routers [122]. Routing protocols can broadly be classified into three different categories:

i) Distance-vector routing, ii) Link-state routing, and iii) Inter-domain routing.

Following our AS-centric approach, we can divide these three categories into Interior Gateway Protocols (IGP) with distance-vector and link-state routing protocols as representatives and Exterior Gateway Protocols (EGP) with interdomain routing. In the following, we briefly highlight these three protocols:

Distance-Vector Routing In distance-vector routing, for example, implemented by the Routing Information Protocol (RIP) [123], each router has direct knowledge only of the cost to reach its direct neighbors. This information is sent to its corresponding neighbors. Using these gathered costs, each node calculates the minimal distance to every other node. In this way the information is iteratively propagated throughout the network.

Link-State Routing Link-state routing can be split into two phases, i) flooding of local information and ii) path calculation. The goal is to make the routing information globally available, but in a straightforward manner. An example protocol implementation for link-state routing is Open Shortest Path First (OSPF) [124]. The advantages of OSPF over RIP are its faster convergence and better scalability.

Inter-Domain Routing A famous example for interdomain routing is the Border Gateway Protocol (BGP) [125]. Interdomain or path-vector routing exchanges reachability information between ASs of multiple organizations. This improves the scalability and allows us to route between millions of routers. The problem caused by this type of networking is the lack of flexibility, as any change in protocol would require a massive hardware roll-out. In addition, the criteria for route selection by BGP is inflexible as well and considers neither demand nor capacity. This often causes congestion as described by Schlinker et al. in [126] and at IETF 104. If such a BGP route is misconfigured, then it can impact major parts of the Internet, resulting in significant outage levels. This happens frequently and affects even major service providers, such as *Cloudflare* [127].

The interplay of internal routing configurations and external interconnections, as well as planning for potential outages and fallback routes can be quite challenging. Additionally, consistency of network configurations is an important issue and needs to be considered as well, leading to the requirements of organized network administration.

6.2 The road to SDN

To ease the administration of networks, the concept of SDN was developed. In this section, we introduce the concept and explain multiple example use cases, which are enabled by SDN.

6.2.1 What is software-defined networking?

A centralized controller was proposed by the *Ethane* project [128] to manage the switches, countering the common decentralized approaches that dominated the Internet routing. The ideas of *Ethane* influenced the development of *OpenFlow* [129], which provides a standardized protocol for the communication between the controller and the switches/routers in a network. This approach is one of the key principles of SDN according to the Open Networking Foundation (ONF) [130]: providing a *centralized management*, which has a global view of the network. An additional key principle is a *directly programmable* network control by decoupling the forwarding or data plane and the control plane. This softwarization improves *agility*, since the network traffic can be dynamically adjusted (see Section 6.1).

SDN, in contrast to the ideas of the ARPANET, tries to centralize the management of the network and can dynamically program its behavior. In SDN, software plays a central role in the operation of networks by introducing abstraction for the data plane, separating it from the control plane [54]. To achieve such a separation, three types of components are needed:

i) a (logically) centralized SDN controller, ii) SDN-capable switches, and iii) a management protocol such as *OpenFlow* [129] or *Network Configuration Protocol (NETCONF)* [131].

Software-defined networks can be *programmatically configured*, that is, network administrators can write their own SDN programs to configure, manage, secure, and optimize network resources via automated scripts. For this, *open and vendor neutral standards* are needed, as we outlined in Chapter 2. The benefit of open application programming interfaces (APIs) is that a vendor lock-in is avoided. Through this abstraction, it does not matter which hardware is used, similarly to personal computers.

6.2.2 Architecture

The core of SDN is the separation between the control plane and the data plane, illustrated in Fig. 6.1. In addition, there is an application plane, which communicates its need to the control plane. This separation approach is similar to the OSI stack, where it helps to develop and test new protocols and hardware without changing the whole stack. Each layer provides an interface to its neighbors. The southbound interface, which is used by the controller to program the data plane, is standardized by the *OpenFlow* or *NETCONF* protocols. For the northbound interface, there is currently no standard, but, for example, a REpresentational State Transfer (REST) API could be used to let applications communicate their requirements to the network.

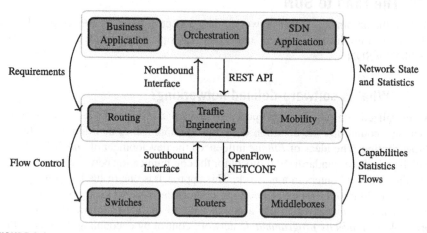

FIGURE 6.1

SDN architecture overview.

A misconception about SDN is that it cannot scale, because every packet is sent to the controller. However, this is not required, since the switches are only programmed by the controller. Additionally, there can be multiple controllers to improve scalability. Thus the data plane can be partitioned, and SDN *federations* can be built. With centralized control, we could assume that the reliability is worse than with decentralized control. Differently put, the SDN Controller is a single point of failure, but here, too, the remedy can be found in the form of several redundant or load-balanced controllers. In the next sections, we provide a more detailed discussion of *OpenFlow*, the most notable southbound protocol, as it has been widely adopted in industry and research. This, however, does not mean that SDN is only possible with the usage of *OpenFlow*. There are many alternatives, such as *NETCONF* or *P4* [132]. Furthermore, the legacy protocols can and are still used as well.

6.2.3 SDN use cases

We now briefly highlight some additional motivations to employ SDN in today's networks. We note that these use cases are nonexhaustive but represent some of the most common application scenarios and reasons to employ SDN.

6.2.3.1 Maintenance dry-out

Consider the case where a router has to be scheduled for maintenance and thus is known to be unavailable for a certain time. In most cases, there are redundant connections in the Internet. With SDN, the alternative routes can be easily configured to serve as backup routes during the maintenance without impacting the existing traffic. Although such configuration could be performed ex-ante for most modern routing equipment, it is worth noting that these configurations are typically static and require other failover systems to be reachable at known endpoints. With SDN, the failover points can be configured dynamically and are significantly more convenient to manage.

6.2.3.2 Traffic scheduling and predictability

The scheduling capability of SDN is also useful for predictable traffic peaks. Normally, routers select paths with the lowest cost metric that satisfies demands, for example, for latency or bandwidth. If the demand cannot be satisfied, then problems can arise. We can readily think of typical spikes in demands during peak times, for example, in the evening when people start watching *Netflix*. In such a situation, SDN can be used to route the excess over alternative routes, which may not have been selected from the standard routing protocols. This approach has been implemented, for example, by Schlinker et al. at *Facebook* [126]. An approach to realize this with the help of Machine Learning (ML) will be discussed later in Chapter 16.

6.2.3.3 Service function chaining

Another application for SDN is Service Function Chaining (SFC). In Service Function Chaining (SFC), new network functions are added to a flow of packets, for example, compression, coding, or a firewall, to name a few. The functions are provided at the intermediate hops in virtual environments attached to the switches. Using the capabilities of SDN, the traffic is redirected through the SFs. A practical implementation of an SFC is demonstrated later in Chapter 19.

6.2.3.4 User handover

The on-demand routing capabilities of SDN can be also exploited for dynamic scenarios, for example, a user handover. Consider the following case where we have an end device (User Equipment, UE), which can be a cell phone or an autonomous car moving from base station B1 toward B2, as illustrated in Fig. 6.2.

The UE requires a service provided by host H1 or H2. The UE, which was previously communicating via base station B1 to H1 (1), is automatically connected to B2 (dashed line). This is managed by the radio network. In addition, the traffic of the UE is still routed via B1 (2). An SDN controller can identify this flow and trigger a mi-

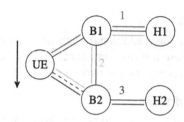

FIGURE 6.2

Handover scenario.

gration of the service from H1 to H2. In this way, both the latency between the UE and the service is lowered and the load of the radio backbone network (link between B1 and B2) is reduced. This feature is especially useful for the implementation of the MEC, as explained in Chapter 4.

6.2.3.5 Network access control

With Network Access Control, resources of a network can be restricted for different users. These resources can represent either a certain consumption or a service. Traditionally, access control is handled by firewalls or Virtual Private Networks (VPNs). With SDN, this can be more agile configured by an interface between the SDN controller and the application. For example, an application could register to the controller using a REST API and Hypertext Transfer Protocol (HTTP). In this way, providers have a fine-grained control over each service. In addition, this would require no configuration at the client side, as VPN does.

6.3 Technologies and standards

In this section, we describe different technologies required to enable SDN, that is, i) various controller and switch implementations, ii) OpenFlow as major standard, and iii) alternative protocols such as *NETCONF* and *P4*.

6.3.1 SDN controllers

For the control plane, there are multiple open-source controller implementations available, for example, *Floodlight* [133], *Opendaylight* [134], *Faucet* [135], *Ryu* [136], *Beacon* [137], *Pox* [138], and *Nox* [139]. Table 6.1 provides an overview of the various controllers, their primary domain of application, and how they are configured.

The reference controller implementation for the *OpenFlow* protocol was *Nox*, which is C++ based, and its sibling *Pox*, which is Python based. Both implementations were research-oriented and are now deprecated. However, they deliver many examples, which can be modified and adapted to different needs. In contrast, industry-

Table 6.1 Comparison of different SDN controller implementations.

Controller	Domain	Configuration type
Nox	Research & Industry	C++ application
Pox	Research	Python application
Beacon	Research	Java *bundles*
Floodlight	Industry	REST API, Java modules
Opendaylight	Industry	REST API, YANG data modeling
Faucet	Industry	YAML-based configuration file
Ryu	Research & Industry	Python application

oriented implementations, such as *Floodlight*, *Opendaylight*, and *Faucet*, often provide a REST API and are configured via a configuration file in Yet Another Next Generation (YANG) or YAML Ain't Markup Language (YAML) format, rather than actual programming. *Ryu* is comparable to *Pox*, but still under active development and supports the up-to-date *OpenFlow* versions.

6.3.2 SDN switches

SDN compatible switches can be implemented in hardware and software. Many manufacturers already offer *OpenFlow* compatible switches. The bandwidths range from Gigabit Ethernet for common business purposes with up to 64K flow table entries to 100 Gb switching capacity with 1000K table entries for edge-to-core applications [140]. In the scope of this book, we limit ourselves to software switches, which later form the basis for our *ComNetsEmu* emulator, introduced in Chapter 13. One of the most popular implementations for virtualized infrastructures and data centers is Open vSwitch (OVS) [141].

6.3.3 OpenFlow

A widely supported and deployed protocol for the southbound interface in SDN is *OpenFlow*. It will be used in the examples throughout this book, and therefore a detailed description is provided here.

6.3.3.1 Flow table

Each switch has a flow table, in which all the routing rules are stored. The components of such an entry are provided in Table 6.2.

6.3.3.2 Classifiers and actions

A basic component of an *OpenFlow* capable switch is the classifier used to match packets. The classifier consists of multiple subsequent match fields applied to the packet header. It starts with the physical layer, that is, the information on which port the packet has arrived. Afterwards, the Ethernet header (layer 2) is parsed, for example, for basic switching operation, the source and destination MAC addresses are

Table 6.2 Structure of flow table entries.

Match Fields	Identify packets matching this flow
Priority	Evaluation order in relation to other flow entries
Counters	Number of matched packets
Actions	Instruction for handling the packet (e.g., routing)
Timeout	Idle timeout - flow gets erased if no packet is received Hard timeout - flow gets erased after fixed time
Cookie	Chosen by controller to identify the flow. Usage depends on controller application

evaluated. For the network and transport layer, dependencies have to be met. It is not possible to directly match the IPv4 destination address by implicitly assuming that the packet under consideration is an IPv4 packet. Therefore the Ethernet frame-type field has to be evaluated, as depicted in Fig. 6.3. If the Ethernet type matches, then the IP (layer 3) header can be next evaluated. The same approach has to be followed for the transport layer, whereby the IP protocol number has to be first specified. An excerpt for some examples of match fields is listed in Table 6.3.

Table 6.3 Excerpt of *OpenFlow* match fields.

OSI layer	Examples	Description
Layer 1	`IN_PORT`	Switch input port
Layer 2	`ETH_DST, ETH_TYPE`	Ethernet header fields.
Layer 3	`ARP_OP, IPV4_SRC`	IP addresses and ARP
Layer 4	`TCP_SRC, UDP_DST`	Port information and control messages

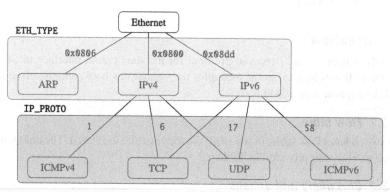

FIGURE 6.3

Dependencies of classifiers per layer.

If one packet matches a particular flow table entry, then an action is performed on the packet. Table 6.4 lists some possible actions.

Table 6.4 Excerpt of *OpenFlow* actions.

OSI	Proto.	Action	Header	Example application
Layer 1	Output	Forw.	Port ID	Drop, flood, or forward packet
	Queue	Set	Queue ID	Bandwidth shaping
Layer 2	Ethernet	Set	VLAN ID	Manipulate VLAN tags
Layer 3	IPv4	Set	Src./Dst.	Network address translation
Layer 3	IPv4	Decr.	TTL	Decrement Time-To-Live
Layer 4	TCP	Set	Port	Port address translation

A basic action is the actual rule to which port a packet will be forwarded to. This can be either a single port or all ports (flooding), which would be already sufficient for a MAC address-learning switch. In addition, multiple ports can be defined to realize multicast communication. For this, Internet Group Management Protocol (IGMP) packets can be intercepted by the controller. The controller can *snoop* [142] into Internet Group Management Protocol (IGMP) messages and determine which ports have interested listeners connected to them. The default behavior would be flooding, which, in turn, would cause a significant amount of extra traffic, especially in large networks. Another action that *OpenFlow* switches perform is dropping packets. This approach is used, for example, by firewalls to only let packets of certain connections pass and stop everything else from being forwarded. Bandwidth shaping to ensure a certain QoS for different traffic types is possible by forwarding packets into different queues as well.

These examples already show how extensible an *OpenFlow* compatible switch can be. It can be used as a simple MAC address-learning switch, but also with multicast and firewall support. Besides these basic functionalities, *OpenFlow* actions support manipulation of the headers of higher layers. In layer 2, this can be applied to manipulate the VLAN-ID, which can be used to split a network into multiple logical ones. Another classical task performed in this context is Network Address Translation (NAT) and decreasing the Time to Live (TTL) of a packet. Furthermore, Port Address Translation (PAT) can be performed on layer 4 using *OpenFlow*. To conclude, all today's network devices can be replaced by programmable and *OpenFlow* compatible switches, softwarizing and paving the way for centralizing the former decentralized networks.

6.3.3.3 Workflow of OpenFlow

Next, we take a look at the deployment of an actual flow rule in the network. A very basic event in networking is the deployment of a new communication path, which can be separated into the flow initialization and the flow continuation parts in an *OpenFlow* context. Utilizing the example illustrated in Fig. 6.4, consider the case where host H1 wants to communicate with host H2 over a network consisting of switches S:

Flow Initialization Initially, a packet is sent from host H1 to switch S1 (1). The switch evaluates its flow table entries looking for a match. If it is the first packet

of this flow, then it is most likely that there is none, and thus the switch asks
the controller what to do with it (2). The controller replies with a new rule,
for example, forwarding the packets to switch S3 (4). Upon receipt, the switch
stores the rule in its flow table (3). This rule is subsequently applied to all
incoming packets from host H1 to switch S1. The same procedure is repeated
for the succeeding switch S3 (5, 6). Finally, the packets are sent to the desired
destination H2 (7).

Flow Continuation   As mentioned earlier, it is a common misconception that all
packets would be sent to the controller to request what do with them, resulting
in significant inefficiencies. Once the initial communication between the switch
and the controller have configured the rule set, and they are stored in the switch
flow table, no further communication between the switch and its controller(s) is
needed. The switches initially search for a matching entry in their flow tables.
If found, then they will not send the packet to the controller and instead apply
the actions associated with the entry.

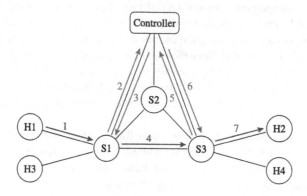

FIGURE 6.4

Example topology for the *OpenFlow* workflow.

6.3.4 **P4**

OpenFlow was one of the first widely used protocols for enabling SDN. However,
OpenFlow also has disadvantages, for example: i) Internet service providers cannot
choose specific functions, although not all functions are always needed, and thus
costs are high; and ii) each new *OpenFlow* version requires new hardware. Therefore
Bosshart et al. [132] proposed *P4*, a data plane programming language. In contrast
to *OpenFlow*, the data plane functionalities described with *P4* are not fixed by the
hardware. Vendors provide a abstraction model of the networking devices (the *P4
Architecture Model*) and a target specific *P4* compiler, as illustrated in Fig. 6.5.

Network engineers can then write *P4* programs for a specific architecture. Since
multiple targets can have the same architecture, programs can be ported. This approach is similar to general-purpose Central Processing Units (CPUs), which share

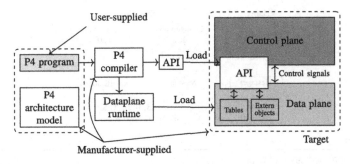

FIGURE 6.5

Programming of network devices with *P4* according to P4$_{16}$ language specifications [143].

the same architecture, for example, *x86*. A C program compiled for the *x86* architecture would run on any other *x86* CPU, independently of the vendor. If a *P4* is compiled, then two artifacts are produced: i) a data plane configuration, which is loaded onto the target and implements the actual forwarding logic given by the *P4* program; and ii) an API that can be used by the control plane for managing the data plane. The API is target-specific, in contrast to the *OpenFlow* approach, where the API is defined by the protocol itself. To summarize, *P4* has multiple benefits, which include:

i) it is easy to add support for new protocols, in contradiction to *OpenFlow*; ii) data plane bugs are easier to fix compared to Application-Specific Integrated Circuit (ASIC)-based solutions, and iii) *P4* reduces complexity, since it allows us to remove unnecessary features and therefore make use of the target resources more efficiently, which in addition reduces costs.

6.3.5 NETCONF

NETCONF [131] is another management protocol used to configure network devices. Its operations are realized on top of a Remote Procedure Call (RPC) paradigm. The NETCONF protocol uses an Extensible Markup Language (XML)-based data encoding for the configuration data and the protocol messages. The *NETCONF* protocol can be separated into four layers as represented in Fig. 6.6. The functions of the four layers are the following: i) The Secure Transport Layer provides the actual communication between client (controller) and server (network device) and can be any transport protocol that ensures certain requirements, such as authentication and security. ii) The Message Layer defines the encoding of the Remote Procedure Calls (RPCs) and notifications. iii) The Operations Layer defines a set of basic operations, for example, `get-config` and `edit-config`, which can be used to retrieve or edit a configuration, respectively. iv) The Content Layer is not defined by the actual *NETCONF* standard. However, the *YANG* data modeling language has been proposed by [144] as a candidate.

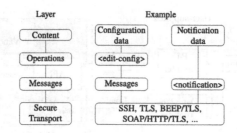

FIGURE 6.6

NETCONF protocol layers [131].

In contrast to *OpenFlow* and *P4*, *NETCONF* is designed as a generic configuration protocol. It could be used, for example, to configure future Time-Sensitive Software-Defined Networks (TSSDNs) as described by Nayak et al. [145]. Time-Sensitive Networking (TSN) will be introduced in Chapter 25. TSN is a set of standards for real-time capable communication needed, for example, by the *Tactile Internet*. Because it is not possible to configure TSN with *OpenFlow* or *P4*, general purpose configuration protocols, such as *NETCONF*, are needed. The protocols could complement each other, for example, by *OpenFlow* taking care of the individual network flows and *NETCONF* configuring the real-time communication.

Network function virtualization

Riccardo Bassoli
Technische Universität Dresden, Dresden, Germany

If you can dream it, you can do it.
Walt Disney

7.1 Introduction

SDN and NFV represent the main paradigms employed to realize software-based virtual networks [146]. The concept of NFV originated in 2012 when a group of network operators, such as AT&T, BT, Deutsche Telekom, Orange, Telecom Italia, Telefonica, and Verizon, selected the ETSI to be the responsible entity for the ISG for NFV, labeled ETSI ISG NFV. The main reasons for operators to focus on the virtualization of network functions were the necessity to reduce their CAPEX[1] and OPEX[2] paired with the necessity to increase network flexibility and facilitate network upgrades and modifications. Moreover, NFV can simplify the structure of radio sites by avoiding the need for additional rooms or cooling systems.

Network function virtualization represents a logic abstraction of physical networks and resources. The two main objectives of the NFV paradigm are the decoupling of software-based network functions from hardware-based physical network equipment and the deployment of network functions flexibly on demand. The former implies a centralized network management and control of network functions (which can be run locally or remotely) independent of the underlying hardware and solely dependent on service requirements. Moreover, such decoupling can enable easier maintenance of hardware (which now can be general-purpose computing hardware and no longer needs to be specialized hardware) and software. A flexible network function deployment can ensure easier management and assignment of network resources to specific services by adapting flexibly assigned resources according to their

---

[1] *Capital expenditure* or *capital expense* is the group of expenses related to infrastructures and fixed assets of a company, such as buildings, lands, or equipment.

[2] *Operating expense*, *operating expenditure*, *operational expense*, *operational expenditure* is the group of expenses related to running the activities and systems of a company.

current needs. This approach can also increase the capabilities of dynamically scaling network sizes and characteristics.

Among the pros NFV can provide, there is the efficient use of network infrastructure via its software-based VNFs. Next, NFV can increase the degree of freedom in the creation, deployment, and management of network services without worrying about configuration of vendor-specific network equipment. In fact, VNFs consist of software running on general-purpose hardware (i.e., servers). Moreover, NFV can create chains where individual functions are set up and deployed together to automatically produce compounded network services only when needed. In terms of cost and investments, NFV can rapidly adapt to technological innovation and thus, provide a better long-term Return On Investment (ROI). Whereas the ability of rapid adaptation increases in importance as product life-cycles are becoming shorter, it also enables companies to provide effective support for newly deployed network services.

Fig. 7.1 depicts the idea of NFV applied to current 4G cellular networks (on the left). Current 4G networks are composed by three main parts: the RAN, the core (operator) network, and the IP network (internet). The RAN is connected to the core network via backhaul/fronthaul links, which can be either wireless (e.g., microwave) or wired (e.g., fiber). The RAN includes end users wirelessly connected to the BS, which are constituted by RRH and Baseband Unit (BBU): the first is the antenna/radio equipment, and the second is the hardware devoted to baseband signal processing and FEC. Next, an operator's network is composed of a broad variety of connected devices. However, the principal ones are: Packet data network GateWay (P-GW), Serving-GateWay (S-GW), Mobility Management Entity (MME), and PCRF. By applying network virtualization these specific devices can be implemented as software-based virtual network functions into either VMs or containers. Thus it becomes immediately clear that general-purpose hardware and especially data centers and servers (so-called cloud computing resources) become the core infrastructure of future generation networks in the context of NFV. Furthermore, *softwarization* of network functions can open various and more efficient implementations of legacy hardware-based network equipment, such as what has happened for the BBU over the last decade.

Fig. 7.1 (on the right) also illustrates an example of a BBU virtualization and split. The baseband unit can be divided into subfunctions [147], such as radio processing, Fast Fourier transform (FFT), modulation/demodulation, FEC, and hybrid automatic repeat request (HARQ). These subfunctions can also be performed remotely, in separate servers, or VMs. This chapter has the following structure. In Section 7.2, we describe the concept of NFV by explaining its logical architecture and main characteristics. In Section 7.3, we discuss the proposed logic architectures for a joint NFV-SDN system. Next, in Section 7.4, we introduce the concept of a programmable protocol stack. Finally, in Section 7.5, we present the concept of BBU virtualization and split in detail.

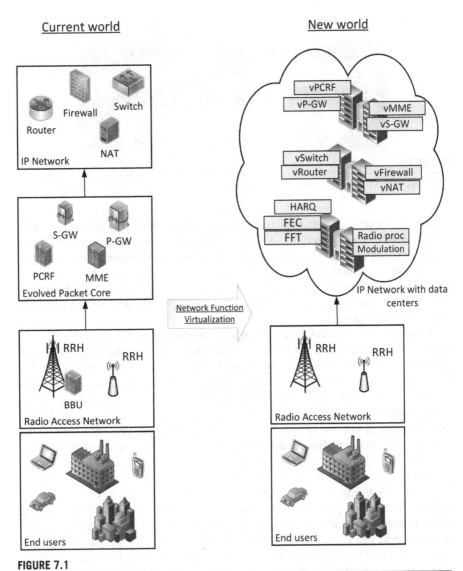

FIGURE 7.1

Legacy 4G wireless cellular network (on the left) and its logical representation after virtualization of network functions (on the right).

7.2 Network function virtualization

Fig. 7.2 shows the logic architecture of NFV. The upper layer contains all virtual network functions, which represent the services. Next, these functions rely on the virtual resources that are dynamically assigned to them. These resources can be grouped into computing, storage, and network resources. Whereas the first two groups are

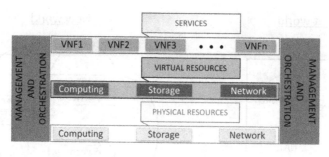

FIGURE 7.2

Logic structure of network function virtualization.

focused on making a single VNF running properly, the third kind of resources permits inter-VNF communications and collaborations. Virtual resources represent a projection/mapping of physical resources onto virtualization layers, whereby the former provide computation, storage, and network communications at hardware level. Finally, a vertical layer is responsible for management and orchestration of how hardware resources are mapped into virtual ones and how VNFs communicate with each other and collaborate.

Running virtual network functions on a server should generally result in keeping the hosting server continuously on, even if the full resources of the hardware are not all necessary. That characteristic would have led to an infrastructure challenge called *server proliferation*, mainly caused by increasing numbers of servers used very inefficiently, together with significantly growing power consumption for usage as well as cooling and augmenting expenses to buy infrastructures to host the servers. Virtual machines helped initially to avoid that upcoming significant challenge since it became possible to run multiple functions on the same server using a technique called *consolidation*. Virtual machines are mainly composed of three parts: i) the hosting Operating System (OS), ii) the hypervisor, and iii) the guest operating system. The first is the OS directly installed on the hardware. The second is a software hosting different VMs and responsible for resource management, monitoring, and managing VMs via coordination with the underlying hosting OS. There can be two kinds of hypervisors, type I and type II. The former is a hardware-based hypervisor, which does not need any host OS, because it directly communicates with the hardware resources. The latter is a software-based hypervisor, which requires a host OS, because it runs on top of the supported OS as an additional layer that interacts with the underlying hardware. Inside the VM, a guest OS runs all the virtual services. Fig. 7.3 illustrates a comparison between three aforementioned logic architectures to realize NFV.

As it is possible to observe in Fig. 7.3, virtual machines, containers, and unikernels are not the only ways to design and implement NFV. In 2014 an additional solution was proposed by Amazon, called serverless. *Serverless computing* [148] is a paradigm that allows developers of services to neglect certain aspects, such as server management and provisioning of resources, which becomes the responsibility of the provider of the platform. The common architecture for serverless solutions is

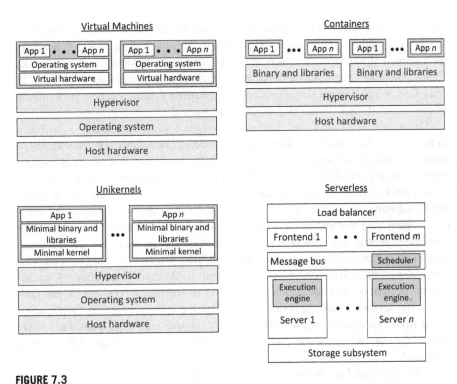

FIGURE 7.3

Comparison among different logic architectures to realize network function virtualization.

mainly composed of five components. First, the *storage subsystem* is the layer where states or data are made persistent to be shared by different functions (applications). Second, the *execution engine* is an element that runs on each server that deals with incoming requests: it addresses them by launching respective runtime environments (e.g., a container), with its required libraries, for the lifetime of the function. These containers are classified into *cold* and *warm*: the former is a container launched for each incoming request, whereas the latter is a container already active and can be reused by other functions. The deployment of warm containers was implemented to reduce latency due to startup. Third, the *message bus* and the *scheduler* constitute the interface responsible to forward messages from front ends to execution engines. Finally, the *front end* represents the interface for developers and for their applications. Multiple front ends can run behind a load balancer to improve scalability.

Serverless computing has various advantages, such as no need for server and resource management by application, resource efficiency, lower costs, and higher scalability. On the other hand, the main drawback for applications in specific 5G verticals is the significant startup latency, which makes current serverless computing ineffective for low-latency communications. Since the usage of VMs still incurs some overhead to simulate hardware inside a virtual environment, a lighter virtual package was created with low-level isolation and a shared Kernel OS, called *container*. Virtualization based on containers is more efficient because containers use lightweight APIs instead of hypervisors, which are the elements that introduce major overhead. Finally, unikernels are single-address space machine images constructed by using library OSs, which can run single processes. Unikernels achieve the best performance when compared to the solutions previously mentioned.

The properties of NFV can be divided into three main categories: i) attributes, ii) threats, and iii) means. *Attributes* were defined as *availability* (i.e., probability of readiness) and *partial availability* (i.e., availability in respect to a subset of requirements or users), *reliability* (i.e., probability of service continuity), *survivability* (i.e., system-level reliability), and *maintainability* (the ability to maintain and to repair functional units). *Threats* were grouped into *fault* (i.e., cause of system error), *error* (i.e., system state that can cause a failure), and *failure* (i.e., deviation of the service from the expected requirements). Faults were classified as *physical* (i.e., hardware-based fault), *transient* (i.e., temporary fault), *intermittent* or *sporadic* (i.e., recurrent fault), *design* or *logical* (i.e., human-based fault made during definition of specifications, design, or implementation), *interaction* or *operational* (i.e., accidental fault happening during human interactions with system), *environmental* (i.e., faults caused by environment where the system is located), *excessive load* (i.e., faults due to load greater than system capacity), and *malicious attack* (i.e., faults caused by external attackers).

When a system fails, it has to recover by going back to its original state. The recovery phase can be classified as repair (i.e., fix the component that is under failure) or replacement (i.e., substitute the failed component). Moreover, the process of repairing is performed through the stages of i) detection, ii) localization, iii) isolation, and iv) repair/replacement. The transformation of current hardware-based networks into virtual network based on virtual network functions is expected to significantly increase failure consequences at a low failure frequency, which will be a new important challenge to solve.

7.3 NFV-SDN architectures

Whereas SDN and NFV emerged as independent paradigms, during the last decade, research and industrial communities have been focused on merging them into a single architecture to make SDN and NFV work together as a unique stack/system. By analyzing the literature it is possible to identify three main SDN-NFV architectures, with the one proposed by ETSI, one of the most popular. Fig. 7.4 illustrates the ETSI SDN-NFV MANO architecture. This architecture consists of four main foundational blocks, which can contain other logical subblocks:

Network Function Virtualization Infrastructure (NFVI) contains the set of software-based (virtual) and hardware-based (physical) resources, which are nec-

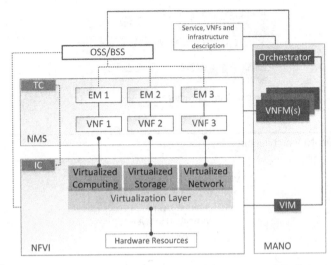

FIGURE 7.4

ETSI SDN-NFV MANO architecture.

essary to the internal functionalities of VNFs and to communications among VNFs.

MANO manages tasks and coordinates and automates the NFV architecture. This block also includes the VNF manager (VNFM), the VIM, and the Resource Orchestrator (RO) of VNFs:

> **RO** manages the resources of the NFVI, which are used by VIM(s).
> **VNFM** manages and handles configurations of the domain and the lifecycle of virtual network functions.
> **VIM** uses and controls the resources of NFVI.

Network Management System (NMS) manages the virtual network. It contains Element Management (EM), TC, and various VNFs:

> **EM** handles all information and events that are referred to VNFs such as configuration, performance monitoring/analysis, security, and failures.
> **TC** can be a VNF itself or a part of NMS in general. It is normally located in the tenant's domain.

Operation/Business Support Scheme (OSS/BSS) represents the set of applications that belong to the Internet Service Provider (ISP). These system-level and management applications are used by the ISP to provide specific network services.

Figs. 7.5 and 7.6 illustrate two additional SDN-NFV architectures, as further described in [149–151]. Fig. 7.5 depicts two main layers, network infrastructure and compute infrastructure. The former includes SDN switches (stateless processing),

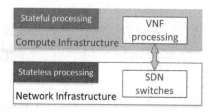

FIGURE 7.5

Two-layer SDN-NFV architecture.

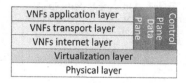

FIGURE 7.6

Pile-based SDN-NFV architecture.

which are devoted to provide connectivity to the VNFs. The latter represents the compute infrastructure, which is comprised of the VNFs and the related stateful processing.

Fig. 7.6 illustrates the pile-based SDN-NFV architecture. It is worth noting that there are two ways of interpreting this architecture: the first (horizontal) approach highlights the VNFs, which perform operations at each layer of the stack, whereas the second (vertical) approach is focused on the management of connectivity at both data and control planes of the SDN network. The virtualization layer is responsible to map resources of equipment at the physical layer into the virtual network layers. VNFs can virtualize internet layer or transport layer functions or services on the application layer.

7.4 Programmable protocol stack

The concepts of a programmable protocol stack and a wireless network operating system are additional important concepts in the context of network virtualization and virtualization of network functions. Legacy network protocols, as they are, have become increasingly less effective and less efficient in satisfying certain QoS conditions (such as latency), especially when considering the path toward future generation networks. The satisfaction and prediction of specific QoS levels have become increasingly difficult, especially when considering combinations of multiple protocols. Coordination across different protocols has similarly increased in difficulty, especially given the lack of a unified architecture. The upcoming realization of a unique

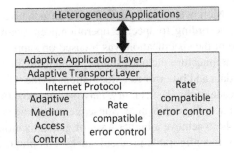

FIGURE 7.7

Adaptive protocol stack (AdaptNet) proposed in [153].

SDN-NFV architecture represents the fertile ground for the growing emphasis on developing reconfigurable protocol stacks.

This trend leads to the question of *What exactly is a programmable protocol stack?* This paradigm represents the implementation of a software-based environment that supports flexible and adaptive management of protocols and network layers. Reconfiguration refers to actions (such as parameter reassignment, service updating, and replacement of functionalities) according to user/network/environmental requirements.

The idea of programmable protocols, and consequently programmable stacks, is derived from preliminary works at the end of the last century [152]. Given the emergence of applications for multimedia content distribution through networks, the research community started thinking of adaptive/programmable transport protocols, which could have better answered to the requirements – in terms of greater QoS – of end users. This *protovirtual* system featured an abstraction layer to manage and remotely control the signaling system.

In the very beginning of the current millennium, society enormously increased its mobile internet population together with great evolution of wireless networks. These historical factors increased the degree of heterogeneity along several dimensions, such as the access technology, network model, device, and application requirements. Such an increasingly complex context enforced the idea behind the need of different nature of a network protocol stack, capable to adapt its different layers dynamically to the varying operating environment. A solution called AdaptNet [153] proposed a network protocol stack where different layers (such as application, transport, and link layers) contained adaptive protocols. The system proposal only left the network layer with IP unchanged to facilitate easy deployment while maintaining existing routing infrastructure.

Fig. 7.7 illustrates the adaptive architecture proposed in [153]. The blocks in the stack with gray background are the software-based ones, which can adapt according to network/application variations. The link layer includes an adaptive MAC to seamlessly change MAC characteristics without requiring any additional changes in the existing network infrastructure. The transport layer has an adaptive structure according to mobile hosts. Moreover, this layer includes an adaptive congestion control algorithm, adjusting according to specific operational environments. The application layer at the time of the specification was focused on supporting real-time video streaming. Thus the architecture includes source- and channel-adaptive coding to effectively handle the data and bit error rate fluctuations of the wireless channel.

After the first decade of the 21st century, network virtualization implementations became a reality and actually deployable. Subsequently, the idea of a protocol stack virtualization was able to achieve a higher level of generalization. In 2012 a model-driven framework for reconfigurable protocol stacks was proposed [154]. The realization of a programmable stack capable of supporting various kinds of real-time applications and protocols required the characterization and modeling of the complete system structure, including its traffic classification and constraints. The design includes interfaces for data to real-time applications, effectively providing a real-time intertask communication channel, which is capable to carry reconfigurations of protocols and layer logic blocks. Between 2017 and 2018, SDN and NFV paradigms started to become mature, which influenced the evolution of architectures and practical implementations of fully virtualized protocol stacks, as it is possible to see in newer proposals [155,156].

Fig. 7.8 depicts the logical architecture of Wireless Network Operating System (WNOS). The majority of research community and industry efforts have focused on SDN and NFV since the main interest has been virtualization of routing, network resources, and network functions. Nevertheless, the control problems in wireless networks can require further elements to be considered to optimally allocate resources. Specifically, allocation decisions should take the multilayer characteristics of the network protocol stack into account.

The *network abstraction framework* is the interface through which the targeted network control problem can be designed according to specific aims of the end-to-end applications. This logic block provides the characterization of network behavior and the centralized definition of the network control problem. The objectives can be defined via APIs to target throughput maximization, low latency, and so on. Furthermore, constraints should be included as well to take the characteristics of the physical network into account.

Next, the *automated network control problem decomposition* considers the definition of wireless network behavior to divide the targeted network control problem into distributed subproblems with their specific characteristics. According to the network and network control problem structures, decomposition can take different forms and can imply different overhead and complexity. It is important to mention that a network control problem and its decomposition can involve subsets of network layers and protocols, without necessarily applying modifications to all the layers of the stack.

Finally, the programmable protocol stack is a software-based stack that inputs information from the higher logical layers and configures various parameters at each layer, repeating the procedure at each network device. This update of parameters is dynamic according to network changes and end-to-end service requirements. The

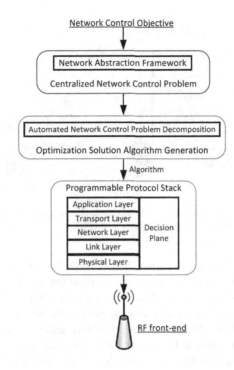

FIGURE 7.8

Architecture of wireless network operating system proposed in [155].

adaptive nature of the programmable protocol stack also involves the physical layer via the deployment of SDRs to optimize spectrum allocation and wireless resource management.

Fig. 7.9 illustrates the system layout of the Software-Defined Protocol (SDP) system. This system consists of SDP controllers and servers, which contain SDP blocks. In particular, SDP blocks perform processing of the paths of packets. New connections send SDP requests to the SDP controller, which establishes all the functionalities and characteristics (and eventually aggregations of multiple data flows) to release the on-demand protocol stack to satisfy the required QoS. In fact, an SDP controller maps the existing SDP requests onto the available SDP servers. The number of functional blocks involved in the processing path mainly depends on the end-to-end latency requirements. Furthermore, an SDP controller makes decisions on processing procedures for specific traffic flows, on configuration of flow table in the switches and on function blocks in SDP servers.

Fig. 7.9 also showcases the internal logic structure of an SDP server. This entity consists of four main logical blocks: i) control agent, ii) SDP block pool, iii) switch module, and iv) lower-layer interfaces. The *control agent* receives the control commands from the SDP controller, which are translated into rules for the functional blocks in the switch module. Moreover, the controller also updates the flow tables to

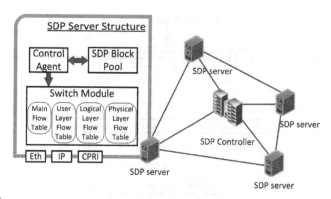

FIGURE 7.9

Software-defined protocol architecture proposed in [156].

assign the packets to their specific flows. The *SDP block pool* contains different kinds of functions to be performed. The processing carried out by these functions can also be subject to decomposition (flow tables are designed to support this feature).

Next, the packets belonging to a flow are sent to the necessary flow tables they request: i) the main flow table, ii) the user layer flow table, iii) the logical link layer flow table, and iv) the physical layer flow table. The *main flow table* sorts incoming data packets. Next, the *user layer flow table* and *logical link flow table*, respectively, categorize data packets according to the different users and logical links/services/applications they belong to. Finally, the *physical layer flow table* aims at forwarding the packets to their specific physical channels and interfaces. In Fig. 7.9 the considered interfaces are Ethernet (Eth), Common Public Radio Interface (CPRI), and IP. In this way, SDP servers can connect to other servers or controllers via different kinds of physical channels and also via RRHs.

7.5 Virtualization of RAN and BBU splitting

Cloud RAN is a network function virtualization paradigm with the specific scope of virtualizing baseband procedures. Fig. 7.10 depicts a schematic comparison between the legacy 4G RAN and the future cloud RAN. The legacy 4G RAN mainly consists of base stations, which are connected to a baseband unit. The baseband units are located at each radio site. Base stations and relative BBUs are connected via fibers supporting the CPRI standard. In current 4G cellular networks, baseband processing refers to processing of all the lower layers performed within the 4G protocol stack. Specifically, a BBU includes physical layer processing equipment (e.g., ASICs, DSPs, microcontrollers, and FPGAs), smart antennas, multiuser detection (for interference reduction), modulation/demodulation, error correction coding, radio scheduling, encryption/decryption of Packet Data Convergence Protocol (PDCP) communication (both, downlink and uplink), and multi-carrier modulation (MCM).

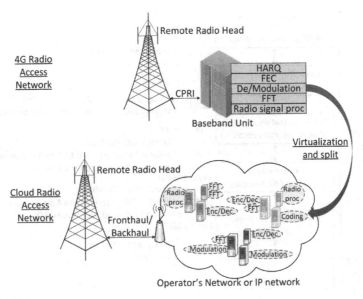

FIGURE 7.10

Components of the legacy 4G radio access network (top) and future virtualization of the radio access network functions (bottom).

Subcarriers are created and recovered at receiver and transmitter by using Fast Fourier transform (FFT) and inverse Fast Fourier transform (IFFT).

However, the current 4G implementation of a RAN is neither scalable nor efficient for future large heterogeneous scenarios, such as those emerging for 5G. Thus the objective of cloud RANs is the virtualization of BBUs (v-BBUs) to achieve higher flexibility in access network management and configuration. This enables baseband processing to be detached from standard hardware and location by transforming BSs into pure RRH. Subsequently, baseband processing can be moved to servers in data centers. Generally, the bandwidth of a cloud RAN can vary according to aggregate carrier bandwidth, cell load, the number of sectors, the number of antennas, modulation scheme, and error-correcting codes for FEC.

Fig. 7.11 illustrates the most commonly proposed architecture for this BBU split. The legacy 4G baseband unit can be decomposed into five layers [157] above the radio frequency equipment (RRH) devoted to analog-digital/digital-analog conversions. Layer 1 Low removes the cyclic prefix and performs FFT/inverse Fast Fourier transform (IFFT) on the signal. Layer 1 High maps/demaps resources. Next, Layer 2 Low is responsible for detection, equalization, modulation/demodulation, and precoding the information. Layer 2 High applies FEC to the user data. Finally, Layer 3 performs MAC and HARQ. Between each couple of these logical layers, a so-called split is possible. However, the lower the split in the stack, the more stringent the requirements and limitations in terms of throughput and latency. As one example,

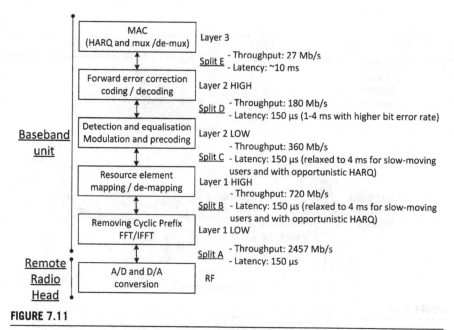

FIGURE 7.11

Virtual baseband unit splitting into virtual subfunctions with specification of requirements in terms of throughput and latency for each specific split.

consider the potential splits and requirements in Fig. 7.11. Whereas *Split E* (between Layer 2 High and Layer 3) requires 27 Mb/s and ≈10 ms delay, *Split A* asks for guaranteed minimum throughput of 2457 Mb/s with 150 μs latency.

Innovation track

Outline

As computing in communication networks breaks new ground for rapid deployments of disruptive innovations, we highlight three potential innovations, namely machine learning, network coding, and compressed sensing, as interesting candidates.

Machine learning

8

Riccardo Bonetto, Vincent Latzko

Technische Universität Dresden, Dresden, Germany

I propose to consider the question, Can machines think?
Alan Turing

8.1 Introduction

In 1950, Alan M. Turing wrote the opening quote of this chapter as the incipient of his seminal paper on artificial intelligence *Computing Machinery and Intelligence* [158]. He then proceeded to recognize that since the definitions of *thinking* and *machine* are far from being unambiguous, a new question should be asked instead. This question is known as the *imitation game* and marks the very dawn of the systematic research on artificial intelligence and, more specifically, on machine learning.

Machine learning is a wide umbrella term encompassing a plethora of heterogeneous theories and algorithms (e.g., statistical learning, Bayesian networks, self-organizing maps, etc.) developed during a time span of 70 years. The goal of every machine learning strategy, however, can be formulated as follows. Given a (possibly, partial) observation of the state of a system and a parametric model that outputs predictions based on the observed system state, find the best parameters set for the given model to maximize the prediction accuracy with respect to the task at hand.

This definition raises a number of questions: *What is, exactly, the observed state of a system? How should the parametric model be chosen? How is the prediction accuracy measured and improved?* This chapter is meant to provide answers to these questions.

As shown in Fig. 8.1, machine learning algorithms can be grouped into three macrocategories, namely *unsupervised learning*, *supervised learning* (Section 8.2), and *reinforcement learning* (Section 8.4). Briefly, unsupervised learning refers to the set of algorithms that aim at extracting information from unlabeled data (i.e., data for which no specific classification is known or needed). An example of an unsupervised learning task is *Clustering*. Clustering refers to grouping data into a number of (nonoverlapping) sets based on some similarity measure. Supervised learning in-

Computing in Communication Networks. https://doi.org/10.1016/B978-0-12-820488-7.00021-9

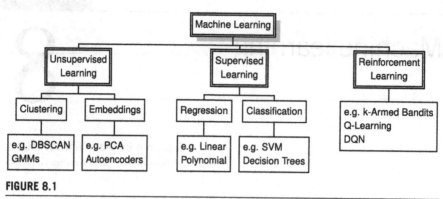

FIGURE 8.1

The machine learning fields and some example algorithms.

cludes the set of algorithms that, given some input, aim at predicting some of its characteristics (e.g., given the picture of a digit, predict the actual digit the picture represents). Lastly, reinforcement learning algorithms address the problem of teaching an agent (e.g., a car) to interact with an environment (e.g., the road) based on an observation of its current condition (e.g., the traffic and the traffic lights state). In this book, we focus on supervised and reinforcement learning.

8.2 Supervised learning

Supervised learning is the field of machine learning dealing with the approximation of unknown functions for which a set of input and output pairs is available. Two main tasks can be identified in this context. Namely, approximating a function whose codomain is continuous or one whose codomain is discrete. In the first case the learning task is called *regression*. In the latter the learning task is called *classification*.

8.2.1 Problem formulation

The supervised learning problem lays its foundation on the availability of a (possibly, large) collection of multidimensional data called *data set*. In such a collection, each element is called a *datapoint*. Each datapoint lies in an invariant M-dimensional space, called the *feature space*, and defined as the Cartesian product of individual feature domains. Each datapoint element is called a *feature*. Feature domains may be continuous or discrete. In the latter case the elements of such domains are called *classes*. Formally, a data set is defined as

$$\mathcal{D} = \{\boldsymbol{d}_i, \; i = 0, \dots, N-1 : \boldsymbol{d}_i \in \mathcal{F}_0 \times \cdots \times \mathcal{F}_{M-1}\}. \tag{8.1}$$

A convenient way to represent a data set is a matrix (or table)

$$D = \begin{matrix} & \begin{matrix} \mathcal{F}_0 & \mathcal{F}_1 & \cdots & \mathcal{F}_{M-1} \end{matrix} \\ \begin{matrix} i=0 \\ i=1 \\ \vdots \\ i=N-1 \end{matrix} & \begin{pmatrix} d_{0,0} & d_{0,1} & \cdots & d_{0,M-1} \\ d_{1,0} & d_{1,1} & \cdots & d_{1,M-1} \\ \vdots & \vdots & \ddots & \vdots \\ d_{N-1,0} & d_{N-1,1} & \cdots & d_{N-1,M-1} \end{pmatrix} \end{matrix}. \tag{8.2}$$

Let $\mathcal{F} = \{\mathcal{F}_0, \ldots, \mathcal{F}_{M-1}\}$ be the set of M individual feature domains. Then, depending on the task at hand, $\mathcal{F}$ is partitioned into $\mathcal{F}_p, \mathcal{F}_t \subset \mathcal{F}$: $\mathcal{F}_p \cup \mathcal{F}_t = \mathcal{F}$ and $\mathcal{F}_p \cap \mathcal{F}_t = \emptyset$; $\mathcal{F}_p$ are called *predictors* (i.e., these are the features used to make predictions), whereas $\mathcal{F}_t$ are called *targets* or *labels* (i.e., the features that are predicted), depending on whether the task is classification or regression, respectively. In the following, when the context bears no ambiguity, we will refer to both targets and labels as targets.

We assume that a function mapping predictors to targets (or labels) exists; see Eq. (8.3). However, this function is unknown.

$$f: \underset{\mathcal{F}_i \in \mathcal{F}_p}{\LARGE\times} \mathcal{F}_i \rightarrow \underset{\mathcal{F}_j \in \mathcal{F}_t}{\LARGE\times} \mathcal{F}_j \tag{8.3}$$
$$x \mapsto y.$$

Nonetheless, as shown in Eq. (8.4), we can define a parametric approximation of f:

$$\hat{f}: \left(\underset{\mathcal{F}_i \in \mathcal{F}_p}{\LARGE\times} \mathcal{F}_i \right) \times \Theta \rightarrow \underset{\mathcal{F}_j \in \mathcal{F}_t}{\LARGE\times} \mathcal{F}_j \tag{8.4}$$
$$(x, \theta) \mapsto \hat{y},$$

where Θ is called the *parameters space*, and $\theta \in \Theta$ is a specific parameter vector.

By using these definitions we can formally define the supervised learning problem: given an approximation function $\hat{f}$ called a *model*, find the optimal parameter vector θ^* such that a specific distance measure between y and $\hat{y}$ is minimized for all the pairs $(y, \hat{y})$ such that y is included in $\mathcal{D}$. The process of finding such an optimal parameter vector is called *training*. The selection of a specific model (i.e., a function $\hat{f}$) is referred to as *model selection*.

In the rest of this chapter, we describe state-of-the-art models, distance measures, and training strategies.

8.2.2 Supervised learning workflow

As already mentioned, to obtain meaningful predictions from a model, we need to train it (i.e., search for a parameter vector that minimizes some distance measure between the predictions and the targets). To do that, it is customary to split the data set $\mathcal{D}$ into three distinct subsets, namely *Training set* ($\mathcal{D}_{\text{train}}$), *Validation set* ($\mathcal{D}_{\text{val}}$), and *Test set* ($\mathcal{D}_{\text{test}}$). The split is done in such a way that $(\mathcal{D}_{\text{train}} \cup \mathcal{D}_{\text{val}}) \cup \mathcal{D}_{\text{test}} = \mathcal{D}$ and $(\mathcal{D}_{\text{train}} \cup \mathcal{D}_{\text{val}}) \cap \mathcal{D}_{\text{test}} = \emptyset$. Moreover, let D_{train}, D_{val}, and D_{test} be the matrix representations of the training, validation, and test sets, respectively. The training set is used to search for the optimal parameter vector, the validation set is used to monitor the prediction accuracy during the training phase, and the test set is utilized to perform a final assessment of the trained model performance. In Fig. 8.2, we show a block diagram of the generic training process. Given a model $\hat{f}$, a set of data points $x_0, \ldots, x_{N-1}$, and the corresponding targets $y_0, \ldots, y_{N-1}$, the data points are fed (one by one or in batches) to the model that, based on a parameter vector θ_i, produces predictions $\hat{y}_0, \ldots, \hat{y}_{N-1}$. These predictions are compared to the actual targets according to an error measure (also called objective function), thus producing the errors $e_0, \ldots, e_{N-1}$. In turn, these are utilized to improve the model parameters through the process of minimizing the objective function. This yields a new (and hopefully improved) parameter vector θ_{i+1}. The training process stops when the accuracy of the model when exposed to the validation set does not improve anymore (i.e., when we cannot find any other better minima of the objective function). It is worth noting that the model should *never* be exposed to the test set during the training phase. Intuitively, this is due to the fact that we want the model to perform well on previously unseen data.

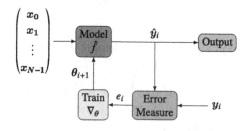

FIGURE 8.2

Block diagram explaining the supervised training workflow.

If we used the test set to tune the training process (e.g., we train our model, evaluate it on the test set and, if we are not satisfied with the results, perform additional training steps), then this would result in a model that has been tuned to perform well on a specific test set. However, this will give us little to no information regarding the performance of our trained model when exposed to new and unseen data. Additional explanation of this phenomenon will be given in Section 8.2.2.5, where we will introduce the underfitting and overfitting phenomena, and we will provide some general recipes for preventing them. Once the model is trained on the available training set (i.e., the validation error has been minimized, or it is acceptably small), the final performance can be assessed on the test set. This step eventually determines the actual performance of the model. Once a model has been trained and tested, it can be deployed and start serving the application purposes for which it was designed. As the model is exposed to new data, its parameters might need to be updated (especially

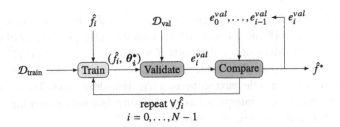

FIGURE 8.3

Block diagram explaining the model selection workflow.

in the case in which the initial training and test sets did not contain enough data to extract an accurate representation of the world of interest). To determine whether the model parameters need such an update, the model performance needs to be monitored carefully, even when deployed. An increased prediction error usually means that new properties of the world of interest emerged, but the model is not able to grasp them (as during the initial training process, these properties were still unknown). Hence the training phase might be performed periodically, or as soon as enough (depending on the model and the measured error) new data are available.

Until now we have taken the model $\hat{f}$ for granted. However, different models may have drastically different performance on the same data set. For this reason, it is important to select the right model for the task at hand. The process of selecting such a model is called *model selection*.

In Fig. 8.3, we introduce the model selection process through a block diagram. Given a number of potential models $\hat{f}_0, \ldots, \hat{f}_{N-1}$, each model is trained, and the validation error is computed, resulting in the validation errors $e_0^{val}, \ldots, e_{N-1}^{val}$. Each model is then compared to the others in terms of the respective validation errors. The best model $\hat{f}^*$ is then selected. Eventually, the performance of $\hat{f}^*$ is assessed with respect to D_{test}.

8.2.2.1 Feature encoding

Besides being discrete, some feature domains are not numerical (e.g., the eye color on passports is classified into *Brown*, *Blue*, and *Green*). For most models, it is convenient (if not required) to define a bijective mapping between a nonnumerical domain and a (suitably chosen) numerical one. This mapping is called *feature encoding*. There are two main feature encoding strategies, *label encoding* and *one-hot encoding*. It is also worth mentioning that instances exist for which it is beneficial to perform feature encoding even for numerical categorical feature domains. However, this applies in particular to the latter encoding strategy, one-hot encoding.

Label encoding

Label encoding is perhaps the most intuitive feature encoding strategy. Given a discrete and nonnumerical feature domain $\mathcal{F}_k = \{F_0, \ldots, F_K\}$, assign to each possible value F_i its index i: $F_i \iff i$. According to this encoding, a feature $d_k : d_k \in \mathcal{F}_k$

is replaced by the corresponding class encoding $\tilde{d}_k \in \tilde{\mathcal{F}}_k$. If, for example, the feature domain is $\mathcal{F}_k = \{$Brown, Green, Blue$\}$, then the corresponding label encoding is $\tilde{\mathcal{F}}_k = \{0, 1, 2\}$. Moreover, a datapoint $d = [d_0, \ldots, \text{Green}, \ldots, d_M]^T$ becomes $d = [d_0, \ldots, 1, \ldots, d_M]^T$. This strategy is memory-efficient and easy to compute; however, as we will see in the next sections, it has some drawbacks. The most impacting one is that it cannot be interpreted as a probability distribution over the likelihood of a feature belonging to a specific class.

One-hot encoding

Given a discrete feature domain (in this case, the transformation applies no matter whether the domain is numerical or not) $\mathcal{F}_k = \{F_0, \ldots, F_K\}$, a feature $d_k : d_k \in \mathcal{F}_k$ such that d_k belongs to class F_i is replaced by a vector v such that $v_j = 0$ if $i \neq j$ and $v_j = 1$ otherwise. For example, given a feature domain $\mathcal{F}_k = \{$Brown, Green, Blue$\}$, a feature $d_k = \text{Green}$ becomes $d_k = [0, 1, 0]$. With respect to memory usage, this strategy is less efficient than label encoding (each feature is encoded using a vector of the same size as the cardinality of the corresponding categorical feature domain). However, it generates a probability distribution over the likelihood of a feature belonging to a specific class. For labels, this distribution has exactly one nonzero value (the class the feature belongs to). Soon we will see that by obtaining predictions of the same probability distribution we can compute the confidence that our model has when assigning a datapoint to a specific class.

8.2.2.2 Commonly used distance measures

Depending on the learning objective and on the specific nature of the features being predicted, different indicators of the distance between a prediction and the respective target exist. These measures are usually called *loss* functions, and so they will be addressed in the following. Generally, a loss function is defined in the equation

$$\begin{aligned} L: \ &\mathbb{R}^k \times \mathbb{R}^k \to \mathbb{R} \\ &(\hat{y}, \ y) \mapsto c . \end{aligned} \tag{8.5}$$

Here we will introduce the most common two loss functions, the *Mean Squared Error* (MSE, often used in regression) and the *Categorical Cross-Entropy* (CCE, often used in classification).

Mean squared error

Given n predictions $\hat{y}_0, \ldots, \hat{y}_{n-1} : \hat{y}_i \in \mathbb{R}$, and the corresponding targets $y_0, \ldots, y_{n-1} : y_i \in \mathbb{R}$, the Mean Squared Error (MSE) is computed as

$$\text{MSE} = \frac{1}{n} \sum_{i=0}^{n-1} (\hat{y}_i - y_i)^2 . \tag{8.6}$$

Categorical cross-entropy

Let $\hat{y} = [\hat{y}_0, \ldots, \hat{y}_{m-1}]^T \in \mathbb{R}^m$ be a prediction, and let $y = [y_0, \ldots, y_{m-1}]^T \in \mathbb{R}^m$ be the corresponding target. Moreover, assume that both the target and the prediction are probability distributions. Then the Categorical CrossEntropy (CCE) can be computed as

$$CCE = -\sum_{k=0}^{m-1} y_k \log(\hat{y}_k).$$

(8.7)

It is worth noting that to compute the CCE, the targets need to be one-hot encoded, as detailed in Section 8.2.2.1, whereas the model predictions need to be probability distributions. MSE and CCE are not the only possible choices; indeed, many other loss functions exist. For a thorough discussion on the topic, see [159–161].

8.2.2.3 Error minimization: gradient descent

Now that we can compute the cost of a prediction, we have all the elements to train a model. The training process is, in fact, the minimization of the loss function L with respect to the model parameters. Hence the parameter vector we are looking for is

$$\theta^* = \underset{\theta \in \Theta}{\operatorname{argmin}} \left(L(\hat{y}, y) \right) = \underset{\theta \in \Theta}{\operatorname{argmin}} \left(L(\hat{f}(y, \theta), y) \right),$$

(8.8)

where the rightmost part of Eq. (8.8) highlights the dependence of the prediction $\hat{y}$ on the model parameters θ.

Eq. (8.8) tells us that training a model is, in fact, an unconstrained (often non-convex) optimization problem [162]. Here we will introduce the Gradient Descent (GD) algorithm along with its main variations, Stochastic Gradient Descent (SGD) and Mini-batch Gradient Descent (MGD).

Gradient descent

Fig. 8.4 shows an iteration of GD for a real-valued function $f: \mathbb{R} \to \mathbb{R}$. Given a point $x_i \in \mathbb{R}$ in the domain of f, a new point x_{i+1}, closer to the (possibly local) optimum x^*, is found by computing

$$x_{i+1} = x_i - \eta \frac{\partial f}{\partial x}(x_i),$$

(8.9)

where η is a parameter called the *learning rate*. As can be noticed in Fig. 8.4, if f is decreasing in x_i, then its derivative is negative (proportionally to the slope of the tangent to the graph of f in x_i) and the new point x_{i+1} will be to the right of x_i (i.e., $x_{i+1} \geq x_i$, with equality only when either the derivative or the learning rate is 0). If, on the other hand, f is increasing in x_i, then its derivative is positive, and the new point x_{i+1} will be to the left of x_i. An additional important item to notice is that the choice of the learning rate might have a nonnegligible impact on the convergence rate (and on the convergence at all) of GD. In particular, if η is chosen too small, then GD will take a long time to converge to an optimum point. However, we will be ensured

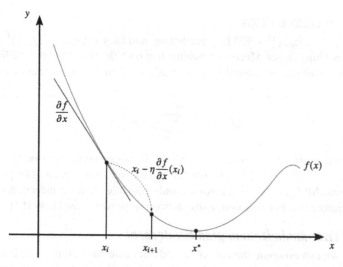

FIGURE 8.4

Graphical example of one gradient descent iteration.

that the search for such a point will be thorough. On the other hand, if η is chosen too large, then GD will move rapidly on the x-axis. However, the possibility exists that the optimal point x^* lies in between the current point x_i and the next one x_{i+1}, so that a too large learning rate might cause the algorithm to miss x^*.

The following equation shows the update rule for minimizing a generic multidimensional loss function L with respect to the model parameter vector $\theta \in \Theta$:

$$\theta_{i+1} = \theta_i - \eta \nabla_\theta \left[L(\hat{y}, y) \right] = \theta_i - \eta \nabla_\theta \left[L(\hat{f}(y, \theta_i), y) \right]. \qquad (8.10)$$

In this case, ∇ is the gradient operator, and the rightmost side of Eq. (8.10) renders the dependency of the loss from the model parameters explicit.

In many scenarios the loss function is a cumulative measure of the error for every prediction in the training set. For example, the MSE is the average of the squared errors of every target-prediction pair in the training set; see Eq. (8.6). Similarly, the CCE is usually accumulated by averaging the individual values over all the training set:

$$\text{CCE} = \frac{1}{n} \sum_{i=0}^{n-1} \left(- \sum_{k=0}^{m-1} y_k^{(i)} \log(\hat{y}_k^{(i)}) \right), \qquad (8.11)$$

where $y_k^{(i)}$ ($\hat{y}_k^{(i)}$) is the k component of the ith target (prediction) in a series of n target-prediction pairs. As a consequence, for each GD iteration, the model has to process the entire training set, compute the individual losses, and then average them. This, in turn, results in a computational workload that is proportional to the size of the whole training set. To mitigate this, two GD variations exist, SGD and MGD.

Stochastic gradient descent

SGD is based on the assumption that a single target-prediction pair is sufficient to approximate the aggregate loss function, for example:

$$\text{MSE} = \frac{1}{n}\sum_{i=0}^{n-1}\left(\hat{y}_i - y_i\right)^2 \approx \left(\hat{y}_k - y_k\right)^2 , \qquad (8.12)$$

where k is uniformly distributed in $0, \dots, n-1$ (i.e., $k \sim U\left(\{0, \dots, n-1\} \subseteq \mathbb{N}\right)$). If the approximation of Eq. (8.12) holds, then SGD only needs to evaluate the loss function with respect to a single target-prediction pair. For each optimization step, this noticeably reduces the computational workload required for the optimal parameter vector search.

Minibatch gradient descent

Despite its computational efficiency, SGD suffers from the fact that the loss function estimate tends to be of poor quality, especially in the early training stage. To mitigate this issue, MGD was introduced. MGD utilizes a randomly sampled subset of the training set called *Minibatch*, instead of approximating the loss function using a single, uniformly sampled, target-prediction pair. The size of the minibatch determines the quality of the approximation. For each optimization step, a minibatch of size b is sampled (without repetitions) from the training set. When the full training set has been used, a training *epoch* has been performed, and the sampling process starts over. Additional parameters of MGD are the minibatch size b, and the number of epochs E (i.e., the number of times the training process goes through the entire training set). For example:

$$\text{MSE} = \frac{1}{n}\sum_{i=0}^{n-1}\left(\hat{y}_i - y_i\right)^2 \approx \frac{1}{|\mathcal{D}_i|}\sum_{y\in\mathcal{D}_i}\left(\hat{y} - y\right)^2 , \qquad (8.13)$$

where $\mathcal{D}_i$ is the ith minibatch. MGD mitigates the intrinsic SGD estimation instability by evaluating the loss over a larger portion of the training set. This, however, comes at the expense of an increased computational cost per optimization step. This, in turn, requires to carefully tune the minibatch size.

8.2.2.4 Predicting probability distributions: SoftMax

As pointed out in Section 8.2.2.2, to compute the cross-entropy loss, the prediction vector $\hat{y} = [\hat{y}_0, \dots, \hat{y}_m]^T$ has to be a probability distribution (i.e., $\hat{y}^T \cdot \mathbb{I}_m = 1$, $\mathbb{I}_m \in \{1\}^m$, and $\hat{y}'_i \geq 0$, $i = 1, \dots, m-1$). Moreover, in Section 8.2.2.3, we noted that optimizing the model parameters requires the ability to compute the gradient of the loss function, which, in turn, depends on the model output. Hence, the model output and the loss function need to be differentiable. A differentiable function returning as output a probability distribution is the so-called *SoftMax* presented in the equation

$$\text{SoftMax}(i, \hat{y}) = \frac{e^{\hat{y}_i}}{\sum_{j=0}^{m-1} e^{\hat{y}_j}}, \tag{8.14}$$

where $\hat{y} \in \mathbb{R}^m$. An equivalent formulation is the vectorized one, shown in the equation

$$\overline{\text{SoftMax}}(\hat{y}) = \begin{bmatrix} \dfrac{e^{\hat{y}_1}}{\sum_{j=0}^{m-1} e^{\hat{y}_j}} \\ \vdots \\ \dfrac{e^{\hat{y}_{m-1}}}{\sum_{j=0}^{m-1} e^{\hat{y}_j}} \end{bmatrix} = \begin{bmatrix} \text{SoftMax}(1, \hat{y}) \\ \vdots \\ \text{SoftMax}(m-1, \hat{y}) \end{bmatrix}. \tag{8.15}$$

8.2.2.5 Overfitting vs. underfitting

As introduced in Section 8.2.2, to monitor the training process, a dedicated part of the data set, called validation set, is used. The loss function computed on the output of the model when presented with the validation set is called the *validation error*, which measures the ability of the model to *generalize*. The validation process has three possible outcomes: (i) low training error and low validation error, (ii) high training error and high validation error, or (iii) low training error and high validation error. In the first case the model is properly trained, and it is time to move to the testing phase. In the second case, either the training process is not completed (i.e., if more optimization steps are performed, the training error decreases, and so does the validation error), or the model is said to *underfit*. In the third case the model is said to *overfit*.

Underfitting

Under fitting is a consequence of the model not being complex enough to learn the relationships between the input features that are needed to correctly predict the output. Two solution strategies exist (and often they need to be used in combination): utilizing a larger data set (either by collecting more data or by artificially augmenting the available ones) or (if possible) selecting a more complex model.

Overfitting

When a model overfits, it learns to correctly predict all the training targets but fails when predicting validation or test targets. In other words, it fails to apply the knowledge gained during the training phase to data it has never seen before (i.e., it fails to generalize). We can spot overfitting by monitoring the training and validation errors. As soon as the validation error starts increasing (or remaining constant), whilst the training error continues decreasing, we can safely conclude that the model is overfitting. To mitigate this issue, we have three main options (or a combination thereof):

(i) utilize a larger data set (again, either by collecting more data or by artificially augmenting the available ones), (ii) lower the model complexity, or (iii) use a regularization term in the loss function.

Here we will introduce three general regularization techniques: $L1$ regularization, $L2$ regularization, and *early stopping*. It is worth noting that other regularization techniques exist (e.g., $L0$ regularization or *dropout*). These techniques, however, are either strictly related to a specific goal (e.g., $L0$ regularization is used to enforce the sparsity of the object being regularized [163]) or to a specific model (e.g., dropout is specific for Artificial Neural Networks; see Section 8.2.6). Hence here we will not discuss them in greater detail.

$L1$ regularization

Given a parameter vector $\boldsymbol{\theta} \in \Theta \subseteq \mathbb{R}^k$ and a loss function $L: \mathbb{R}^m \times \mathbb{R}^m \to \mathbb{R}$, the $L1$ regularized loss $L^{(L1)}: \mathbb{R}^m \times \mathbb{R}^m \times \Theta \to \mathbb{R}$ is defined as

$$L^{(L1)}(\boldsymbol{y}, \hat{\boldsymbol{y}}, \boldsymbol{\theta}) = L(\boldsymbol{y}, \hat{\boldsymbol{y}}) + \lambda \sum_{i=0}^{k-1} |\theta_i|, \tag{8.16}$$

where $\lambda \in \mathbb{R}_+$ is a parameter weighting the impact of the regularization term on the loss, and $|\cdot|$ is the absolute value function. $L1$ regularization aims at forcing as many elements θ_i of $\boldsymbol{\theta}$ to 0, thus enforcing sparsity in the parameter vector. This pushes the model to consider only the features that are strictly necessary to perform the task at hand.

$L2$ regularization

Given a parameter vector $\boldsymbol{\theta} \in \Theta \subseteq \mathbb{R}^k$ and a loss function $L: \mathbb{R}^m \times \mathbb{R}^m \to \mathbb{R}$, the $L2$ regularized loss $L^{(L2)}: \mathbb{R}^m \times \mathbb{R}^m \times \Theta \to \mathbb{R}$ is defined as

$$L^{(L2)}(\boldsymbol{y}, \hat{\boldsymbol{y}}, \boldsymbol{\theta}) = L(\boldsymbol{y}, \hat{\boldsymbol{y}}) + \lambda \|\boldsymbol{\theta}\|_2 = L(\boldsymbol{y}, \hat{\boldsymbol{y}}) + \lambda \sqrt{\sum_{i=0}^{k-1} \theta_i^2}, \tag{8.17}$$

where $\lambda \in \mathbb{R}_+$ is a parameter weighting the impact of the regularization term on the loss. $L2$ regularization aims at keeping the magnitude of the elements of $\boldsymbol{\theta}$ as small as possible. Pushing all the parameter vector components close to 0 results in the model not assigning a very high importance to any specific feature.

Early stopping

When using early stopping, the training process is interrupted as soon as the generalization error (i.e., the error on the validation set) starts increasing (or stops decreasing for a certain number of optimization steps). The main advantage of early stopping is that it neither modifies the loss function nor constrains the parameter vector.

8.2.3 Linear and logistic regression

Linear models are among the simplest machine learning models, yet efficient if applied to the right class of problems. Here we first introduce the *Linear Regression* model as a regression technique predicting targets based on a linear combination of the input features. We subsequently modify the linear regression model in such a way that it can perform (binary) classification, thus obtaining the *Logistic Regression* model.

8.2.3.1 Linear regression

The linear regression model requires the input datapoints to be fixed size real-valued vectors $d = [d_0, \ldots, d_{n-1}]^T \in \mathbb{R}^n$, and the parameter space to be $\mathbb{R}^n$ (note that the parameter vector dimension coincides with that of the input vector). According to the notation introduced in Section 8.2.1, the model is defined as

$$\hat{f}: \ \mathbb{R}^n \times \mathbb{R}^n \to \mathbb{R}$$
$$(d, \theta) \mapsto d^T \cdot \theta . \tag{8.18}$$

Usually, linear regression is associated with the MSE loss. Let the training set be defined as a real-valued matrix $D_{\text{train}} = [d^{(0)}, \ldots, d^{(m-1)}]^T \in \mathbb{R}^{m \times n}$, and let $y = [y_0, \ldots, y_{m-1}] \in \mathbb{R}^m$ be the target vector such that $(d^{(i)}, y_i)$ are the training input-target pairs. The MSE loss is then computed according to the equation

$$\text{MSE}(\hat{y}, y) = \frac{1}{m} \|D_{\text{train}} \cdot \theta - y\|_2^2 = \frac{1}{m} \sum_{i=0}^{m-1} \left[\left(\sum_{j=0}^{n-1} d_j^{(i)} \theta_j \right) - y_i \right]^2 , \tag{8.19}$$

where $d_j^{(i)}$ is the jth element of datapoint $d^{(i)}$.

Optimal solution

Optimizing the linear regression model with respect to the MSE loss (i.e., finding θ^* such that the MSE loss is minimized) is a well-known convex optimization problem (see [162]), and it can be solved analytically. Recall that

$$\|D_{\text{train}} \cdot \theta - y\|_2^2 = \sum_{i=0}^{m-1} \left(d^{(i)T} \theta - y_i \right)^2 . \tag{8.20}$$

Then the gradient of the MSE loss with respect to the parameter vector θ is

$$\nabla_\theta \left[\text{MSE}(y, \hat{y}) \right] = 2 \sum_{i=0}^{m-1} \left(d^{(i)T} \theta - y_i \right) d^{(i)T} = 2(D_{\text{train}} \theta - y) D_{\text{train}} . \tag{8.21}$$

This leads to

$$\nabla_\theta \left[\text{MSE}(y, \hat{y}) \right] = 0 \iff \theta = (D_{\text{train}}^T D_{\text{train}})^{-1} D_{\text{train}}^T y , \tag{8.22}$$

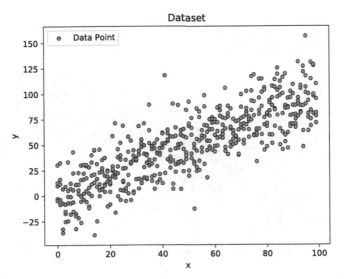

FIGURE 8.5

Example of a data set exhibiting a linear trend.

and we can conclude that

$$\boldsymbol{\theta}^* = \underset{\boldsymbol{\theta} \in \mathbb{R}^n}{\text{argmin}} \ \left(\text{MSE}(\boldsymbol{y}, \hat{\boldsymbol{y}})\right) = (D_{\text{train}}^T D_{\text{train}})^{-1} D_{\text{train}}^T \boldsymbol{y} \quad \text{Q.E.D.} \tag{8.23}$$

The rightmost side of Eq. (8.23) is called the *Normal Equation*.

Despite being able to analytically find the (unique) optimal parameter vector, the normal equation is seldomly used in practice, and MGD iterative optimization is preferred. This is due to the fact that the normal equation requires to operate on the whole training set (posing memory management challenges) by performing two matrix products (recall that for $n \times n$ matrices, the matrix product has a computational complexity of $O(n^{2.373})$ [164]). Fig. 8.5 shows a data set that exhibits a linear trend. Data sets with this characteristic are particularly suited for the linear regression algorithm. In Fig. 8.6, we can see the result of the optimization of the linear regression parameters applied to the data set of Fig. 8.5.

8.2.3.2 Logistic regression

Logistic regression models utilize a linear combination of an input datapoint to solve a binary classification problem (i.e., there are only two possible classes). Using the notation introduced in Section 8.2.3.1, the logistic regression model is defined as

$$\hat{f}: \ \mathbb{R}^n \times \mathbb{R}^n \to \mathbb{R}$$
$$(\boldsymbol{d}, \boldsymbol{\theta}) \mapsto \sigma(\boldsymbol{d}^T \cdot \boldsymbol{\theta}), \tag{8.24}$$

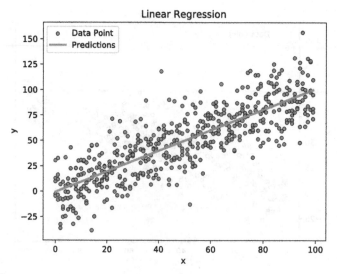

FIGURE 8.6

Example of a data set exhibiting a linear trend and the corresponding linear regression-based predictions.

where

$$\sigma: \mathbb{R} \to \mathbb{R},$$

$$x \mapsto \frac{1}{1 + e^{-x}}, \tag{8.25}$$

is the so-called *logistic function* (other widely used names, especially in the context of artificial neural networks, are a *sigmoid function* and *logit*). By taking a close look at Eq. (8.25) we can notice that its output is bounded in the interval $[0, 1]$. In turn, it can be interpreted as a probability distribution. In particular, given two classes $\mathcal{A}$ and $\mathcal{B}$, the output of the logistic function can be interpreted as the probability $0 \le p \le 1$ of a data point to belong to class $\mathcal{A}$. As the problem under consideration is a binary classification one, the complementary event is that the data point belongs to class $\mathcal{B}$, and this event has probability $1 - p$. Hence the class prediction for a given data point d and parameter vector θ is based on the following rule:

$$\begin{cases} \sigma(d^T \cdot \theta) \le 0.5 \implies P[d \in \mathcal{A}] \le 0.5 \implies d \in \mathcal{B}, \\ \sigma(d^T \cdot \theta) > 0.5 \implies P[d \in \mathcal{A}] > 0.5 \implies d \in \mathcal{A}, \end{cases} \tag{8.26}$$

where $P[E]$ denotes the probability of event E happening. As mentioned in Section 8.2.2.2, CCE is a suitable loss function for classification tasks. We can subsequently define the logistic regression model loss as

$$L(\boldsymbol{y}, \hat{\boldsymbol{y}}) = -\frac{1}{m} \sum_{i=0}^{m-1} \left[y_i \log{(\hat{y}_i)} + (1 - y_i) \log{(1 - \hat{y}_i)} \right]. \qquad (8.27)$$

Once again, this is a convex function, and hence the optimal parameter vector $\boldsymbol{\theta}^*$ is unique. To find it, we need to compute the gradient of the loss function. To this end, note that the derivative of the logistic function is $\sigma'(x) = \sigma(x)(1 - \sigma(x))$. Then the gradient of the loss function with respect to $\boldsymbol{\theta}$ is

$$
\begin{aligned}
\nabla_{\boldsymbol{\theta}} \left[L(\boldsymbol{y}, \hat{\boldsymbol{y}}) \right] = &-\frac{1}{m} \sum_{i=0}^{m-1} \left[y_i \frac{1}{\sigma(\boldsymbol{d}^{(i)T}\boldsymbol{\theta})} \sigma'(\boldsymbol{d}^{(i)T}\boldsymbol{\theta})\boldsymbol{d}^{(i)} \right. \\
&\left. - (1 - y_i) \frac{1}{1 - \sigma(\boldsymbol{d}^{(i)T}\boldsymbol{\theta})} \sigma'(\boldsymbol{d}^{(i)T}\boldsymbol{\theta})\boldsymbol{d}^{(i)} \right] \qquad (8.28) \\
= &\frac{1}{m} \sum_{i=0}^{m-1} \left[(\hat{y}_i - y_i)\boldsymbol{d}^{(i)} \right].
\end{aligned}
$$

Since Eq. (8.28) cannot be solved analytically, we can apply MGD to iteratively find $\boldsymbol{\theta}^*$ such that $\nabla_{\boldsymbol{\theta}} \left[L(\boldsymbol{y}, \hat{\boldsymbol{y}}) \right] (\boldsymbol{\theta}^*) = 0$.

8.2.4 Support vector machines

Support Vector Machines (SVMs) are binary classifiers that use hyperplanes to separate data. The result of the training process is interpretable models with comparably small memory footprint and quick execution times. Given a labeled data set, SVMs form a constrained optimization problem with a quadratic formulation. The task is to separate datapoints x, for example, to classify the class C_1 if some threshold $y(x) > 0$ is passed.

8.2.4.1 Linear separation

In the case of linear separation in two dimensions, this corresponds to using a line (a one-dimensional hyperplane) $\langle w^T, x \rangle + b > 0$ with parameters w and b. For calibration of parameters, N data points are used to find the hyperplane that best separates data. Depending on the data, there may be many hyperplanes that perfectly *label* points. However, the best separation *maximizes* the distance between all the points and the hyperplane, which is called the *margin* and is denoted by $\frac{1}{\|w\|}$. The constrained optimization problem is developed as

$$\underset{w}{\text{argmin}} \frac{1}{2} \|w\|^2 \;\; \text{s.t.} \;\; y(w^T x + b) - 1 \geq 0.$$

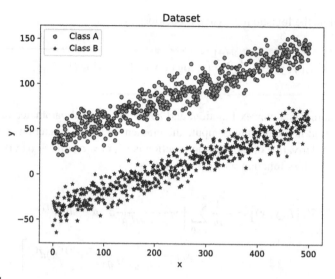

FIGURE 8.7

Example of a data set that can be linearly separated.

Quadratic programming is also able to find the dual problem obtained by constructing the Lagrangian $L = \frac{1}{2}\|w\|^2 - \sum_{i=0}^{N} \alpha_i (y_i (w^T x_i + b) - 1)$. It should be noted that the support of the hyperplane usually is dominated by only a few samples, that is, nearly all α_i are zero. This corresponds to the fact that there are only few support vectors.

8.2.4.2 Linear separation with margin

Even for not strictly linearly separable data, for example, slightly mixed or noisy data, SVMs can be used for classification. Margin classification uses an additional relaxation term in the problem description, yielding a nearly identical formulation. This additional term acts as a penalty to deviations:

$$\underset{w}{\mathrm{argmin}}\, \frac{1}{2}\|w\|^2 + \lambda \sum_{i=0}^{N} \chi_i \quad \text{s.t.} \quad y(w^T x + b) - 1 + \chi_i \geq 0, \chi_i > 0. \qquad (8.29)$$

Here $\lambda \in [0, 1] \subseteq \mathbb{R}$ is a weight to allow for a trade-off between regularization and fit, that is, how much violation of the class boundary to accept.

The numerical value is typically determined through cross-validation. Using this margin violation extension proves always beneficial, since it has been found to counter overfitting. Fig. 8.7 illustrates an example of a data set with two classes of data points (i.e., one class is identified by circular markers, whereas the other is identified by star-shaped markers). This data set can be clearly partitioned by a straight line. Hence it is said to be linearly separable. Fig. 8.8 shows the linear separation obtained by training an SVM to classify the data set of Fig. 8.7. We can see how the

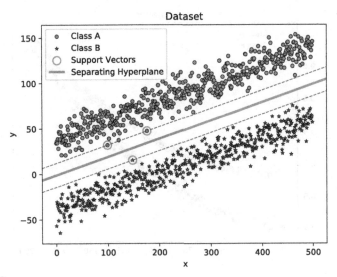

FIGURE 8.8

Example of a data set that can be linearly separated, and the corresponding linear Support Vector Machine (SVM) attempt at separation.

separating hyperplane maximizes the distance between each of the support vectors (identified by magenta circles).

8.2.4.3 Nonlinear separation

If the data set is not linearly separable, as that shown in Fig. 8.9, then it may be such in a higher-dimensional space. The so-called *kernel-trick* applies a suitable transform, which may be a nonlinear map from the raw input, instead of the linear dot product. This generates the feature space in which separation by a linear hyperplane may be possible. The kernel is a function $K(x_i, x_j) = \phi(x_i)^T \phi(x_j)$, which is an operator approximated by a matrix. Notable examples include Gaussian kernels, χ^2-, Pyramid Match-, and Histogram Intersection kernels.

Fig. 8.10 shows the result of applying a linear SVM to the nonlinearly separable data set of Fig. 8.9. As expected, this results in poor classification performance. On the other hand, Fig. 8.11 shows an SVM with a Gaussian kernel applied to the same data set as before. As we can see, the classification performance is improved dramatically. Care should be taken to transform into feature spaces too aggressively, since the kernels always increase the generalization error of the model, except for identity kernels.

8.2.5 Decision trees

Decision trees are widely used classification and regression models. They are based on successive binary splits of subsets of the training set until the final split generates

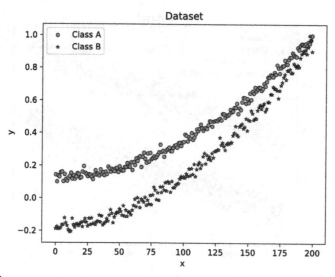

FIGURE 8.9

Example of a data set that cannot be linearly separated.

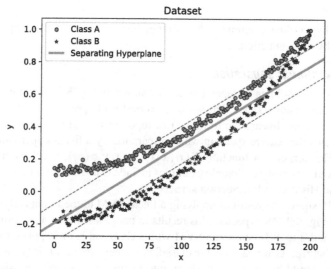

FIGURE 8.10

Example of a data set that cannot be linearly separated and the corresponding linear SVM attempt at separation.

the desired classification output. Each split is based on a single feature and is aimed at generating the purest subsets (with respect to some purity measure).

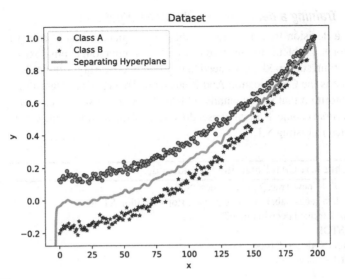

FIGURE 8.11

Example of a data set that cannot be linearly separated and the corresponding SVM separation when using a Gaussian kernel.

Decision trees get their names from mathematical structures called *trees*. A tree is defined as a pair $T = (\mathcal{N}, \mathcal{E})$, where $\mathcal{N} = \{n_0, \ldots, n_N\}$ is called the *node set* (and its elements are called *nodes*), and $\mathcal{E} \subseteq \mathcal{N} \times \mathcal{N}$ is called the *edge set* (and its elements are called *edges*). Trees have the property that if $(n_i, n_j) \in \mathcal{E}$, then there is no $k \in \mathbb{N}$ such that $(n_k, n_j) \in \mathcal{E}$. In this case, we say that n_i (n_j) is the father (child) of n_j (n_i). Moreover, we know that each node has at most one father. If a node has no father, then it is called the *root*. The root of a tree is unique. A node that has no children is called a *leaf*. Trees are organized into levels, with each level containing all the nodes with the same *distance* from the root. This distance is called *depth*, and, for a given node $n_i \in \mathcal{N}$, it corresponds to the number of edges one encounters when traversing the tree from the root to n_i.

In a decision tree, each nonleaf node represents a binary decision with respect to a single feature. Hence each node has exactly two children (making the decision tree a *binary tree*), identified as the *left* child and the *right* child. When a datapoint traverses a decision tree (starting from the root), each binary decision determines whether the next node will be the left or right child. Leaf nodes represent the classes to partition the data set into. When a datapoint completely traverses the decision tree, it ends up in a leaf node, thus being classified.

Unlike other models, trees can be efficiently visualized, and the classification rules can be immediately grasped by an audience with no machine learning specific technical skills. Because of that, decision trees are widely used in business decision making and financial analysis.

8.2.5.1 Training a decision tree: the CART algorithm

Training a decision tree (i.e., finding the best sequence of split rules) is a known NP-Complete problem (for an introduction to computational complexity, see [164]). Hence, to train such models, we need to resort to heuristic algorithms. The most well-known one is the Classification And Regression Tree (CART) algorithm. CART is a *greedy* algorithm that incrementally adds nodes to a decision tree, starting from the root. A recursive, object-oriented pseudocode implementation of the CART algorithm is presented in Listing 8.1.

Algorithm 8.1: CART algorithm pseudocode.

Setup: $T \leftarrow$ new Tree, T.root $\leftarrow$ new Node
Input: data set $\mathcal{D}$, labels $\mathbf{y}$, node T.root, stop condition STOP
Output: a trained decision tree T
if not STOP
   currentNode $\leftarrow T$.root
   split[] $\leftarrow$ new array
   for each $\mathcal{F}_i$:
      split[i] $\leftarrow$ new Split
      split[i].feature $\leftarrow i$
      split[i].rule $\leftarrow$ split threshold t_i w.r.t. $\mathcal{F}_i$ with maximum purity w.r.t. $\mathbf{y}$
      split[i].purity $\leftarrow$ purity of the split w.r.t. $\mathcal{F}_i$
      currentNode.split $\leftarrow$ split with maximum purity in split[]
      currentNode.leftChild $\leftarrow$ new Node
      currentNode.rightChild $\leftarrow$ new Node
      $(\mathcal{D}_L, \mathbf{y}_L) (\mathcal{D}_R, \mathbf{y}_L) \leftarrow$ tNode.split($\mathcal{D}$)
      CART($\mathcal{D}_L$, $\mathbf{y}_L$, currentNode.leftChild, STOP)
      CART($\mathcal{D}_R$, $\mathbf{y}_R$, currentNode.rightChild, STOP)
   return T

However, we note that *how* the purity of a split is measured or *what* the stopping conditions are has not been defined so far.

Split (im)purity

The purity of a split is measured either in terms of the *Gini impurity* or in terms of the *Entropy*. Given k classes $C_0, \ldots, C_k$, a split threshold t_i with respect to a feature $\mathcal{F}_i$, and a data set split $\mathcal{D}^{t_i}$, the Gini impurity of $\mathcal{D}^{t_i}$ is defined as

$$G(t_i, \mathcal{D}^{t_i}) = 1 - \sum_{j=0}^{k-1} \left[\frac{\sum_{d \in \mathcal{D}^{t_i}} \mathbb{I}[d \in C_j]}{|\mathcal{D}^{t_i}|} \right]^2, \qquad (8.30)$$

where $\mathbb{I}[d \in C_j]$ is the indicator function defined as

$$\mathbb{I}[d \in C_j] = \begin{cases} 1 & \text{if } d \in C_j, \\ 0 & \text{otherwise.} \end{cases} \qquad (8.31)$$

On the other hand, the entropy is defined as

$$H(t_i, \mathcal{D}^{t_i}) = -\sum_{j=0}^{k-1} \left[\frac{\sum_{d \in \mathcal{D}^{t_i}} \mathbb{I}[d \in C_j]}{|\mathcal{D}^{t_i}|} \log \left(\frac{\sum_{d \in \mathcal{D}^{t_i}} \mathbb{I}[d \in C_j]}{|\mathcal{D}^{t_i}|} \right) \right]. \tag{8.32}$$

These two measures lead to very similar results. However, the Gini impurity tends to isolate the most frequent class in its own branch of the tree (when the classes are not uniformly distributed).

8.2.6 Artificial neural networks

Artificial Neural Networks (ANNs) are interconnected computational models with a layered structure, consisting of nodes (*artificial neurons*), weighted connections, and functionality. They have gained massive traction in the years following 2012 and continue to evolve at a rapid pace, both in research and applications. Although the actual working mechanism is very far from any biological reference, the name will be briefly explained in the following. Nerve cells like in the cerebellum consist of a cell body, the soma, and a drainer, the axon. Axons from multiple neurons may be connected to a soma via terminals called dendrites. A typical axon in the human brain connects to approximately 10^4 neurons over synapses, electro-chemical connections. Through biochemical processes, voltage gradients built up in the soma and are released along the axon, triggering a nerve pulse arriving at many dendrites. Neurons accepting this pulse tend to increase their potential as a result, eventually also discharging if enough input is accumulated in a specific time frame: *What is wired together, fires together.* Any such firing is followed by a relaxation period where the neuron slowly balances itself against its embedding cerebral tissue and supporting cells by way of ion pumps and channels. This time-dependent behavior is common to all biological neurons, and, consequently, they are referred to as *spiking*. With the average human brain having over $80 \cdot 10^9$ neurons, with an estimated 10^{14} synapses, the connection of ANNs to biological neurons is merely inspirational.

8.2.6.1 Artificial Neural Network (ANN) fundamentals

In ANNs the key idea is to use a simple, parameterized structure in a repetitive manner, where the internal parameters are tweaked during training. Neurons are arranged in layers like an onion, with input data flowing toward some final computational step. This structure is pictured in Fig. 8.12. Each neuron in every layer is associated with two variables, a set of weights and a bias. Every connection a neuron receives (via its *dendrites*) from other neurons one layer before is associated with weight w_j. Only the most simple case of full connectivity is covered here, where every neuron receives signals from *all* nodes in the previous layer, and every layer features N neurons. For every neuron, this corresponds to a weighted sum $\mathbf{w} \cdot \mathbf{x} = \sum_{i=1}^{N} w_i x_i$. The bias is simply an offset to the activation, acting as a threshold.

Finalizing every neuron is its activation function, which yields a scalar value as the neuron output. Critically, this function is required to be nonlinear: any chain of linear functions after the multiplication of input and weights (itself a linear operation) would result again in a linear function, independent of input, dimensionality, or repetition. For reference, a function $f(x)$ is said to be linear if it satisfies the conditions (i) $f(x+y) = f(x) + f(y)$ and (ii) $f(\alpha x) = \alpha f(x)$. Otherwise, it is called *nonlinear*. Althoughthere is no principle limit in terms of choice, the most commonly used nonlinearities include *ReLU* (the *rectified linear unit*), $\sigma(x) = \max(0, x)$ and its more complicated variants, and also well-known mathematical functions such as $\sigma(x) = \tanh(x)$ or the sigmoid $\sigma(x) = \frac{1}{1+e^{-x}}$.

For any layer, all neurons have a vector of weights and a bias, and these parameters are joined into one matrix and one vector, respectively. A single neuron output a, when fed with an N-dimensional input $\mathbf{x}$, can therefore be written as

$$a(\mathbf{x}) = \sigma(\mathbf{w} \cdot \mathbf{x} - b) = \sigma\left(\sum_{i=1}^{N} w_i x_i - b\right). \tag{8.33}$$

This *activation* is a weighted sum of inputs, thresholded by a bias, and followed by a nonlinear operation.

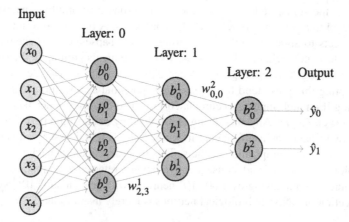

FIGURE 8.12

A feed forward ANN along with the corresponding notation. For example, $w^1_{2,3}$ flows *into* Layer 1, specifically into its neuron 2, coming from neuron 3.

Neuronal networks are universal function approximators, which means that any mathematical function can be approximated by a sufficiently large neuronal network. In turn, this implies that a neural network can act as stand-in for an unknown function, and parameters can be found for any nonlinear mapping. This is an extremely attractive property in the world of big data.

8.2.6.2 Layers

Typically, an ANN is considered to consist of at least three layers: input, processing, and output layers. Layers that have weights are also referred to as hidden layers.

The more the layers between input and output, the deeper the neural network. It has been found that deeper nets tend to perform better than a wide network with the same number of parameters (i.e., a network with 100 neurons in total may perform better if the neurons are distributed in multiple smaller layers instead of just one wide layer). The reasoning is that a network is pushed by optimization to extract useful patterns from the data that allow informed decisions to be made. This corresponds to combining features from lower levels into a composite pattern, each consisting of again lower-level features in some nonlinear way. With N neurons in a layer l, the N weight vectors are joined into a matrix W. The dimensions obviously depend on the size of each vector, which in turn depends on the number of neurons in layer $l-1$ and the connection scheme (however, for now, this technicality is set aside). Assuming that all layers have the same number of neurons N, the weight matrices are of shape $N \times N$ and can be thought of as an operator $W : \mathbb{R} \to \mathbb{R}$. This results in the sequence of matrices $w^l_{j,k}$ with $j, k \in [1, N]$ and the superscript l indicating the layer. The notation is pictured in Fig. 8.12. For simplicity, $z^l = w^l \cdot a^{l-1} - b^l$ is introduced, which is simply the sum of all active connections in layer l before the nonlinearity. The activation a^l is then composed of the activations a^l_j of individual neurons,

$$a^l_j = \sigma \left(\sum_{k=1}^{N} w^l_{j,k} a^{l-1}_k + b^l_j \right), \tag{8.34}$$

arranged as a vector. Consequently, this activation is fed into the next layer of neurons with a weight for each connection.

In various machine learning frameworks, the layers also accomplish certain other tasks, such as normalizing inflow data, reducing the dimensionality of feature maps, regularizing the model, or link to other layers not directly adjacent. Therefore the individual implementations must be checked.

8.2.6.3 Training with backpropagation

ANNs tend to have significantly more free parameters than the methods described earlier. This leads to both higher expressivity of the models and larger optimization spaces. The latter have proven to be prohibitive in the past, until the arrival of massive data sets, affordable computation (both CPU and GPU as well as, lately, specialized accelerators), and stimulating competition in the form of challenges, most notably the Large Scale Visual Recognition Challenge (ImageNet). ANNs with their layered structure, where each layer is shielded by a nonlinearity, pose a big problem for classical optimization techniques.

Already invented in 1960 by Kelley, the backpropagation algorithm is the de facto standard way of training neuronal networks, as it is able to find both function values and gradients. It first passes data through the network toward the final layer (the so-called *forward pass*), measures the discrepancy versus the label via some loss function $L(\hat{y}, y)$ (often, MSE or cross entropy), and computes the gradient of the error with respect to the parameters using the chain rule. This is done per layer, and this gradient is propagated backward toward the input layer (hence the name *backward pass*).

For a single-layer single-output network, with loss L and nonlinearity σ, the loss of the network for an input x is given by $L\left(\hat{y}, y^*\right)$. This is calculable right after the forward pass through the network. Note that $\hat{y} \equiv a \equiv \sigma(z)$. Next, the gradient $\nabla_w L$ of the loss is calculated by

$$\frac{\partial L}{\partial w_i} = \frac{\partial L}{\partial a} \frac{\partial a}{\partial z} \frac{\partial z}{\partial w_i}, \tag{8.35}$$

consisting of three components:

1. $da = \frac{\partial L}{\partial a}$, which directly depends on the chosen loss function,
2. $dz = \frac{\partial L}{\partial z}$, which includes the derivative of the activation, and
3. $dw = \frac{\partial L}{\partial w}$, which depends on the previous layer or, in this case, the input x.

Additionally, the term $db = \frac{\partial L}{\partial b}$ per layer can be updated right away. It can alternatively be written as

$$\frac{\partial L}{\partial w_i} = L'(\sigma)\sigma'(z)x. \tag{8.36}$$

Incidentally, this one-layer neural network exactly corresponds to the logistic regression case.

In a more complicated network with multiple layers and multiple neurons, the weight vector becomes a matrix, the bias and the activation scalars become vectors, and every parameter is associated with its layer $l \in [1, N]$, where the 1 has been dropped in the above example for simplicity. This approach is then repeated until the input layer, as w_{ij}^N are connected via the activations of layer $N - 1$ to previous layers and, ultimately, to the input. The weights are then changed by

$$w_{ij}^l = w_{i,j}^l + \eta \Delta w_{i,j}^l = w_{i,j}^l - \eta \frac{\partial L}{\partial w_{i,j}^l} \tag{8.37}$$

with η the learning rate as in prior sections. However, performing these calculations for a single example has proven to be inefficient due to overfitting. In contrast, performing these calculations on the whole data set at once is intractable for most of today's data sets. As a remedy, MGD is applied as described in Section 8.2.2.3.

8.2.6.4 Best practices, new trends

In the years following 2012, a multitude of new best practices began to emerge due to an immense increase in research, both scientifically and academically, as well as industry-driven. One of the first problem spaces addressed was the initialization of parameters. Whereas random values work, it poses a harder problem for the MGD algorithm due to mismatched expectation and variance and therefore is not a common choice. Different approaches for initial values have been proposed (Xavier, variance scaling, etc.) but have been rendered largely irrelevant with the introduction and widespread adoption of *batch normalization* (or BatchNorm for short). This technique forces the data to take on a unit *Gaussian* distribution at the beginning of the training process, effectively including a differentiable preprocessing at every layer of the network.

Regularization in the form of dropout has significantly increased generalization performance. Here specific layers are added into the network architecture (the dropout layers), which exclude a certain amount of neurons for a training batch. In essence, neither the forward pass nor the backpropagation step is applied to all nodes, but only a fraction. This forces the network to make due with lower expressive power.

Also, architecture search found some very specific helpful approaches. Residual Networks revolutionized the training and deployment of *deep networks*, adding a simple identity map that effectively skips a layer and its nonlinearity. This helps the gradients flow unhindered in the backward pass, which hugely stabilizes and speeds up the training process. A plethora of additional tricks exist, and they are evolving rapidly. Automatic differentiation and usage of a tape to record values during forward pass sped up training significantly, allowing the gradients to be recorded simultaneously. Replacing fully connected layers with fully convolutional ones reduces the amount of memory needed enormously.

8.2.7 Convolutional neural networks

One specific architecture of neural networks, Convolutional Neural Networks (CNNs), has revolutionized the domain of computer vision and image processing. These models use the mathematical operation that is well known from signal processing, where convolution is often used in conjunction with Fourier transform and the analysis of linear time-invariant systems. In this field, convolution is typically considered in one dimension only; the mathematical operation however is more general. For two functions $f(t)$ and $g(t)$, the operation

$$f(t) * g(t) = \int_{-\infty}^{\infty} f(t)g(t - \tau)d\tau \qquad (8.38)$$

is called the convolution.

Just like in the case of discrete time cyclical convolution, the integral can be approximated by a finite sum. The resulting discrete time convolution is predominantly used in the calculation of the fast Fourier transform

$$(f * g)[n] = \sum_{j=0}^{N-1} f[k]g[n - j]. \qquad (8.39)$$

An intuitive way for internalizing is to flip either but exactly one function horizontally and *slide it over* the other function domain, multiplying for any shift $d\tau$ the function values and keeping a running sum. Critically, this concept generalizes to higher dimensions, where visualization and intuition are much harder, if at all possible.

In Convolutional Neural Networks (CNNs), one of the two functions is some (typically, high-dimensional) input, be it an image with multiple color channels or some other representation, whereas the second function is a filter, also called a kernel, in a nod to the mathematical background of functional analysis. These filters are typically compact and much smaller than the input. CNNs are mostly used in tasks involving image recognition, but in fact their structure allows for arbitrary patterns to be matched. The result of any such convolution is a map of responses to the filter, exactly like in signal processing.

8.2.7.1 Convolutional layers

The main motivation for convolutional elements is connected to the sliding window that is the kernel. Convolving a filter with an image essentially multiplies for every step j that is taken across the image, resulting in j filter responses that are passed to the next layer (or rather, the next operation, such as a multiply–accumulate or nonlinearity). These responses again form an image (see Fig. 8.13), albeit transformed by

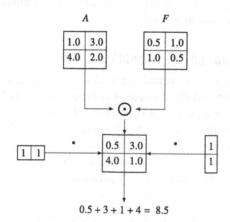

$$0.5 + 3 + 1 + 4 = 8.5$$

FIGURE 8.13

Example of the convolution between a filter F and a matrix A. $\odot$ represents the Hadamard product operation, whereas $\cdot$ represents the inner product operation.

the filter. Intuitively, this achieves translational invariance for pattern matching, since the kernel is applied to the whole input domain. One of the most powerful properties of CNNs is that not only a single, but rather k kernels are simultaneously convolved with the input map, yielding k responses. This means that a multitude of feature maps are created in every convolutional layer.

This specifically includes rotated kernels, yielding rotational invariance. Other kernels result in scale invariance, but in general this depends on the input data and the training process.

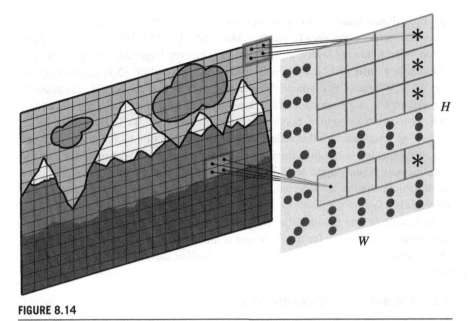

FIGURE 8.14

Example of a convolutional layer along with its parameters.

In addition, the kernels may move in steps larger than one (and therefore skip some input data), resulting in *strided* convolutions that give different dimensionality output maps. This gives a general formula for the output dimension of a convolutional layer:

$$H \times W = \left(\frac{H_{\text{in}} - F_H + 2P}{S} + 1 \right) \times \left(\frac{W_{\text{in}} - F_V + 2P}{S} + 1 \right), \qquad (8.40)$$

where W_{in} and H_{in} are the input dimensions, F_H and F_V are the filter sizes (not restricting to symmetrical filters), P is an optional padding at the borders of the input, and S is the stride. Again, one such map is created for every filter. In real implementations, instead of a simple two-dimensional map, a volume is typically used that has resulted from prior filter responses or various input channels. In Fig. 8.14, a 2×2 filter and its corresponding feature map of size $H \times W$ are shown.

By far the most powerful part of CNNs is the fact that the kernels are learned in the sense that backpropagation is applied *into* the kernels. The result is a data-driven extraction of features that are optimized for the task the neural network is designed and trained for. This element allowed CNNs to surpass traditional methods in the ImageNet Large Scale Visual Recognition Challenge (ILSVRC).

8.2.7.2 Pooling layers

Pooling layers are used for dimensionality reduction and, as a by-product, allow networks to achieve scale invariance for detected features on earlier layers. Images have very high dimensionality in comparison to other problems; for example, a simple 384×384 pixel image already yields a $384 \cdot 384 \cdot 3 = 442,368$-dimensional input vector. In the following layers, this number may be even larger depending on the kernels used (e.g., using 512 kernels with stride 1 would result in 1536 feature maps of size 384×384 *each*). For typical classification or detection challenges, this needs to be condensed to, for example, 1000 classes (typical ImageNet challenge) or a four-tuple for location and size of a detection bounding box (e.g., YOLO).

Pooling layers achieve this reduction by various selection techniques, which are – as all approaches in machine learning – constantly evolving. We concentrate on the so-called *max pooling* approach. For a feature map, a certain area (e.g., 2×2) and the maximum number are selected. This number is then placed at in the output, and the result is a feature map with its dimensionality reduced by a factor of $2 \cdot 2 = 4$.

In the case of max pooling, the operation corresponds to a nonmaximum suppression. Intuitively, filtering for the dominant response gives a sort of scale invariance. Other pooling approaches exist, where average pooling and global pooling are among the most commonly used.

8.2.7.3 Residual (skip) connections

Residual connections have been hinted as before and have revolutionized deep learning. They allowed the training of much deeper networks by addressing the problem of vanishing gradients. Conceptually, shortcuts are introduced that *skip* intermediate layers and directly introduce one layer response into a later layer. However, instead of simply concatenating the results, a subtraction is carried out, and the difference (*the residual*) is formed and used. The connection therefore is equivalent with multiplication with the negative identity matrix. This, in turn, allows the gradient to flow backward during backpropagation, since there always exists a path toward the input. The reasoning behind this is the realization that deeper elements in a network represent higher-level features.

8.3 Intermission

Neural networks are nonlinear high-dimensional mappings from some problem-dependent input space to some output space. The expressive power and performance lies in the amount of parameters that are tractable with current processing power. Their black-box behavior and combination of data points during training acts as a wrapper, often shielding users from exactly following the flow of data or gradients. This often leads to less interpretability, which has been widely criticized, since at the same time explainability suffers. Still, ANNs have revolutionized the fields of pattern matching, recognition, and modeling, which is especially true for CNNs. Their adaptability to a multitude of different problems due to their function approximation power has made them a convenient and competent tool for a variety of fields.

8.4 Reinforcement learning

As discussed throughout the previous sections, the supervised learning problem can be summarized as *given a set of input–output pairs of an unknown function, find the best approximation of such a function.*

Reinforcement learning, on the other hand, addresses the problem of teaching an *agent* (e.g., a robot) to interact with a (possibly dynamic) system by sensing its *state* and, based on this information, taking an *action* that results in a *reward* signal and possibly triggers a *state transition* in the system. The goal is to find a *policy* (i.e., a probability distribution on the available actions given the current state) that maximizes the expected cumulative reward. The state transitions, actions, and reward signals are all assumed to be stochastic. Thus the reinforcement learning problem is a particular instance of stochastic programming [165,166].

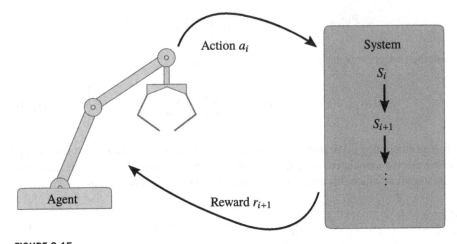

FIGURE 8.15

A typical reinforcement learning scenario: A robot (agent) interacting with a system through actions triggering state transitions and receiving reward signals.

Fig. 8.15 illustrates an example of the scenario we just introduced. The robot arm (identified as *agent*) interacts with a system by performing an action a_i that triggers a transition in the system state, namely $S_i \Rightarrow S_{i+1}$. This transition generates a reward signal r_{i+1} sampled from an (unknown) probability distribution that depends on the triple (S_i, a_i, S_{i+1}). This generates a repeating sequence of state observation, action, reward, state transition: $S_0, a_0, r_1, S_1, a_1, r_2, \ldots, S_n, a_n, r_{n+1}, \ldots$.

8.4.1 Finite Markov decision processes

A mathematical framework particularly suited to model the problem introduced in Section 8.4 is called *finite* Markov Decision Process (MDP). Given a specific system (or environment) and an agent, let S be the set of all possible system states (i.e., the *state space*), let S_i, $S_{i+1} \in \mathcal{S}$ be random variables representing consecutive states of the system, and let $\mathcal{A}_s$ be the set of all possible actions that can be taken by the agent when the system is in state s (in the following, we will lighten the notation of the action space by omitting the state information, thus identifying the action space as $\mathcal{A}$). Then, in accordance with the notation introduced in [167], we define the probability of transitioning to state $S_{i+1} = s'$ and obtaining reward $R_{i+1} = r$, given that the system was in state $S_i = s$ and the agent performed action $A_i = a$, as

$$p: \mathcal{S} \times \mathcal{A} \times \mathcal{S} \times \mathcal{R} \rightarrow [0, 1] \subset \mathbb{R}$$
$$(s', r \mid s, a) \mapsto P[S_{i+1} = s', R_{i+1} = r \mid S_i = s, A_i = a]. \tag{8.41}$$

From Eq. (8.41) (among other measures) we can compute the state-transition probabilities

$$P[S_{i+1} = s' \mid S_i = s, A_i = a] = \sum_{r \in \mathcal{R}} p(s', r \mid s, a) \tag{8.42}$$

and the expected reward for a given state-action pair

$$E[R_{i+1} = r \mid S_i = s, A_i = a] = \sum_{r \in \mathcal{R}} r \left[\sum_{s' \in \mathcal{S}} p(s', r \mid s, a) \right]. \tag{8.43}$$

As mentioned in the previous section, the ultimate goal of an agent is maximizing the expected cumulative reward, called *Return* and denoted as G_i (where i denotes the instant at which this quantity is computed). The return is defined as

$$G_i = \sum_{j=0}^{\infty} \gamma^j R_{i+j+1} = R_{i+1} + \gamma G_{i+1}, \tag{8.44}$$

where $\gamma \in [0, 1] \subset \mathbb{R}$ is the so-called *discount factor*. The discount factor serves two purposes. On the one hand, when $\gamma < 1$, it models the uncertainty in the future (i.e., the farther away the reward is in the future, the less weight it is assigned). On the other hand, it prevents the return going to infinity when the task the agent is performing does not have a *terminal state*.

To maximize the return, an agent needs to be able to evaluate the *quality* of a state-action pair. This is done by means of a *value function* measuring the *action value*. Such functions are defined as the expected return given the current state and action pair. Since the return depends on the future state transitions, which in turn depend on the future actions, to compute the action value, we need to know the probability distribution of the actions given a specific current state. Such a probability distribution is called *policy* and is defined as

$$\pi(a \mid s) = P[A_i = a \mid S_i = s]. \tag{8.45}$$

Given a policy π, the *value* of a state-action pair (s, a) is defined as

$$q_\pi : \mathcal{S} \times \mathcal{A} \to \mathbb{R}$$
$$(s, \, a) = E_\pi[G_t \mid S_i = s, A_i = a]. \tag{8.46}$$

The goal in reinforcement learning is finding the best action value function

$$q^*(s, a) = \max_\pi q_\pi(s, a). \tag{8.47}$$

8.4.2 Q-learning

Unfortunately, Eq. (8.47) cannot be analytically solved. As a matter of fact, the state transitions and the reward probability distributions are unknown. Thus the action value function can only be approximated. To this end, one of the most known algorithms is *Q-learning* [168]. Q-learning is based on an iterative update rule

$$Q_k(S_i, A_i) = Q_{k-1}(S_i, A_i) + \alpha \left[R_{i+1} + \gamma \max_{a \in \mathcal{A}} Q_{k-1}(S_{i+1}, a) - Q_{k-1}(S_i, A_i) \right],$$
$$\tag{8.48}$$

where k is the current iteration, and $\alpha \in [0, 1] \subset \mathbb{R}$ is a parameter called *Step Size*. Q-learning is an *off policy* method. This means that it converges to the optimal action value function q^* irrespective of the policy with which actions are selected during the iterative update process. Despite its proven effectiveness, Q-learning suffers from a crucial weakness. As a matter of fact, to update the Q function (i.e., the approximation of q^*), all the state-action pairs need to be accounted for in a tabular data structure (that is why Q-learning is said to be a *tabular* algorithm). Thus the memory footprint of the algorithm grows exponentially with respect to $|\mathcal{S}|$ and $|\mathcal{A}|$. Although this might be acceptable for small-sized problems, for many real-life tasks, as, for example, reinforcement learning based routing (see Chapter 16), the Q-table also needs to be approximated. To this end, deep learning based strategies, such as Deep Q-Learning [169], have been developed.

8.4.3 The exploration vs. exploitation dilemma

In Section 8.4.2, we described how to obtain state-action values and how to update these values to obtain the optimal policy for the finite MDP at hand. To do that, however, a rule is needed to select the actions *while* updating the Q-Table. One option could be using the Q-values determined so far as policy and selecting the next action accordingly. This approach, however, is likely to lead to local optima, thus preventing the algorithm from finding the best possible policy. Another approach could be selecting random actions, regardless of the Q-values learned so far. This guarantees that every action is treated *equally*, thus mitigating the risk of getting stuck in local optima, but it may also severely slow down the algorithm convergence rate. This is known as the *exploration* (i.e., choosing random actions) *vs exploitation* (i.e., selecting the best action) *dilemma*. To address this issue, several approaches have been proposed. Here we introduce the two most common ones, the ϵ-*greedy* policy and the Upper Confidence Bound (UCB).

8.4.3.1 The ϵ-greedy policy

The ϵ-greedy policy is based on a parameter $\epsilon \in [0, 1] \subseteq \mathbb{R}$ that determines the degree of exploration that should be applied to the algorithm. As shown in Eq. (8.49), at any given time t, selection of the best action requires awareness of the state the system is currently in (i.e., $\text{argmax}_{a \in \mathcal{A}} Q(S_t, a)$) with probability $1 - \epsilon$ and, with probability ϵ, a random action is chosen by drawing from a uniform distribution $\mathcal{U}\{\cdot\}$ over the action space $\mathcal{A}$.

$$A_t = \begin{cases} \underset{a \in \mathcal{A}}{\text{argmax}} \; Q(S_t, a) & \text{with probability } 1 - \epsilon, \\ a \sim \mathcal{U}\{\mathcal{A}\} & \text{with probability } \epsilon. \end{cases} \quad (8.49)$$

This strategy allows tuning the exploratory behavior of the algorithm by selecting different ϵ parameters. The closer ϵ is to 0, the more this approach exploits current knowledge. On the other hand, the closer ϵ is to 1, the more exploration of options. A common approach is to make ϵ decay (linearly or exponentially) proportionally to the number training episodes. In this manner the approach starts by aggressively exploring options and, as the training progresses, moves to exploiting the acquired knowledge more and more.

8.4.3.2 The upper confidence bound policy

One of the drawbacks of the ϵ-greedy policy is that whenever choosing a random action, we completely discard all the knowledge we acquired so far. The UCB policy aims at mitigating this drawback by taking into account how many times a state-action has been visited before and how far into the training we are. This results in the action selection policy shown in the equation

$$A_t = \underset{a \in \mathcal{A}}{\text{argmax}} \left[Q(S_t, a) + c \sqrt{\frac{\log t}{N_t(a)}} \right], \quad (8.50)$$

where $N_t(a) \in \mathbb{N}$ is the number of times action $a \in \mathcal{A}$ has been selected, and $c \in \mathbb{R}_+$ is a parameter weighting the importance of the exploration.

8.4.4 Deep Q-learning

Deep Q-Learning (DQN) [169] is a reinforcement learning algorithm that exploits the universal approximation property of ANNs (see Section 8.2.6.1) to approximate the Q-table and break the dependency of the memory complexity of the learning process from the size of the state-action space.

The rationale behind DQN is that we can use an ANN to predict the Q-values for each action $a \in \mathcal{A}$ based on the state that is fed as input to the network. To obtain meaningful predictions, the ANN (called the *Q-network*) needs to be trained, and in turn a loss function needs to be defined. However, since here we are not considering a supervised learning problem, we have no any *label* identifying the correct Q-value for any given state-action pair. Hence, defining a suitable loss function is not straightforward. We start by initializing two identical ANNs with the same parameters. Using the notation introduced in [169], let these networks be Q and $\hat{Q}$, and let the parameters be $\boldsymbol{\theta}$ and $\boldsymbol{\theta}^-$ ($\boldsymbol{\theta} = \boldsymbol{\theta}^-$), respectively. To train the Q-network Q, an action $a_t \in \mathcal{A}$ is initially selected according to a user-defined policy (e.g., the ϵ-greedy one) and then applied, the state transition (i.e., from state S_t to state S_{t+1}) is observed, and the corresponding reward r_t is collected. The target update (i.e., the value with respect to which the loss function will be evaluated) is defined as

$$y_t = \begin{cases} r_t & \text{if } S_{t+1} \text{ is terminal,} \\ r_t + \gamma \max_{a \in \mathcal{A}} \hat{Q}(S_{t+1}, a; \boldsymbol{\theta}^-) & \text{otherwise.} \end{cases} \tag{8.51}$$

The target y_t is compared to the output of the Q-network by means of an MSE loss, as shown in the equation

$$L = (y_t - Q(S_{t+1}, a; \boldsymbol{\theta}))^2 \,, \tag{8.52}$$

and the parameters $\boldsymbol{\theta}$ are updated accordingly by means of a gradient descent step. Every K training steps (where K is user defined), the parameters $\boldsymbol{\theta}^-$ of the target network $\hat{Q}$ are updated by setting $\boldsymbol{\theta}^- = \boldsymbol{\theta}$.

Whereas DQN allows us to apply a variation of Q-learning to problems that could not be tackled with the traditional tabular approach, its convergence may be difficult to achieve. To mitigate this issue, several approaches have been proposed as, for example, the *replay memory* introduced in [169].

Network coding

Juan A. Cabrera G., Morten V. Pedersen, Frank H.P. Fitzek
Technische Universität Dresden, Dresden, Germany

Why don't we just call plans what they really are: guesses.
Jason Fried and David Heinemeier

9.1 Interflow network coding – the basics

The concept of network coding originated in the seminal paper of Ahlswede et al. in 2000 [170]. The goal of this work was characterizing the coding rate region for the general multicast problem in point-to-point communication networks without noise. This type of network is commonly encountered, for example, in the case of computer applications. In these networks, information is typically sent within atomic and non-modifiable units called *packets*, which are obtained from the fragmentation of larger contents generated at an information source. A network is modeled as a graph where the nodes represent its physical elements and the edges represent the corresponding connections between those elements. The packets are sequentially sent through the nodes o reach an intended destination, which is an information sink. In the general multicast problem, multiple information sources generate packets that have to be conveyed to multiple sinks. A convention in this type of networks was that intermediate nodes in the network can either forward or replicate packets in a specific order before being conveyed to the next node. This concept would be later known as *routing*.

In [170], to address a solution to the general multicast problem, the authors narrowed the problem to the case of one source and various sinks with a multicast requirement. It was found that the admissible coding rate region is defined by the max-flow min-cut solution of this problem viewed from a graph-theoretical perspective (we will describe this solution later in this chapter). Deriving this fundamental result, the authors in [170] noted that allowing operations on and modifications of the packets coming to the input edges of a node would permit to achieve the max-flow min-cut capacity. This core idea represented a major difference to methods considered at the time and former routing-based solutions. Thus modifying the information coming into a network node prior to its transmission was defined as *Network Coding*.

One important aspect of network coding is that it can be implemented at any layer of the OSI model, from physical to the application layer. In this chapter, we describe

Computing in Communication Networks. https://doi.org/10.1016/B978-0-12-820488-7.00022-0

the fundamental types of network coding at the packet level, given that it is also the most well-understood form in the literature.

The initial research of network coding was based on inter-flow network coding, where coding is based on an *opportunistic coding* approach. Following this strategy, coded packets are generated in a defined region of a network to serve nearby nodes with different packet requirements. This increases the throughput, since the total amount of transmissions required to serve the sinks is reduced. To achieve this, the encoder node creates coded packets by mixing the original packets from *various* network flows. Since the coding opportunities depend on the network topology to be considered, we present some network examples where this type of coding is suitable.

9.1.1 The butterfly network

In [170] the authors introduced the butterfly network illustrated in Fig. 9.1 as an example to show the potential gains of NC. This network consists of a source node S and two sink nodes R_1 and R_2. In this network, the source wants to deliver two packets (represented as blue [dark gray in print version] and yellow [light gray in print version] packets in Fig. 9.1) to both sink nodes. We consider that both packets have the same arbitrary length in bits. The network also includes two relay nodes C and D that allow the source to convey these packets to the sinks. In this example, let us consider that the capacity of each link between a pair of nodes is one packet in a given unit of time and that there are no sources of noise that could corrupt the information in any packet.

Fig. 9.1A presents the case of using routing at the relay node C. In this case, this node has to choose between the blue and yellow packets for transmission over the link that connects itself with D. If it chooses packet blue, then D forwards it to both sinks. Then R_2 receives only one packet duplicated. If, instead, it chooses the yellow packet, then a similar problem occurs, and R_1 receives a duplicated packet. However, when using network coding as shown in Fig. 9.1B, a new *coded* packet is created as a bitwise XOR of the other two packets. The coded packet is represented as a green (mid gray in print version) packet to illustrate that it is the combination of the blue and yellow packets. The resulting coded packet has the same length as the original packets. In real networks, we may drop the constraint for the blue and yellow packets to have the same length, since a shorter packet could be zero-padded before the XOR operation takes place. The padding could later be removed as postprocessing. After sending the coded packet through D, R_1 *decodes* the blue packet by performing the XOR of the green packet and the yellow packet; R_2 decodes the yellow packet in a similar manner.

In linear network coding (i.e., the type of network coding we discuss in this chapter) the modifications applied to the received packets at a network node are linear over a finite field. Differently put, the outgoing packet is created multiplying the incoming packets by coding coefficients and adding them together with the addition and mul-

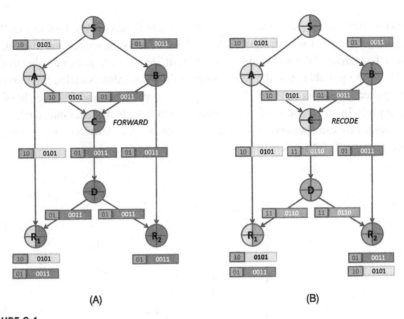

(A) (B)

FIGURE 9.1

Forwarding and coding in the butterfly network. (A) Routing in the butterfly network. The throughput is 1.5 packets per time unit; (B) Network coding in the butterfly network. The throughput is 2 packets per time unit, which is the min-cut max-flow.

tiplication operations defined within the finite field (the details of the operations are explained later in this chapter). In network coding, to differentiate the packets at the destinations, some type of signaling must be appended to a coded packet to identify which coefficients were used to create it. A representation of the coefficients is added to the packet (constituting a packet overhead) to achieve this goal. With knowledge of the coefficients, each destination identifies the packets received and performs the algebraic operations necessary to obtain the required original packets. In Fig. 9.1 the coding coefficients are represented as an overhead in pink, prepended at the beginning of the packets. Therefore 10 refers to the data of the packet containing zero times packet blue and one time packet yellow. Similarly, the coding coefficients 11 represent that the data packet is the combination of one time packet blue plus one time packet yellow.

When the butterfly network uses a routing approach, we see that a bottleneck occurs at C, since only one packet can be forwarded per time unit. This leads to an overall throughput of 1.5 packets per time unit. However, using network coding permits to achieve a throughput of two packets per unit of time, which is the maximum throughput that can be achieved in this network according to the min-cut max-flow theorem.

The min-cut max-flow theorem states that the maximum flow that can pass from a source to a sink is given by the total weight of the edges of the network in the minimum cut. For example, consider the cuts of the butterfly network shown in Fig. 9.2. A cut in the network removes some edges and disconnects the source and the sink. If we add the weight of the removed edges in the cut, then we can assign a value to that cut. Of all the possible cuts, the one with the minimum value is called the min-cut. In the butterfly network in Fig. 9.2, each edge has a weight of one. The weight of an edge is given by the capacity of the communication link. In this example, each communication link can transport one packet per unit of time. Therefore the minimum cut of the butterfly network is two.

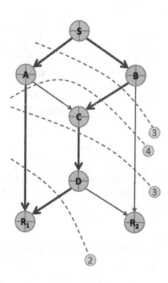

FIGURE 9.2

Different cuts of the butterfly network. The min-cut is two.

To achieve potential gains through this representation, network coding introduces processing at any network node with three operations: encoding, decoding, and recoding. The first two are common to any form of coding, since they indicate how to map the original information and recover it, whereas recoding is a distinctive feature of network coding only. This feature enables a node to create new valid coded packets from packets that have been encoded before, but *without* requiring to decode them. A condition necessary for recoding to be ideal is that a recoded packet is indistinguishable from a packet obtained from encoding, as we will discuss later

Network Coding can achieve the min-cut max-flow capacity of any multicast network [170–172]. For example, in Fig. 9.3, it does not matter how the relay nodes are connected within the network. As long as the min-cut of the network is three, it is possible to transmit three packets to the multicast sinks by enabling encoding, recoding, and decoding operations at the nodes.

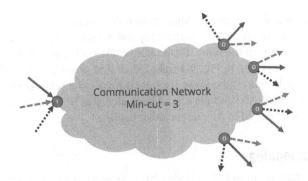

FIGURE 9.3

Network coding can achieve the min-cut max-flow capacity of any multicast network.

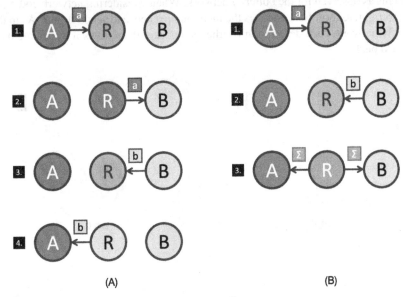

(A)   (B)

FIGURE 9.4

Intersession Network Coding in the Alice and Bob Topology. (A) Routing; (B) Coding.

9.1.2 Alice and Bob topology

Fig. 9.4 illustrates the Alice and Bob topology, which occurs in a region of a network where two unicast opposite flows in the same path cross at a relay. For this scenario, node A (Alice) wants to send packet a to node B (Bob) through the relay R, forming a flow from A to B. Similarly, Bob wants to send packet b to Alice through the same relay as well, creating a flow from B to A.

When using a routing approach in a time-slotted packet network without losses or delays, we require four time slots to transmit both flows, as presented in Fig. 9.4A.

We require two time slots to send a, Alice to the relay and the relay to Bob; we require two more time slots to send b, Bob to the relay and the relay to Alice. However, there is a coding opportunity at the relay. If we allow coding, then we reduce the required transmissions as illustrated in Fig. 9.4 right. Alice first sends a to the relay, keeping a copy of it, then Bob does the same for b, and finally the relay sends $\Sigma = a \oplus b$. To obtain the originally desired packets, Alice calculates $b = a \oplus (a \oplus b)$, and Bob calculates $a = b \oplus (a \oplus b)$ to decode the other packet.

9.1.3 The X topology

This topology may be regarded as a simplified version of the butterfly network as shown in Fig. 9.5. In this scenario, A wants to send packet a to C, and B wants to send packet b to D. Using a routing approach, shown in Fig. 9.5 left, will result in the same problems as observed for the butterfly network. When considering network coding in Fig. 9.5 right, relay R both encodes the incoming packets and sends $A \oplus B$. As in the butterfly network, each destination overhears a packet that allows it to decode their intended packet.

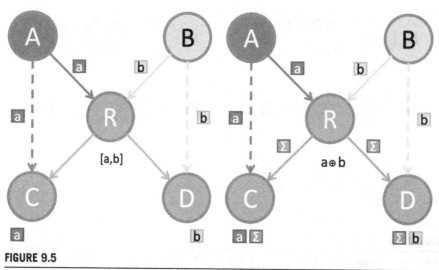

FIGURE 9.5

Network Coding in the X Topology.

9.1.4 The cross topology

This topology can be considered as a more complex variant of the X topology we described previously. This scenario is also referred to as an extension of the Alice and Bob topology. Here the two flows of the X topology are bidirectional. Subsequently, there are four unicast flows traversing the relay R, as illustrated in Fig. 9.6. We have also differentiated two cases of possible overhearing connections (which, e.g., could represent a wired and a wireless case). Fig. 9.6 left considers the case of no overhearing, and the flows connect only through the relay. Fig. 9.6 right includes overhearing links between each node and its closest neighbors.

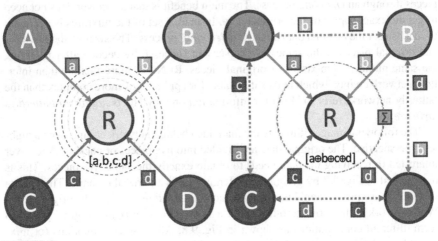

FIGURE 9.6

Cross topology.

Regardless of the overhearing case, there is no direct connectivity between a node and its diametric opposite. For example, a packet from A cannot reach D without going through the relay B or C; similar examples apply to all other nodes.

Consider relaying in the topology without overhearing in a time-slotted system, it would require eight time slots for all the packets to reach all the nodes. Four transmissions are required to send a distinct packet from each node, and four more are required to broadcast each packet from the relay to its three intended nodes. If we name the packets from each node respectively as a, b, c, and d, using network coding will require only six time slots. After sending the four uncoded packets, the relay broadcasts $a \oplus d$ to A and D and then $b \oplus c$ to B and C. From these transmissions each node may recover its intended packets. Moreover, if we allow overhearing as in Fig. 9.6 right, then using network coding only requires five transmissions. After the four original transmissions, each node has received a packet from its neighboring nodes, but not from its diametrical opposite. Therefore the relay only needs to send $\Sigma = a \oplus b \oplus c \oplus d$ to satisfy all the nodes. This network topology can be generalized to have N nodes and a relay. We can see that when more nodes are present, the gains of coding in the relay become greater.

9.2 Intraflow network coding – now it gets interesting

In the previous section, we focused on interflow network coding, that is, the network nodes would code the packets of different network flows. In this section, we address intraflow network coding and particularly Random Linear Network Coding (RLNC) [173]. The main idea behind intraflow network coding is dividing the original data from a single network flow into smaller pieces and creating mixtures of these pieces through an *encoding* process. The main benefit is that a receiver does not need to get the exact copies of the original data. Instead, a set of the mixtures is sufficient to reconstruct the original data through a *decoding* process. The sizes of the mixtures and original pieces is the same. A receiver can successfully decode after receiving the same number of mixtures or original pieces. RLNC is more flexible than interflow network coding, which makes it suitable for applications more complex than the butterfly network (refer to [174] for information on why NC *is not about butterflies anymore*).

The following images illustrate the main idea behind network coding from a high-level perspective. The original data are divided into five pieces or packets. A receiver interested in receiving the data needs to obtain exactly the same five pieces. This is represented in Fig. 9.7 by giving the different pieces a particular shape. This means that only those pieces would match with each other and reconstruct the original data. With network coding, on the other hand, it is possible to mix the original pieces to form different combinations as shown in Fig. 9.8. All these combinations (or mixtures) have the same size, and they are formed by mixing the different *colors* of the original pieces (encoding) or by mixing two or more mixtures together (recoding) as illustrated in Fig. 9.9. The size of a particular color within a mixture is used in Figs. 9.8 and 9.9 to represent that each color can be mixed with a particular *weight*. There are two novel aspects of RLNC over traditional forms of coding: i) each mixture, that is, the weight of each color used, is chosen randomly, and ii) it is possible to create a new valid mixture combining two other mixtures through a *recoding process*.

data broken into five pieces

data conveyed / stored on five different paths / clouds

data retrieved from five different paths / clouds → success

FIGURE 9.7

Legacy – The receiver needs an exact copy of the five original pieces.

The difference between sending mixtures and original pieces in this example is that a receiver needs a set of five mixtures to reconstruct the original data, as illustrated in Fig. 9.8. From the set of all possible mixtures not all subsets of five mixtures

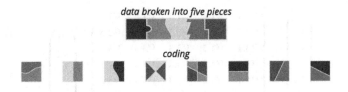

FIGURE 9.8

Coding – Different mixtures are created from the original pieces.

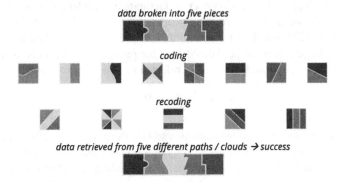

FIGURE 9.9

Recoding – From mixtures it is possible to create new and valid mixtures.

will suffice to reconstruct the original data. Later in this chapter, we will describe the limitations on the mixtures needed at the receiver to enable decodability in greater detail.

The idea of the mixtures and colors is beneficial to help understanding the underlying operating principle of RLNC. However, what does it mean, mathematically, to produce a mixture out of the original data? To produce a coded packet (i.e., a mixture), we must linearly combine the original pieces. For example, if **P** is the vector containing five original packets $\{p_1, \ldots, p_5\}$, then each coded packet c_i is the sum of all the original packets multiplied by a coding coefficient (or weight). With RLNC, it is possible to produce as many coded packets as needed out of the five original packets. The encoding process is shown in a matrix form in the following equation, where the five original packets of the example are used to produce n coded packets:

$$
\begin{pmatrix} c_1 \\ c_2 \\ c_3 \\ c_4 \\ c_5 \\ \vdots \\ c_n \end{pmatrix} = \begin{pmatrix} \alpha_{1,1} & \alpha_{1,2} & \alpha_{1,3} & \alpha_{1,4} & \alpha_{1,5} \\ \alpha_{2,1} & \alpha_{2,2} & \alpha_{2,3} & \alpha_{2,4} & \alpha_{2,5} \\ \alpha_{3,1} & \alpha_{3,2} & \alpha_{3,3} & \alpha_{3,4} & \alpha_{3,5} \\ \alpha_{4,1} & \alpha_{4,2} & \alpha_{4,3} & \alpha_{4,4} & \alpha_{4,5} \\ \alpha_{5,1} & \alpha_{5,2} & \alpha_{5,3} & \alpha_{5,4} & \alpha_{5,5} \\ \vdots & \vdots & \vdots & \vdots & \vdots \\ \alpha_{n,1} & \alpha_{n,2} & \alpha_{n,3} & \alpha_{n,4} & \alpha_{n,5} \end{pmatrix} \cdot \begin{pmatrix} p_1 \\ p_2 \\ p_3 \\ p_4 \\ p_5 \end{pmatrix} . \tag{9.1}
$$

In RLNC, all the coding coefficients $\alpha_{i,j}$ are chosen randomly from a uniform distribution. If the coefficients were real numbers and the operations of multiplication and addition were also performed over real numbers, then it would be difficult to implement the encoding and decoding operations. Furthermore, the size of the coded packets would increase when more original packets were added together. Instead, each coefficient $\alpha_{i,j}$ is chosen randomly according to a uniform distribution from the Galois field $GF(q)$ of size q. Moreover, to facilitate the implementation of the operations in widely spread hardware, the Galois fields used are binary extension fields of the form $GF(2^h)$ with $h \geq 1$. Later in this chapter, we will explain in detail how the finite field operations are performed over the packets. Eq. (9.1) can be rewritten in general for g original packets (usually referred to as *generation*) as

$$
\underset{n \times 1}{\mathbf{C}} = \underset{n \times g}{\mathbf{G}} \cdot \underset{g \times 1}{\mathbf{P}} , \tag{9.2}
$$

where $\mathbf{C}$ is the vector of the coded packets, and $\mathbf{G}$ is the matrix containing the random coefficients. A receiver needs to receive enough linearly independent coded packets in order to decode the original information. The decoding is performed by means of Gaussian elimination. Once the receiver collects g linearly independent coded packets $\overline{\mathbf{C}}$, it can reconstruct $\mathbf{P}$ by solving the equation

$$
\underset{g \times g}{\overline{\mathbf{G}}^{-1}} \cdot \underset{g \times 1}{\overline{\mathbf{C}}} = \underset{g \times 1}{\mathbf{P}} , \tag{9.3}
$$

where the matrix $\overline{\mathbf{G}}$ is a square submatrix of $\mathbf{G}$ with g rows instead of n, and g columns. The matrix $\overline{\mathbf{G}}$ is available to the recoder because this matrix contains the coding coefficients used to generate the coded packets of $\overline{\mathbf{C}}$, and this information is appended to the coded packets produced by the encoders or recoders. Previously, we mentioned that not all sets of g mixtures are sufficient to decode the original data. The reason is that if the g received packets are not linearly independent, then the matrix $\overline{\mathbf{G}}$ cannot be inverted, and Eq. (9.3) would constitute an underdetermined system of linear equations – the original data could not be decoded.

The idea of the colored mixtures is useful to illustrate some advantages of RLNC in different applications. In the area of multipath transport and multicloud storage, RLNC allows the transmission of the mixtures without any particular order or planning. The mixtures can arrive at the receiver in any order, from any path and from any cloud storage, as illustrated in Fig. 9.10. A traditional system, without coding, could also achieve multipath and multicloud communications with an optimized scheduling algorithm. The problem is that optimal scheduling becomes increasingly difficult with the number of paths and clouds involved. Planning is even more complex in the presence of heterogeneous and dynamic links with different capacities. For example, imagine a user who would like to retrieve a certain large file from two storage clouds A and B (for example, two network nodes caching information) with homogeneous links. Assume, furthermore, that both clouds contain the whole file stored. To maximize the download throughput, the clouds could schedule the transmission. For example, cloud A may send the file from the beginning as a data stream, whereas cloud B would send the data stream from the end of the file. This approach guarantees the maximum download throughput. The problem becomes more difficult when there are three clouds A, B, and C. In this case, clouds A and B could still send data from the beginning and the end of the file correspondingly, whereas cloud C would start sending a packet from the middle of the file, then one packet above the middle, one packet below the middle, and so on, until it reaches a packet already sent by cloud A or B. At this point, the three clouds would continue with their scheduling scheme from the part of the file that has not been send yet. This scheduling maximizes throughput, but it requires that the clouds communicate with each other to identify when certain parts of the file have been transmitted. The problem of scheduling three clouds is complex but still manageable. If now we consider four or more clouds and also consider heterogeneous and dynamic links, then the scheduling becomes extremely difficult, and the signaling overhead between clouds becomes nonnegligible.

Nevertheless, it is possible to guarantee maximum throughput without any complex scheduling if the system uses RLNC. All the clouds have to do is to create random mixtures of the original data and transmit them to the receiver as illustrated in Fig. 9.10. The receiver only needs enough linearly independent coded packets, independently of the path used or the cloud that sent them. Therefore, without any signaling, all clouds can contribute to achieve the fastest transmission of information, because every coded packet will be linearly independent to the receiver with high probability if the Galois field used is big enough (in practice, $GF(2^8)$). This chapter covering the theory of network coding is complemented by hands-on exercises highlighting the practical usage of network coding for transport in Chapter 20 and for distributed storage in Chapter 21.

RLNC also offers an inherent benefit in terms of security and the possibility of adding a light-weight encryption mechanism. Individual mixtures do not provide much information about the original fragments of the data. Since all the original colors are mixed, a receiver can decode the original information after enough mixtures are collected. Eq. (9.3) shows that without enough coded packets (i.e., linear equations), an eavesdropper has an underdetermined system of linear equations, that is, more unknowns than equations. The benefits are shown in Fig. 9.11. If we want to protect the communication between a sender and a receiver, we can transmit multiple mixtures through different paths. For example, in Fig. 9.11 the sender transmits four mixtures over one path and one mixture via the other path. An eavesdropper needs to have access to both channels to retrieve enough mixtures for a successful decoding.

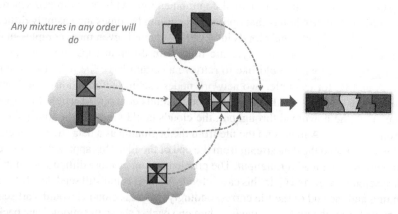

Any mixtures in any order will do

FIGURE 9.10

It is possible to use multiple clouds and multiple paths to transmit and store information. The receiver only needs enough mixtures in any order.

The main idea is not to transmit enough mixtures over any particular path, or not to store enough mixtures at any particular cloud storage, to allow an eavesdropper to have a determined system of linear equations. With more paths or cloud storages, the inherent security of RLNC may be increased by transmitting different mixtures over different paths. The assumption is that breaking into more channels or cloud services becomes more difficult for an eavesdropper. In the context of distributed public cloud storage, it is not uncommon to read in the news about data breaches and information leaks on private servers. To maintain data security even in the scenario of data leaks, we can distribute the mixtures among multiple clouds. This guarantees that the original data cannot be decoded, unless an attacker manages to break into enough public clouds, which should be a harder task. This is illustrated in Fig. 9.11, where four mixtures are stored in one cloud, and the last mixture is stored in the other. Similarly to multipath communication, more clouds can enable a higher level of data protection against attackers.

Furthermore, to increase security against attackers even more, it is possible to use RLNC to achieve a lightweight encryption. In traditional systems the totality of the data is encrypted. This can be computationally expensive. However, in RLNC, one can encrypt only the *recipe* of the mixtures, that is, the coding coefficients or weight of each color used. In contrast, this is computationally less demanding than encrypting all the data.

9.2.1 How to create coded packets

All the operations involved in the encoding, recoding, and decoding of packets are performed over finite fields. For example, if we are operating with a field of the form $GF(2^8)$, then all the operations are performed over 256 elements (from 0 to 255),

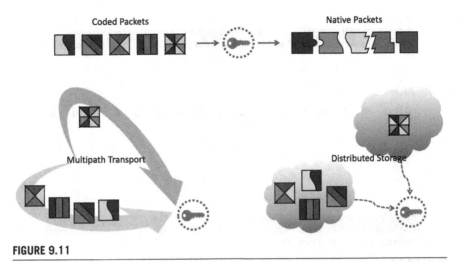

FIGURE 9.11

An attacker needs to break into enough clouds or eavesdrop enough paths to decode.

and each element is 8-bits long. All the mathematical operations performed will not yield numbers that are larger than 255. Similarly, if we operate with a field of the form $GF(2)$, then all the elements will be one-bit long. In the literature, it is usually mentioned that a newly coded packet is a combination of other packets. Similarly, it is often noted, for example, that a coded packet is the sum of the first original packet and three times the second original packet. However, the size of a packet exceeds the size of the elements of a Galois field; for example, for a Galois field $GF(2^8)$, each packet is usually longer than 8 bits. For example, the Maximum Transmission Unit (MTU) of an Ethernet frame is 1500 bytes. Therefore we now shift to an explanation of how to encode and add packets together.

A note on practical hands-on in Python

As part of the offered testbed, the reader can find a set of Jupyter notebooks that will guide him through the topics discussed in this chapter. The notebooks are inside the folder `app/network_coding_introduction/` of the provided vagrant virtual machine. Inside that folder, in the `README.md` file the reader can find the instructions to set up the environment for tinkering with and trying network coding examples. These notebooks contain a guided introduction to the Kodo library for RLNC. This software library is also used as part of other examples accompanying this book.

9.2.1.1 *Coding a packet with a binary field size*

We illustrate the process of encoding and decoding packets with the following examples. Initially, we assume that we have two original packets, that is, our generation size is 2, and the packets look similar to those in Fig. 9.12. In this figure, besides the original data packets, we show how much memory is allocated for the storage of the coded packet. The size of the coded packet in bits will be the same as the size of each

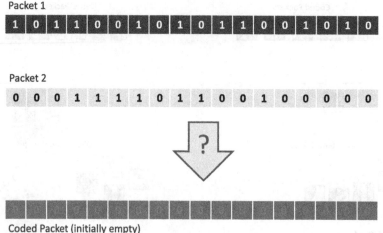

FIGURE 9.12

The original two packets of the example and the memory allocated for the coded packet.

original data packet. This is one of benefits of working with finite field operations, as the size of the coded packets does not increase, no matter how many packets we add together.

For the initial example, assume that we are operating with a binary field size, that is, $GF(2)$. In this Galois field the defined operations for the addition and multiplication are the same, and it is the logical exclusive or operation (XOR). (To understand how the operations are defined in binary extension fields, we refer the interested reader to [175,176].) If we use RLNC to encode these packets together, we would generate two coefficients chosen randomly with uniform distribution from the elements of the finite field. For this example, furthermore assume that the randomly generated coding coefficients $\{\alpha, \beta\}$ are equal to $\{1, 1\}$. This refers to the coded packet being comprised of packet 1 plus packet 2, that is, we will add the two original packets together to generate a coded packet. However, since the operations involved are defined over bits, we cannot really add those two packets together. Instead, we add, one by one, all the symbols of the packets, as shown in Fig. 9.13. A symbol is constituted by the number of bits required to represent an element in the Galois field (i.e., 1 bit per symbol in the binary field or 8 bits per symbol in the field $GF(2^8)$). An encoder or a recoder will typically append the coding coefficients used to generate the coded packet to the packet. In Fig. 9.13 the coding coefficients appended are of one bit each (again, each coding coefficient is as long as the number of bits needed to represent an element in the Galois field), and they are $\{1, 1\}$, represented as pink in the figure. As shown in the example, coding two or more packets together with a binary field size is simply performing a bitwise XOR operation over the data packets.

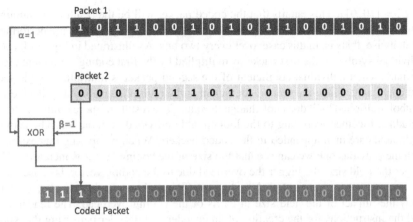

FIGURE 9.13

Adding two packets together refers to performing the addition via XOR of all the symbols of the packets, one by one. The coding coefficients are appended to the coded packet before a transmission.

9.2.1.2 Coding a packet with a larger field size

Even with a larger Galois field, the process is similar to the previous outline. Assume that our original data packets are the same as in the previous example, but this time we are operating with a Galois field of four elements, that is, $GF(2^2)$. In this example, each element of the field is a number in the range $\{0, 1, 2, 3\}$, and it is represented by two bits. Just as in the previous example, all the operations performed over the packets will be performed over symbols of the Galois field. In this case, each symbol has two bits instead of one bit as in the binary field. The addition and multiplication operations are different for this example, and they are summarized in Table 9.1. (We refer the interested reader to [175,176] for a more in-depth discussion of the operations and how Table 9.1 was created.)

Table 9.1 Lookup tables for operations of a Galois field of the form $GF(2^2)$.

	00	01	10	11		00	01	10	11
00	00	01	10	11	00	00	00	00	00
01	01	00	11	10	01	00	01	10	11
10	10	11	00	01	10	00	10	11	01
11	11	10	01	00	11	00	11	01	10
	(A) Addition					(B) Multiplication			

Again, in RLNC, the coding coefficients are chosen randomly with a uniform distribution from the elements of the Galois field. Since we consider two packets, two random coefficients are chosen. This time, assume that the Random Number Generator (RNG) chose the coding coefficients $\{\alpha, \beta\} = \{2, 1\}$ or, in binary representation,

$\{\alpha, \beta\} = \{10, 01\}$. This entails that the coded packet will be the result of combining two times packet 1 and one time packet 2. Once again, the operations are performed symbolwise, thats is, in this case, over every two bits. As illustrated in Fig. 9.14, each individual symbol of the first packet is multiplied by the first coding coefficient $\{10\}$. Similarly, each individual coefficient of the second packet is multiplied by the second coefficient. Since the second coefficient is the multiplicative identity $\{01\}$, each symbol multiplied by it does not change its value. The results of these multiplications are added together according to the lookup table provided in Table 9.1. The coding coefficients are then appended to the coded packets. When comparing this example with the previous one we can see that the size of the coding vector is increased. The bigger the field size, the bigger the overhead due to the coding vector, because more bits are needed to represent each coding coefficient. Later in this chapter, we will study the impact of the field size in terms of the coding overhead. The reader will find the instructions for the creation of an encoder and decoder pair using the Kodo library in the Jupyter notebook called `Kodo_python_getting_started.ipynb`.

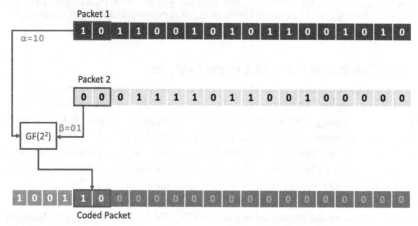

FIGURE 9.14

The coding operations are performed symbolwise. The data symbols are multiplied by the coding coefficients and then added together.

9.2.1.3 Recoding coded packets

The overall process of creating a recoded packet is similar to the individual packet coding. The only difference is that the same operations performed over the data symbols must be performed over the appended coding coefficients (similar to network protocol encapsulations). Since all the operations involved are linear, consistency is guaranteed. It is also important to note that the recoding operations do not increase the size of the coding vectors (different from network protocol encapsulations). Furthermore, as mentioned in the previous sections, a recoded packet is indistinguishable from an encoded packet. The recoding of two coded packets is illustrated in Fig. 9.15, where the original data symbols of the first and second coded packets are denoted as

f_i and e_i, respectively. The coding coefficients of each encoded packet are denoted as $C_{i,j}$, and they are displayed as appended to the coded packets. Coefficients chosen by the RNG of the recoder are represented as d_1 and d_2. The new coded data are then produced similarly as explained before, that is, by multiplying each symbol in the data packet by the random coding coefficient assigned to that packet and adding the results of these multiplications together (symbolwise). The appended coding vectors of the recoded packet (i.e., C_{new} in Fig. 9.15) are generated by applying the same operations performed for the data symbols, that is, for coding the coefficients of the coded packets. In practical applications, the data packets transmitted over the network use a certain protocol with its particular header appended to the payload. These headers are represented as $H_{i,j}$ in Fig. 9.15. If we use NC in the network, then we may need to combine the headers of the coded packets to be consistent with the underlying protocols.

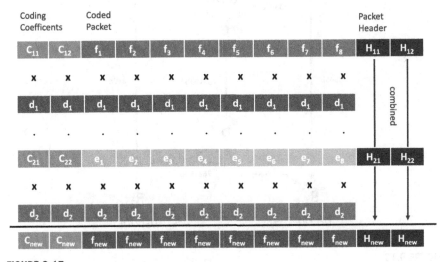

FIGURE 9.15

The recoding process: All the operations performed over the coded symbols are also performed over the coding coefficients.

9.2.2 **RLNC and the butterfly**

After describing the operation of RLNC, we can visualize how the butterfly in Fig. 9.1 can be generalized. Fig. 9.1B shows the operation of a planned, not random, network coding over a binary Galois field. The relay encoding the packets chooses the {1, 1} as the coding coefficients. However, the source and the relay could use RLNC. It can be seen in Fig. 9.16 that the source node encodes the original packets it transmits to all the nodes. When a node, in this case the relay, receives two coded packets, it recodes them using RLNC and transmits the recoded packets to the sinks. Once each sink collects enough linearly independent packets, the original packets can be decoded. By increasing the field size used (in practice, $GF(2^8)$ is large enough) the throughput of the system can be made arbitrarily close to the min-cut max-flow of the network as desired.

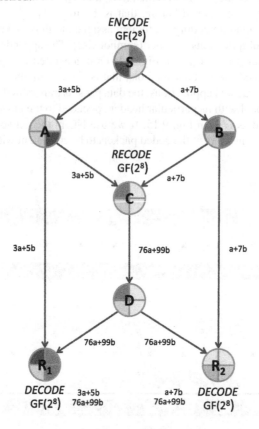

FIGURE 9.16

The butterfly example extended to use RLNC and a higher field size: The source encodes the original packets, and the relay stores and recodes the received packets.

9.2.3 Impact of the coding parameters

There are two main parameters that can be controlled when implementing network coding, the field size and the generation size. Their values have an impact on the coding overhead and on the computational complexity of the encoding, recoding, and decoding of the information. The coding overhead added by network coding originates from two sources. On one hand, the coding coefficients are appended to each coded packet. On the other hand, the randomness of the chosen coefficients leads to the transmission of linearly dependent packets, which are not useful for the sinks and therefore are considered overhead.

9.2.3.1 Overhead due to linear dependencies

Depending on the field size (and even not considering any packet losses), the amount of transmissions required to collect g linearly independent coded packets may vary. This is due to linear dependency, which refers to the amount of linearly dependent coded packets that are generated during the encoding and recoding processes. As more linearly independent coded packets are received during the transmission process, linearly dependent coded packets are generated more frequently. Toward the end of the transmission, the number of linearly dependent packets increases [177–180]. We illustrate this by means of an example and the following generalization. We initially assume that we employ a binary field. We want to transmit $g = 4$ original packets, and therefore each coded packet is produced with four randomly generated coefficients. The first coded packet is likely useful for the decoder, because almost all the possible combinations of four coding coefficients (i.e., 16 possible combinations) will contain linearly independent information. There is only one combination that will not be useful for a decoder at the beginning of the transmission, that is, if the four chosen coefficients are 0. Now assume that the first coding vector was $cv_1 = \{1, 0, 0, 0\}$. At this stage, the possible combinations that are linearly dependent are, once again, the zero vector and cv_1. We continue the example assuming that a new coded packet was produced with the coding vector $cv_2 = \{0, 1, 0, 0\}$. At this point the state of the decoder can be represented by the coding matrix

$$A = \begin{bmatrix} 1 & 0 & 0 & 0 \\ 0 & 1 & 0 & 0 \\ 0 & 0 & 0 & 0 \\ 0 & 0 & 0 & 0 \end{bmatrix}. \tag{9.4}$$

The rank of A is two. At this stage an encoder producing the coding vectors cv_1, cv_2, the zero vector, and a linear combination of cv_1 and cv_2, that is, $cv_{ld} = \{1, 1, 0, 0\}$ will not yield any new information to the decoder. Once $g - 1$ linearly independent coded packets have been received, the probability of generating a dependent coded packet is $\frac{1}{2}$ for this particular example and $\frac{1}{q}$ for a general Galois field of size q [178, 181,182]. We can define $P_{r \to r}$ as the probability that any randomly coded incoming symbol does not increase the rank of the coding matrix, and we can see that

$$P_{r \to r} = \frac{1}{2^{g-r}}, \tag{9.5}$$

where r represents the current rank of the encoding matrix. If we assume a field of size q instead of a binary field, we can generalize Eq. (9.5) to

$$P_{r \to r} = \frac{1}{q^{g-r}}. \tag{9.6}$$

Eq. (9.6) can be visualized in the Markov chain illustrated in Fig. 9.17. Each state represents the current rank of the decoding matrix, and the vertex represents the

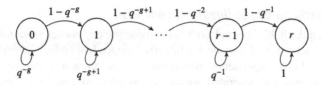

FIGURE 9.17

Markov chain representing the rank of the decoding matrix and the probabilities of moving between states.

probabilities of changing between states. Remaining in the same state r occurs with probability $P_{r \to r}$, and increasing the rank occurs with probability $1 - P_{r \to r}$.

At each step of the transmission (i.e., for each rank), it is possible to produce a certain overhead due to the randomness of the process. We can define E_r as the expected number of packets needed at a receiver for successfully increasing the rank, given that the rank at the receiver is r and the size of the decoding matrix is g. To compute this expectation, we can consider the probability that when encoding i packets, $i - 1$ of them are linearly dependent and do not increase the rank, whereas only the ith packet increases the rank. If we multiply this probability by i and sum over all possible values that i can take, then the expectation is

$$E_r = \sum_{i=0}^{\infty} i \underbrace{(P_{r \to r})^{i-1}}_{\text{Same rank}} \underbrace{(1 - P_{r \to r})}_{\text{Rank increases}} . \tag{9.7}$$

where the overbraces read $i-1$ packets over $(P_{r \to r})^{i-1}$ and ith packet over $(1 - P_{r \to r})$.

Differentiating the geometric power series, we obtain

$$\sum_{i=0}^{\infty} i \, (x)^{i-1} = \frac{1}{(1-x)^2} \qquad \text{for} \quad |x| < 1 , \tag{9.8}$$

and we substitute this result into Eq. (9.7):

$$\begin{aligned} E_r &= \frac{1}{(1 - P_{r \to r})^2} (1 - P_{r \to r}) \\ &= \frac{1}{(1 - P_{r \to r})} \\ &= \frac{1}{1 - \frac{1}{q^{g-r}}} . \end{aligned} \tag{9.9}$$

To compute the total number of transmissions needed per generation E_T, we have to sum the expected number of transmissions per rank (i.e., E_r) over all the ranks of the generation, from the rank $r = 0$ to $r = g - 1$:

$$E_T = \sum_{r=0}^{g-1} E_r$$

$$= \sum_{r=0}^{g-1} \frac{1}{1 - \frac{1}{q^{g-r}}}.$$

(9.10)

If we make the change of variable $r' = g - r$, then Eq. (9.10) becomes

$$E_T = \sum_{r'=1}^{g} \frac{1}{1 - q^{-r'}}.$$

(9.11)

To compute the actual total overhead per generation O_T, we need to subtract g from Eq. (9.11), because a receiver needs at least g packets to decode. Any transmission in addition to g packets is overhead. Therefore

$$O_T = E_T - g.$$

(9.12)

In Eq. (9.12), it is clear that the total overhead due to linear dependencies is a function of the field size q and the generation size g. However, the dependency with respect to g can be neglected. For practical generation sizes $g \geq 5$, we can observe in Fig. 9.18 that the extra number of coded packets needed due to linear dependencies does not vary much as g increases. Using Eq. (9.12), we can see that for $GF(2)$ (i.e., $q = 2$), $g + 1.6$ transmissions on average are required to decode the original packets, as illustrated in Fig. 9.18A, whereas for $GF(2^8)$, this value can be approximated to g for practical purposes, as illustrated in Fig. 9.18B. What do these values mean? Is the overhead of the binary field negligible? The answer depends on the generation size. If the source encodes $g = 100$ packets, then the average overhead per generation due to linear dependencies is of 1.6%, which is a negligible overhead. On the other hand, if the generation size is $g = 6$ packets, then the binary field yields an overhead of 26,7% per generation, which is not negligible anymore.

The plots presented in Fig. 9.18 can be recreated by the reader by using the Jupyter notebook called `Overhead_analytical.ipynb` in the provided environment. Furthermore, in the notebook called `Overhead_practical.ipynb`, we make use of the Kodo library to illustrate, practically, the appearances of the linear dependencies along the transmissions, as well as the 1.6 extra packets in the binary case.

9.2.3.2 Computational complexity

The generation size affects the algorithmic complexity of both encoding and decoding. The computational complexity of encoding RLNC packets scales as $\mathcal{O}(g)$, that is, linearly, since it involves g multiplications and $g - 1$ sums. For the decoding process, Gaussian elimination is of cubic complexity $\mathcal{O}(g^3)$ in principle, given the inversion

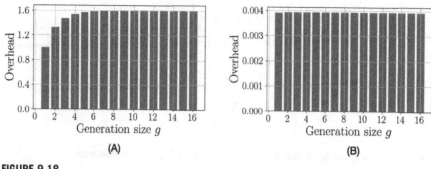

FIGURE 9.18

Extra packets needed due to linear dependencies as a function of the generation size for different field sizes q. (A) $q = 2$; (B) $q = 2^8$.

of a square matrix of size g. However, a structured Gaussian elimination implementation for RLNC can achieve $\mathcal{O}(g^2)$ for $g < 512$ as reported in [180]. Given that the field size effects are diverse, we summarize them in Table 9.2, which shows the effects of the field size for two principal regions, low and high field sizes. The criteria to separate them have to consider a field size higher than 2^8. The table also presents various metrics to evaluate the performance of the code.

The field size complexity accounts for the computational cost and time required for the operations in the Galois field arithmetic required to process the data. Besides algorithmic complexities, the field utilized to operate on the data affects the code performance in terms of time and energy spent on processing. The binary field poses only a low computational burden on the device carrying out the operations, since modulo-2 operations are XOR/AND operations. However, increasing the field size requires defining and operating with new arithmetic approaches, which incur higher processing times. Thus a proper field size for mobile devices is important to ensure a satisfying code processing speeds [177,180]. Information on the encoding and decoding throughput of the state-of-the-art network coding library compiled in AMD and Intel processors with and without Single Instruction Multiple Data (SIMD) support is provided in [183]. Furthermore, we refer the interested reader to [184,185] for information regarding the encoding and decoding throughput of the latest implementations of RLNC in multicore processors.

To reduce the computational complexity, it is possible to use systematic RLNC. In systematic RLNC the original data packets are transmitted uncoded, followed by the coded packets [186]. Systematic coding has the practical effect of filling the coding matrix with zeroes. The coding coefficients of an uncoded packet are a vector of zeroes except for one coefficient equal to 1. The index of this coefficient is the index of the uncoded packet. It is intuitive to think that more zeroes in the coding matrix reduce the computational load. If the reader remembers their courses on linear algebra, the easiest matrix problems to solve were those where the matrix was sparse (i.e., with many zeroes). Based on this intuition, several RLNC techniques were developed where the encoder purposely introduces zeroes to the coding vector to reduce the computational complexity at the expenses of overhead due to linear dependencies. Examples for RLNC approaches following this principle include perpetual RLNC [187,188], sparse RLNC [189], tunable sparse RLNC [190–193], and sliding window RLNC [194,20,195].

9.2.3.3 Overhead due to the coding coefficients

Signaling is interpreted as the amount of bits required to represent the coding coefficients. These bits are appended to each coded packet and for each original packet [179]. The size of the coding vector depends on the number of symbols and the field size. As each coded packet is generated combining g original packets, the coding vector contains g symbols. Each symbol of the coding vector requires $\log_2(q)$ bits to be represented. Therefore the size of the coding vector is $|cv_i| = g \times \log_2(q)$, which grows linearly with g and logarithmically with q. For $GF(2)$, only 1 bit per packet is required to be included in each coded packet to signal the coding coefficients. However, for fields sizes of $q = 2^8 = 256$ or higher, one byte or more are necessary for each original packet to signal its coding coefficients. Consequently, for high generation sizes, high fields can potentially make the amount of signaling much larger than the original packet size. Thus overhead accounts for both effects of the linear dependency and coding coefficients signaling respect to useful data. In Table 9.2, it is specified which effect accounts for most of the total overhead in the specified region. For low fields, most of the overhead originates in linear dependency effects, but for higher field sizes, the signaling from the coding coefficients becomes nonnegligible.

Table 9.2 Field size effects in the code performance.

q	Linear dependencies	Signaling	Overhead	Complexity
$< 2^8$	High	Low	Linear dependencies	Low
$\geq 2^8$	Low	High	Signaling	High

Ideally, parameter configurations are desirable that achieve i) low number of transmissions required to decode, ii) low total overhead, and iii) low computational complexity. However, as previously described, these objectives are conflicting in principle, posing a *trade-off* whenever RLNC is utilized. For more information related on the overhead of RLNC, refer to the in-depth analysis in [179].

9.2.4 The potential of recoding

The statelessness of RLNC enables a unique ability that other, traditional codes do not possess. RLNC, as shown in the previous sections, allows the combination of two coded packets to create a new valid coded packet. This is known as recoding. To illustrate the benefits of recoding, consider the example illustrated in Fig. 9.19, where a source wants to convey four packets to a destination. Since the source and the destination are distant, they need to communicate through a relay. Each communication link in this example has a symmetric erasure probability of 50%, that is, on

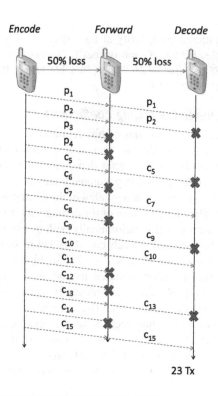

FIGURE 9.19

Forwarding on a lossy medium.

average, half of the packets are lost. In each transmission slot, we assume that both the source and the relay can transmit one packet each. In the example in Fig. 9.19, we additionally assume that the relay implements a simple forwarding scheme. All the packets received from the source are transmitted to the destination. If the source employs a traditional block code, such as a Reed–Solomon code, then we need to add enough redundancy to overcome the erasures of *both* links. With the error pattern of the example, there are 23 packets transmitted in total, and the receiver can decode after 15 time slots. In general, if the erasure probabilities are ϵ_1 for the communication link between the source and the relay and ϵ_2 for the link between the relay and the destination, then a packet is successfully delivered if it is not lost on any link. Successful packet deliveries occur with probability $(1 - \epsilon_1)(1 - \epsilon_2)$. If the source wants to convey g packets, then it needs on average to do Tx_f transmissions, where

$$Tx_f = \frac{g}{(1 - \epsilon_1)(1 - \epsilon_2)}. \tag{9.13}$$

If we consider more relays with erasure-prone communication links, as in a general multihop network similar to those presently encountered in mesh applications, then

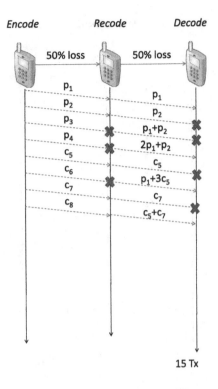

FIGURE 9.20

The potential of recoding in a lossy medium.

the average number of required transmissions at the source increases per link with a factor $\frac{1}{1-\epsilon_i}$. Here ϵ_i represents the erasure probability of the ith communication link.

In the second example presented in Fig. 9.20, RLNC is employed instead. We allow the relay to recombine the received packets from the source before transmitting them to the destination. In this case the number of redundant packets the source needs to transmit is much lower than in the previous example. In this case the source and the relay can add redundancy when it is needed. In this case the relay can also provide linearly independent equations by means of recoding – it does not need to wait receiving a packet to add redundancy! As the figure shows, with the same error pattern, the system needs only 15 transmissions, and the destination can decode after 8 time slots. If we generalize this example, then the required number Tx_r of transmissions by the source when the relay recodes is determined by the *worst link*, that is, $\min\{(1 - \epsilon_1), (1 - \epsilon_2)\}$. In this case,

$$Tx_r = \frac{g}{\min\{(1 - \epsilon_1), (1 - \epsilon_2)\}} . \tag{9.14}$$

Most notably, if we consider a system with multiple hops, then we can see that the number of hops do not affect the total number of packets the source has to transmit.

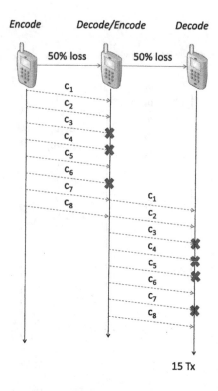

FIGURE 9.21

A proxy-like protocol is as efficient as recoding in terms of channel utilization, but it suffers in latency.

The total number of transmissions in the system is determined only by the worst link, that is, the link with the greatest erasure probability. This makes a multihop network coded system (as those present, e.g., in mesh networks) scalable. Is this finding surprising? If we recall that RLNC achieves the min-cut max-flow capacity of a network (ignoring the linear dependencies), then this result is not a surprise. If we consider all the *cuts* of a multihop network, tthen he minimum cut is determined by the link with the worst capacity. Therefore the maximum flow of the network is determined only by that link, and it is independent of the number of hops.

The third example in Fig. 9.21 illustrates that the same coding efficiency in terms of the total number of transmissions in the system is achieved when using a traditional block code if we allow the relay to use a strategy called decode-and-forward. With this strategy, the source encodes the data and transmits packets to the relay until it decodes the information (i.e., until it receives four linearly independent packets). Once the relay is able to decode, it transmits the information to the destination as well as newly encoded packets. In this case the total number of transmissions in the system is the same as in the recoding example when the same packet erasure pattern is present. This benefit in terms of total transmissions in the network is obtained by means of coding. The difference and advantages of network coding over traditional block codes in this scenario are derived from a reduced latency before the destination is able to decode. Compared with recoding, the decode-and-forward approach in this example needs 14 time slots. Since low latency is such an important characteristic of 5G systems, the potential of recoding is evident.

It is important to note that Eqs. (9.13) and (9.14) show a simplified model valid when the erasure probabilities of the different links are different from each other. When the values of the erasure probabilities are close to each other (i.e., $\epsilon_1 \approx \epsilon_2$), there is an error between the presented equations for the number of transmissions from the source and the values obtained in the measurements. The larger the value of ϵ_i, the larger the error between our model and practical results. The explanation behind this behavior is outside the scope of this book chapter. However, if the reader is interested in delving into the details, then he can find a detailed explanation in Appendix A of [196]. In the bibliography, this error is referred to as the shark fin effect due to the shape of the 3D plot of the error between the values in the measurements and the equations of this chapter plotted against the values of ϵ_1 and ϵ_2. This plot looks like the fin of a shark, and the reader can see it in [196]. For an insightful simulation, the reader can find the Jupyter notebook `Potential_of_recoding`, which will show him the benefits of recoding and the shark fin effect described in this section.

Compressed sensing

Maroua Taghouti

Technische Universität Dresden, Dresden, Germany

> *... Can we not just directly measure the part that will not end up being*
> *thrown away?*
> **Donoho**

10.1 Compressed sensing theory

Compressed Sensing is a novel approach that allows for the efficient acquisition and effective reconstruction of a signal of interest, far below the Shannon–Nyquist sampling rate. It was conventionally assumed in standard disciplines under the condition that the signal admits a sparse representation in a certain domain. This disruptive transform coding[1] was first initiated by Candes et al. [197] in 2005 and later named Compressed Sensing (CS) by Donoho et al. [198]. CS has shown that by fulfilling sufficient conditions a perfect reconstruction of a signal can be achieved from a small number of measurements (samples). Thanks to its performance, CS gained tremendous interest during the last decade.

10.1.1 Problem formulation

From a mathematical point of view, CS finds solutions to underdetermined linear systems of equations using optimization techniques. The CS problem can be summarized as recovering x using

$$\min_{x} \|x\|_0 \text{ subject to } y = Ax, \tag{10.1}$$

where $y = Ax$ is an underdetermined linear system of equations. The conventional wisdom dictates that m, the number of elements of y, must be at least equal to or greater than the number of elements of x, that is, $m \geqslant n$. Otherwise, classical linear algebra solvers classify this problem as underdetermined, which yields an infinite number of solutions (if there exists at least one), unless additional information about

---

[1] The theory that relies on providing a sparse or compressible representation of signals of interest by finding an appropriate frame or basis.

Computing in Communication Networks. https://doi.org/10.1016/B978-0-12-820488-7.00023-2

the system is provided. (See Fig. 10.1 for a visual representation of the dimensions in an underdetermined linear system of equations.)

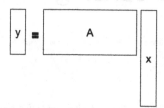

FIGURE 10.1

CS is an underdetermined system of equations to be solved.

This explains the bounds of the Shannon–Nyquist sampling theorem: the sampling rate of a continuous-time signal should be twice as high as the largest frequency to be able to reconstruct it. CS does not contradict traditional, well-known algebra. The central hypothesis behind CS theory is that the signal x is sparse, that is, the majority of its coordinates are zeros. Actually, without this extra hypothesis regarding x, the reconstruction is impossible, as too many solutions exist.

The *magic* of CS relies on the fact that A can be designed such that the sparse signal x can be recovered exactly or at least approximately from the few measurements y. Knowing that the signal of interest $x \in \mathbb{R}^n$ is not sparse for most cases, it is important to note that there exist some transformations that enable the sparsification of x, such as the Fourier transform. Thus x can be represented as $x = \Psi\theta$, where $\theta \in \mathbb{R}^n$ is a sparse representation of x obtained by projection over an orthonormal basis $\Psi \in \mathbb{R}^{n \times n}$. Therefore the compressed sensing problem can also be interpreted as a problem of recovering θ using

$$\min_{\theta} \|\theta\|_0 \text{ subject to } y = A\Psi\theta. \tag{10.2}$$

Both cases are equivalent and yield an underdetermined linear system of equations. On the other hand, solving such a type of problems, based on ℓ_0 optimizations, is challenging and could take considerable amounts of time if the signal is large. Candes et al. proposed to use linear programming based on the ℓ_1 norm instead and proved that it could approximate the ℓ_0 under certain conditions, which we will explain in detail in this chapter. Therefore the problem is reformulated as follows:

$$\min_{x} \|x\|_1 \text{ subject to } y = Ax. \tag{10.3}$$

To summarize, the compressed sensing task is accurate reconstruction of a signal from a minimal number of measurements. The preliminary questions when learning about compressed sensing are i) *What are the conditions on x, A, and m to classify the problem as a compressed sensing one?* and ii) *What kind of algorithms to consider for an accurate reconstruction?*

10.1.2 **Mathematical background**

We assume that the reader has some prior knowledge about linear algebra. Nevertheless, we propose a short but concise review of some definitions needed in this chapter.

10.1.2.1 *Basis and frame of a vector space*

Basis

A basis of a vector space over the real field $\mathbb{R}^n$ is a linearly independent subset $(e_1, e_2, \ldots, e_n)$ of the vector space that spans it. It should satisfy the following two properties:

- *The linear independence property*: For any $\alpha_1, \ldots, \alpha_n$, $\alpha_i \in \mathbb{R}$, if $\alpha_1 e_1 + \cdots + \alpha_n e_n = 0$, then $\alpha_1 = \cdots = \alpha_n = 0$.
- *The spanning property*: For every vector $v \in \mathbb{R}^n$, we can choose $\beta_i \in \mathbb{R}$, $i \in \{1, \ldots, n\}$, such that $v = \beta_1 e_1 + \cdots + \beta_n e_n$.

Example

For the vector space $\mathbb{R}^3$, $\mathcal{B} = (e_1, e_2, e_3)$ is a basis, where $e_1 = (1, 0, 0)$, $e_2 = (0, 1, 0)$, and $e_3 = (0, 0, 1)$.

Frame

Unlike the basis that requires exactly a set of n independent vectors to form a basis in $\mathbb{R}^n$, a frame is a set of $l \geq n$ vectors, $v_1, \ldots, v_l$ that fully represent a basis in $\mathbb{R}^n$. It proposes a stable and redundant way of representing a signal.

10.1.2.2 *Norms*

We consider $\mathbb{R}^n$ as an Euclidean vector space provided with the following ℓ_p norms:

$$\|x\|_p = \left(\sum_{i=1}^{n} |x_i|^p \right)^{\frac{1}{p}}, \quad p = 1, 2, \ldots. \tag{10.4}$$

It is easily recognizable that for $p = 1$, Eq. (10.4) defines the absolute-value norm, and for $p = 2$, it is the Euclidean distance, which gives the ordinary distance from the origin to x.

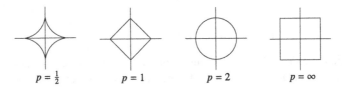

$p = \frac{1}{2}$ $p = 1$ $p = 2$ $p = \infty$

$\| \cdot \|_q$ is a norm only for $1 \leq q \leq \infty$, as it satisfies the q-triangle inequality:

$$\|x + y\|_q \leq \|x\|_q + \|y\|_q, \quad x, y \in \mathbb{R}^n. \tag{10.5}$$

However, it should be noted that when $p = 0$ (also valid for $0 \leq q \leq 1$), the ℓ_0 is not a norm, as it does not satisfy the aforementioned norms rule. It is just an *abusive* term due to its ease of use. It is more accurate to name it a pseudonorm and is defined as follows:

$$\|x\|_0 = \sum_{i=1}^{n} |x_i|^0. \tag{10.6}$$

Knowing that $0^0 = 0$, $\|x\|_0$ can be simply written as[2]

$$\|x\|_0 := Card(supp(x)) := Card(\{i \in \{1, \ldots, n\} : x_i \neq 0\}), \tag{10.7}$$

which means that ℓ_0 serves to count the number of nonzero elements in a vector, based on the support (*supp*), which provides the exact locations of these nonzero elements.

10.1.2.3 Orthogonal matrices

A matrix $A \in \mathbb{R}^{n \times n}$ is said to be orthogonal if and only if it is nonsingular and its inverse is equal to its transpose, that is, $A^{-1} = A^T$, or else if and only if its column vectors $a_1, \ldots, a_n$ are orthogonal to each other, that is, $\langle a_i, a_j \rangle = 0$, $\forall i \neq j$, and of norm equal to one.

10.1.2.4 Matrix decomposition

The Singular Value Decomposition (SVD) is a powerful tool, which is applicable to any matrix, revealing sufficient information about its structure.

Theorem 10.1. *Let $A \in \mathbb{R}^{n \times m}$ ($n \geq m$), be a matrix with s positive singular values. There exist orthogonal matrices $U \in \mathbb{R}^{n \times n}$, $V \in \mathbb{R}^{m \times m}$, and $\Sigma \in \mathbb{R}^{n \times m}$ such that*

$$A = U \Sigma V^T. \tag{10.8}$$

10.1.2.5 Kronecker product

The Kronecker product of two matrices $A \in \mathbf{R}^{m \times n}$ and $B \in \mathbb{R}^{p \times q}$ is a special type of matrix multiplication that involves the multiplication of one coefficient from one matrix by the entire second matrix, one at a time, resulting in a block matrix of size $mp \times nq$, denoted by $A \otimes B$, as follows:

$$A \otimes B = \begin{bmatrix} a_{11}B & \cdots & a_{1n}B \\ \vdots & \ddots & \vdots \\ a_{m1}B & \cdots & a_{mn}B \end{bmatrix}. \tag{10.9}$$

It is a particular case of the tensor product and thus is bilinear and associative. Moreover, it conserves many of the linear algebra properties including the transpose, the determinant, and the inverse, which we do not cover in greater detail here.

---

[2] For example, if $x = (0, 1, 0, 0, 2)$, then $supp(x) = \{2, 5\}$ and $\|x\|_0 = 2$.

10.1.3 Sparse and compressible signals

The sparsity or compressibility of signals is a crucial pillar for CS. Generally, a signal with most of its components being zero is called *sparse*. Moreover, a signal $x \in \mathbb{R}^n$ is called k-sparse if it has $k \in \{1, \ldots, n-1\}$ nonzero elements:

$$\|x\|_0 := Card(supp(x)) := Card(\{i \in \{1, \cdots, n\} : x_i \neq 0\}) = k. \qquad (10.10)$$

If $x \in \mathbb{R}^n$ is a k-sparse signal, then it belongs to the set Σ_k consisting of all vectors that have the same support set, that is, all k-sparse vectors:

$$\Sigma_k = \{x \in \mathbb{R}^n : \|x\|_0 \leqslant k\}. \qquad (10.11)$$

On the other hand, exact sparse representation is not a straightforward task in some real life signals, and the concept of sparsity might be a strong constraint to impose in some cases. Therefore it could be substituted by the weaker concept of compressibility. The signal is not exactly k-sparse, but rather its error of the best k-term approximation $\sigma_k(x)_p$ defined as follows is small:

$$\sigma_k(x)_p := \inf\{\|x - y\|_p, \; \|x\|_0 = k\}. \qquad (10.12)$$

This type of signal is characterized by the fact that the best k-term approximations decay rapidly with k. As a result, the faster the decay happens, the better the x can be approximated. Fig. 10.2 illustrates a basic example of a sparse and compressible signal.

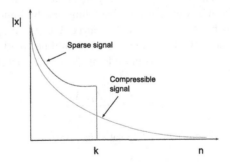

FIGURE 10.2

Compressible and sparse signal (sorted coordinates decay rapidly with power law).

To summarize, it is important to note that the main difficulty in solving sparse linear systems of equations lies in finding the locations of the nonzero components, that is, the support of the signal.

10.1.4 Measurement matrix design

The CS matrix is another foundational component. It must be independent of the signal of interest x. Thus it should be designed in a nonadaptive measurement process, that is, the present measurements do not depend on the previous ones. At the dawn of CS, researchers were highly concerned about designing the right measurement matrix. To ensure that the signal is not corrupted or lost because of the transform, sufficient conditions for sparse recovery, such as the Restricted Isometry Property (RIP), the Null Space Property (NSP), and the Restricted Eigenvalue (RE), have to be fulfilled. Surprisingly, matrices that are drawn from Gaussian, Bernoulli, Rademacher, and some other distributions are able to satisfy the RIP with a very high probability for large n. Thus the burden related to matrix design can be reduced. Unless there is a specific type of a precise CS application, such as operating in finite fields, these matrices represent the easiest and one of the cheapest options.

To see closely the impact of the measurement matrix on the reconstruction, we emphasize the importance of the notion of the measurement itself, denoted by y_i, $i \in \{1, \ldots, m\}$, $m \ll n$, the ith element of the vector y with

$$y_i = \langle x, A_i \rangle, \tag{10.13}$$

where A_i is the ith column of A, and $\langle \cdot, \cdot \rangle$ is the inner product operator. Knowing that the goal is performing the least number of measurements, that is, being able to compress x efficiently where $m \ll n$, the problem can therefore be seen from a different angle as "*How can we build the columns of $A_1, \ldots, A_m$ such that $\forall x \in \Sigma_k$, we can efficiently reconstruct x from y?*".

10.1.4.1 Mutual coherence

The mutual coherence ensures good recovery guarantees. If its value is small, then the performance of compressed sensing algorithms can be improved. To show that the columns, denoted by $A_1, \ldots, A_m$, of a matrix A are well spread in the matrix vector space $\mathbb{R}^{m \times n}$, the mutual coherence of A should be small [199]. It is denoted as $\mu(A)$ and defined as the largest magnitude of the normalized dot product between two columns of A:

$$\mu(A) = \max_{i \neq j} \frac{\|\langle A_i, A_j \rangle\|}{\|a_i\|_2 \|a_j\|_2}. \tag{10.14}$$

Particularly, for full rank matrices, $\mu(A)$ satisfies [200]

$$\sqrt{\frac{n-m}{m(n-1)}} \leqslant \mu(A) \leqslant 1. \tag{10.15}$$

10.1.4.2 Null space property

It gives necessary and sufficient conditions on the reconstruction of sparse signals when using the ℓ_1-relaxation algorithms, such as the basis pursuit. For $T \subset \{1, \ldots, n\}$, we denote by $T^c = \{1, \ldots, n\} \setminus T$ its complement. Moreover, for $\upsilon \in \mathbb{R}^n$, we denote by $\upsilon_T \in \mathbb{R}^{|T|}$ the vector that contains the coordinates of υ on T, and likewise υ_{T^c}. We denote by A_T the $m \times |T|$ submatrix of $A \in \mathbb{R}^{m \times n}$ that contains the columns of A indexed by T. If we define $T = supp(x)$, $x \in \mathbb{R}^n$, then $Ax = A_T x_T$. The NSP was first proposed in [201].

Theorem 10.2. *A matrix A has the NSP of order k if*

$$\|v_T\|_1 < \|v_{T^c}\|_1, \ \forall v \in Ker(A) \setminus \{0\}, \ T \in \{1, \ldots, n\} \ with \ Card(T) \leqslant k.$$

Theorem 10.3. *Given a matrix $A \in \mathbb{R}^{m \times n}$ and $k \in \{1, \ldots, n\}$, every k-sparse vector $x \in \mathbb{R}^n$ is the unique solution of the optimization problem (10.3) if and only if A has the NSP of order k.*

Despite the fact that the NSP is sufficient in recovering the sparse solutions of underdetermined linear system of equations using the basis pursuit, it remains difficult to verify due to its generally high computational complexity.

10.1.4.3 Restricted isometry property

The concept of the RIP was advocated by Candes et al. [197] in 2005. It characterizes nonsquare matrices that are nearly orthonormal when dealing with sparse or compressible signals. It is a finer measure of the quality of the measurement matrix and its suitability for the CS problem when compared to mutual coherence.

Lemma 10.1. *If there exits a constant $\delta_k \in (0, 1)$, called the kth restricted isometry constant of a matrix $A \in \mathbb{R}^{m \times n}$, such that for any k-sparse signal $x \in \mathbb{R}^n$, we have*

$$(1 - \delta_k)\|\theta\|_2^2 \leqslant \|\Phi\theta\|_2^2 \leqslant (1 + \delta_k)\|\theta\|_2^2, \tag{10.16}$$

then Φ is said to satisfy the RIP of order k.

We can also say that A satisfies the RIP when all possible selections of k columns of A are similar to an orthonormal matrix. The restricted isometric constant δ_k tells us how well the matrix A preserves the energy of the compressed signal. However, if a sensing matrix does not obey the RIP, then we cannot ascertain whether the signal can be recovered or not. According to the theorem by Candes [202], if $\delta_{2k} \leq 1$, then problem (P0) has a unique k-sparse solution. As when $\delta_{2k} \leq \sqrt{2} - 1$, the solution to problem (P1) coincides with that of (P0), that is, the convex relaxation is exact. By definition finding the RIP constant requires an exhaustive search over $\binom{n}{k}$ subspaces [203]. In general, the RIP can be computationally intractable [204]. Instead of performing exhaustive searches for finding the right measurement matrices, it was explained that random matrices satisfy the RIP. Specifically, it was shown with an overwhelming probability that Gaussian random matrices have *good* isometry constants [197]; also, Bernoulli and partial Fourier matrices can guarantee the RIP [205].

10.2 Basic reconstruction algorithms

The issue with Eq. (10.1) is based on the fact that such an optimization problem is nonconvex, that is, it requires very high computational complexity to be solved. It depends upon a very high number of searches to find the sparsest solution. Furthermore, it has been proved that the ℓ_0-minimization is NP-hard. The handicap of the

ℓ_0-minimization algorithms is the search within all the sparse sets Σ_k that are candidates to be solutions to the optimization problem (10.1). This suggests a search of $\binom{n}{k}$ possibilities. For example, for a signal of size $n = 1000$ and sparsity $k = 10$, there will be $\binom{n}{k} \geqslant \left(\frac{n}{k}\right)^k = 10^{20}$ possibilities, which would take up to 300 years to solve, provided that 10^{-10} seconds are consumed per iteration. The breakthrough idea proposed by Candes et al. was to substitute the ℓ_0 norm by the ℓ_1 norm, as it is the closest convex norm, as follows:

$$\min_x \|x\|_1 \text{ subject to } y = Ax. \tag{10.17}$$

This is a linear program that is efficiently implemented even for high values of n [197], where the first adopted algorithm to solve it was the Basis Pursuit (BP) [206].

It is most astonishing that the reconstruction algorithms proposed for CS provide some stability. This means that even though perfect reconstruction is not possible in some cases (due to fewer numbers of measurements or errors, vectors are not perfectly sparse, etc.), the algorithm is able to produce approximate solutions with errors that remain under control. However, as witnessed in linear algebra, including coding techniques, the reconstruction is an *all-or-nothing* process.

Since its advent, CS was linked with the ℓ_1 norm-based optimization algorithms. With its expansion to a very wide range of applications, different properties can be taken into consideration. One of the most crucial requirements is the computational complexity of the proposed algorithms, which have a direct impact on the speed of the process of reconstructing the original signals. Additionally, the flexibility, scalability, storage capacity, ease of implementation, and so on are requirements that are more and more claimed in order to keep up with the recent technology progress. As a result, the number of the proposed reconstruction algorithms for the CS problem has noticeably spanned to the point where discerning them has become a difficult task. Furthermore, one of the challenging tasks for researchers is finding the most suitable algorithm among this deluge of CS options for a specific scenario. Their performance can be measured using signal quality assessments such as the SNR or MSE, depending on the application. For some cases, very high accuracy reconstruction is necessary. In the following, we present an overview of major classes of algorithms with emphasis on the most relevant ones.

10.2.1 Convex relaxation

The algorithms proposed in this class aim to solve convex optimization problems through linear programming to reconstruct the signals. Least Absolute Shrinkage and Selection (LASSO) [207], Basis Pursuit De-Noising (BPDN) [206], and Least Angle Regression (LARS) [208] are the common algorithms in addition to the basis pursuit. However, these schemes are still considered very costly, as linear programming is not computationally optimal either. For example, for an image with 10 MP, millions of variables and tens of thousands of constraints have to be involved [209].

10.2.2 Greedy algorithms

This class of algorithms is known to be a simple and intuitive approach for solving optimization problems. Each step comes with an optimal and decisive choice based on a local optimization criterion, as the algorithm attempts to find the optimal global solution of the problem, otherwise a heuristic. Some researchers even consider greedy algorithms as a flexible alternative to convex relaxations.

10.2.2.1 Greedy pursuits

The simplest algorithm within this category is the Matching Pursuit (MP) [210].[3] Most commonly found in the literature are Orthogonal Matching Pursuit (OMP) [212], Compressive SaMPling (CoSaMP), Subspace Pursuit (SP), One-Step Greedy Algorithm (OSGA), and Iterative Hard Thresholding (IHT). They have naturally spanned into other algorithms, which could be less computationally complex or more robust. Examples for these spin-offs include the Lorentzian IHT, which uses the Lorentzian pseudonorm of the residuals in every iteration [213], the Block Orthogonal Matching Pursuit (BOMP), and the Constrained Matching Pursuit (CMP), to name a few.

Orthogonal Matching Pursuit (OMP)

Given $b \in \mathbb{R}^m$, $A \in \mathbb{R}^{m \times n}$, and $n \gg m$, the model $Ax = b$ can be recovered by the OMP algorithm if A and the vector x satisfy the inequality

$$M_A < \frac{1}{2k - 1}, \tag{10.18}$$

where M_A is the mutual coherence of the column vectors of A and k is the sparsity of x. Decisions are made based simply on inner products between the columns of the measurement matrix and a calculated residual. The main and simple algorithm for the OMP is as follows:

Algorithm 10.1: Orthogonal Matching Pursuit algorithm.

Input measurement matrix A, measurement vector y
Result: vector $\bar{x}$
initialization $r_1 = b$, $\Delta_0 = \{\}$
for $j = 1..m$ do
$\quad \lambda_j = \arg\max_j |A_i \cdot r_j|$
$\quad \Delta_j = \Delta_{j-1} \cup \{\lambda_j\}$
$\quad x_j = \arg\min_x \|A_{\Delta_j} x - b\|_2$
$\quad b_j = A_{\Delta_j} x$
$\quad r_{j+1} = r_j - b_j$

---

[3] Also called Pure Greedy Algorithm in approximation theory [211].

This algorithm is employed later in Chapter 22, because it is easy to use and is considered as a fast standard one compared with the other algorithms.

10.2.2.2 Thresholding

Iterative Thresholding algorithms were introduced for CS as an alternative to convex optimization. They are divided into two subclasses, soft (also known as shrinkage, Iterative Shrinkage-Thresholding (IST)) and hard thresholding. The latter, named Iterative Hard Thresholding (IHT), is known to have low computational complexity. In a nutshell, these algorithms start by supposing $x^0 = 0$ and iteratively find the sparsest solution using the main iteration

$$x^n + 1 = H_k(x^n + \Phi^T(y - \Phi x^n)), \qquad (10.19)$$

where H_k is a nonlinear operator that sets its input vector coefficients other than the k largest ones to zero. Fast Iterative Shrinkage-Thresholding Algorithm (FISTA) (an IST variation [214]), Hard Thresholding Pursuit (HTP), Normalized Iterative Hard Thresholding (NIHT), and Compressive Sampling Matching Pursuit with Subspace Pursuit (CSMPSP) (a hybrid algorithm to SP and CoSaMP), to name a few, are algorithms that are inspired by the aforementioned algorithms and rely on iterative thresholding. The comparison of the thorough performance results show that NIHT is typically faster than HTP and CSMPSP due to the greater flexibility in updating the support, which limits unnecessary computation on incorrect support sets [215].

10.2.3 Message passing

These algorithms constitute variants and important modifications of the iterative thresholding algorithms, where the variables are considered as *messages* in a graph that are associated with direct edges [209]. The common algorithms are Approximate Matching Pursuit (AMP), Expander Matching Pursuit (EMP) [216], Sparse Matching Pursuit (SMP), Sequential Sparse Matching Pursuit (SSMP), and Belief Propagation. The latter is one of the most well-known algorithms for message passing and is extensively used as a decoding algorithm for error correction codes or LDPC.

10.2.4 Reconstruction strategies discussion

Additional classes of algorithms exist, such as combinatorial or nonconvex minimization, but they are beyond the scope of this chapter. The straightforward question is which algorithms should be used under which conditions/constraints. So far, there is no complete and thorough classification in the literature. There is also a lack of directions regarding the suitable applications for these classes of algorithms. Nevertheless, it is important to note that the convex optimization algorithms with constrained ℓ_1-minimization, mainly BP, were the first proposed for compressed sensing, and they constitute a solid background for related research. Currently, a vast majority of

Table 10.1 Comparison of computational complexities for reconstruction algorithms in the *KL1p* library.

Algorithms	Complexity	Measurements
BP	$\mathcal{O}(n^3)$	$\mathcal{O}(k\log(n))$
OMP	$\mathcal{O}(k\,m\,n)$	$\mathcal{O}(k\log(n))$
ROMP	$\mathcal{O}(k\,m\,n)$	$\mathcal{O}(k\log^2(n))$
CoSaMP	$\mathcal{O}(m\,n)$	$\mathcal{O}(k\log(n))$
SP	$\mathcal{O}(k\,m\,n)$	$\mathcal{O}(k\log(n/k))$
SMP	$\mathcal{O}(n\log(n/k)\log R)$	$\mathcal{O}(k\log(n/k))$
EMBP	$\mathcal{O}(n\log^2(n))$	$\mathcal{O}(k\log(n))$

works rely on greedy algorithms, especially the popular OMP and its variances. Finally, algorithms could be selected for a specific compressed sensing problem based on their complexities or/and number of measurements required. (See the comparison Table 10.1 for common reconstruction algorithms.)

10.3 Sparse representation

The notion of sparsity has proved its importance and effectiveness in many modern fields. Compressed Sensing came with the premise that if a signal $x \in \mathbb{R}^n$ can have a sparse representation in an orthogonal basis $\Psi \in \mathbb{R}^{n \times n}$ using $\theta \in \mathbb{R}^n$, then only few nonadaptive measurements $y \in \mathbb{R}^m$ are needed to reconstruct the signal. For example, if x is a signal represented in the time domain, and θ is its equivalent representation in the Ψ domain, where Ψ is the inverse of the Fourier transform, then θ is the representation of x in the frequency domain. Based on the structure of the signal of interest, there exists a set of representation systems, which could be adopted to provide a sparse approximation. These systems are expanding and updating based on latest research trends. For instance, it was recently shown that the novel shearlets[4] transform provides optimal sparse representation of most natural images unlike the conventional wisdom regarding the wavelet transform [217].

10.3.1 Well-known transforms

Understanding the transformation from one basis to another, which has a lower dimension, can be simply done by considering a common vector of three dimensions (localization). However, in reality it is lying in a very high dimension, different from its usual three-dimensional coordinates. This means that sparse signals contain much less information then their ambient dimension suggests. In [218], it is explained how

---

[4] Shearlets are a natural extension of wavelets. They differ in the fact that wavelets are isotropic and shearlets enable the encoding of anisotropic features.

to exploit the sparsity properties of signals to process wireless signals in different applications. One of the most known examples is the JPEG2000 coding standard, which uses the sparsity of the wavelet coefficients of natural images [219], and the JPEG standard, which uses the Discrete Cosine Transform (DCT) basis. There exists a large set of sparsifying transforms, including, but not limited to, steerable wavelets, Gabor dictionaries, chirplets, warplets, multiscale Gabor dictionaries, wavelets packets, cosine packets, and so on. Despite the large number of sparsifying transforms, some data unfortunately cannot be sparsely approximated using the aforementioned common bases. The sparse representation of signals based on orthogonal bases in general highly depends on whether the signal characteristics can be matched with the specific basis function. Knowing that it is not an easy task to try these common sparse bases to make the signal fit, we next evaluate approaches that could be generalized to be employed with any nonsparse signal.

10.3.2 Sparsifying dictionary/dictionary learning

Due to the fact that some signals cannot undergo the sparsification step using the previously discussed well-known transforms, the learning dictionaries technique was proposed to provide sparser representations compared with the predefined aforementioned transforms. It is more interesting for CS applications to opt for overcomplete dictionaries. These have no stringent feature of having orthogonal atoms, as they cannot form a basis (see Section 10.1) and thus the scalability and flexibility in representing the data in a richer manner.

The task of finding an overcomplete dictionary requires having a set of training samples, which we denote by $X \in \mathbb{R}^{n \times l}$. The sparsest representation is found by solving either of the following optimization problems:

$$\min_{x} \|x\|_0 \quad \text{s.t.} \quad y = Dx \tag{10.20}$$

or

$$\min_{x} \|x\|_0 \quad \text{s.t.} \quad \|y - Dx\|_2 \leqslant \varepsilon, \tag{10.21}$$

which need optimization algorithms to be solved or approximated.

The most well-known algorithms provided for solving such problems are based on generalizations of the K-means algorithm, with specific differences for each:

- Method of Optimal Directions (MOD) [220];
- Maximum a posteriori probability approach;
- Union of orthonormal bases;
- Maximum likelihood methods;
- K-SVD.

In the following, we present the K-SVD algorithm, because it was proposed at the dawn of CS, and it outperformed the previous schemes [221,222].

On the other hand, selecting the right training samples and reducing the high computational complexity of the dictionary learning algorithms pose obstacles in using dictionaries instead of standard bases. This falsely limits the widespread use of the CS technique in various applications due to the inability of properly sparsifying the signals of interest.

10.3.2.1 K-SVD algorithm

The K-SVD algorithm was proposed for the overcomplete dictionary learning [223]. It is also a generalization of the K-means clustering algorithm. In a nutshell, it uses a two-stage iterative procedure, where it alternates between finding a sparse approximation of the training vectors based on the current state of the dictionary, and then iteratively updates the dictionary atoms to better fit the data.

The objective function in the K-SVD algorithm is defined as

$$\min_{D,X} \left\{ \|Y - DX\|_{\mathcal{F}}^2 \right\} \quad \text{s.t.} \quad \forall i, \|x_i\|_0 \leqslant T_0,$$

where $\|\cdot\|_{\mathcal{F}}$ is the Frobenius matrix norm,[5] defined for a matrix A as $\|A\|_{\mathcal{F}} := (trace(A^*A))^{1/2} = \sqrt{\sum_{i=1}^{m} \sum_{j=1}^{n} |A_{ij}|^2}$, and $trace$ is the sum of the diagonal entries of a square matrix.

10.4 Distributed compressed sensing

The captivating feature of large sets of data types is the fact that they often have a pattern or a structure that can be modeled, thus enabling smart ways of representations and processing. Compressed Sensing is a perfect application for these types of data, as it extends to cover not only the intrasignal correlations, but also the intersignal correlated structures. Distributed Compressed Sensing (DCS) is the theory that combines CS and distributed source coding, therefore allowing the exploitation of inter- and intrasignals. It is linked to the concept of the joint sparsity of the entire set of signals. Consequently, the total number of measurements can be reduced even more than what the standard CS approach would guarantee. DCS opened the door for CS to move from being a signal processing tool to other types of applications, mainly communications thanks to the attention given to the multicorrelation of existing intersignals. WSNs are believed to have the exact type of signals for applying DCS. This is due to its tolerance to noise and quantization and to its robustness to measurement losses. In a nutshell, the basic idea behind DCS is that each node samples

---

[5] Also called the Hilbert–Schmidt norm.

Algorithm 10.2: K-SVD overcomplete dictionary learning algorithm.

Task: Find the best dictionary to represent the data vectors Y as sparse representations by solving

$$\min_{D,X} \left\{ \|Y - DX\|_{\mathcal{F}}^2 \right\} \quad \text{s.t.} \quad \forall i, \|x_i\|_0 \leqslant T_0. \tag{10.22}$$

Result: matrix dictionary D and sparse matrix Γ s.t. $X \approx D\Gamma$
Input: Signal set X, initial dictionary $D^{(0)}$, desired sparsity s, number of iterations k
Initialization: Set $D := D^{(0)}$, set $T = 1$
while *No convergence or number of iterations* **do**

  1. Sparse coding
     Apply a pursuit algorithm to compute γ_i the vectors of Γ for each example x_i, by approximating the solution of
     for $i = 1..n$ **do**
     $$\min_{\gamma_i} \{\|x_i - D\gamma_i\|_2^2\} \quad \text{s.t.} \quad \|\gamma_i\|_0 \leqslant T$$

  2. Codebook update
     For each atom $j = 1, \ldots, m$ of $D^{(T-1)}$ update by

     • Define the group of examples that use the atom,
       $\omega_j = \{i : i \in \{1, \cdots, n\}, \gamma_T^j(i) \neq 0\}$
     • Compute the overall representation error matrix E_j by

     $$E_j = X - \sum_{l \neq j} d_l \gamma_T^l$$

     • restrict E_j by choosing only the columns corresponding to ω_j, and obtain E_j^R
     • Apply SVD, $E_j^R = U\Sigma V^T$

  3. Set $T = T + 1$

---

its readings independently and then transmits the results to a sink node, where all the readings belonging to different nodes are jointly reconstructed.

In the scenario of distributed compressed sensing, we suppose that the number N of nodes (sensors, antennas, etc.) have an ensemble of sparse signals, which can be expressed as $X = [x_1 \ x_2 \ \cdots \ x_N]^T$, where $x_i \in \mathbb{R}^n$ is a sparse signal. The compressed measurements can be written as

$$Y = \Phi X, \tag{10.23}$$

where $Y = [y_1 \ y_2 \ \cdots \ y_N]^T$, and the compressed sensing matrix Φ is represented as

$$\Phi = \begin{bmatrix} \Phi_1 & 0 & \cdots & & 0 \\ 0 & \Phi_2 & 0 & & 0 \\ \vdots & & 0 & \ddots & \vdots \\ 0 & \cdots & & 0 & \Phi_N \end{bmatrix}. \tag{10.24}$$

Following this approach, every individual signal can be expressed as $y_i = \Phi_i x_i$, where $y_i \in \mathbb{R}^m$ and $\Phi_i \in \mathbb{R}^{m \times N}$.

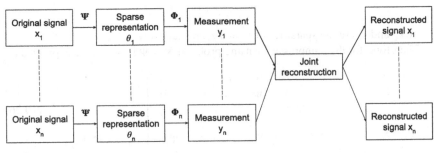

FIGURE 10.3

Distributed CS framework.

Moreover, Fig. 10.3 illustrates the complete DCS framework, where the original signals are not necessarily sparse. It also shows the smoothness of the joint reconstruction and how it is performed using one-step decoding to retrieve all signals involved in the DCS process. This obviously drastically reduces the complexity for reconstruction, thus reducing the delays in delivering the data.

10.4.1 Joint sparsity models

Each signal x_i in these models is generated as a combination of two components:

i) a *common component* denoted by z_c and present in all the signals and
ii) an *innovation component* denoted by z_i, which is unique to every signal.

As a result, the signal x_i can be expressed as

$$x_i = z_c + z_i, \ i = 1, 2, \ldots.$$

As a matter of fact, the correlations of the signals do not have the same characteristics and form and vary according to the type of signals, scenario, and so on. Baraniuk et al. [224] advocated for three main specific models based on the signals characteristics, which exclusively rely on i) and ii), called Joint Sparse Model (JSM)-1, JSM-2, and JSM-3.

10.4.1.1 Sparse common component + innovations (JSM-1)

This model is characterized by the fact that all the signals involved share a common sparse component, but each individual signal has a sparse component, said to be an innovation component:

$$x_j = z_c + z_i, \; j \in \{1, \cdots, n\}, \tag{10.25}$$

where z_c and z_i denote the common component and the innovation component, respectively. They can also be sparsely represented as

$$z_c = \Psi_c \cdot \theta_c, \; \|\theta_c\|_0 = k_c,$$
$$z_i = \Psi_i \cdot \theta_i, \; \|\theta_i\|_0 = k_i,$$

where k_c and k_i are the sparsities of z_c and z_i, respectively.

Therefore, for the compressed sensing problem $X = \Psi\Theta$, Ψ can be expressed as

$$\Psi = \begin{bmatrix} \Psi_c & \Psi_1 & \cdots & 0 \\ \Psi_c & 0 & \Psi_2 & 0 \\ \vdots & 0 & \ddots & \vdots \\ \Psi_c & \cdots & 0 & \Psi_N \end{bmatrix}. \tag{10.26}$$

This is the most common example that models a network of sensors that are located geographically close and simultaneously monitor a natural phenomenon that varies smoothly in time and in space, such as temperature, humidity, light intensity, and so on. These readings contain inter- and intrasignal correlations, related to the spatial and temporal correlations. In the case of temperature monitoring the common component is strongly related to global natural factors, such as the wind or the sun. As for the innovation component, it results from local factors, such as the shadow or the proximity of beings.

10.4.1.2 Common sparse supports model (JSM-2)

All signals in this model can be formed from the same sparse basis, but with different coefficients, as

$$x_i = \Psi\theta_i, \; i \in \{1, \ldots, n\}, \tag{10.27}$$

where Ψ is the sparse basis, and θ_i are the k-sparse coefficients.

This is the most suitable model to characterize the ensemble of signals in WSNs. A common practical application of this model is the case where different sensors acquire the same signals, but with phase shifts. Additionally, the JSM-2 is an efficient approach for acoustic signals, multilead ElectroCardioGram (ECG) signals [225], or MIMO communication [226].

10.4.1.3 Nonsparse common component + sparse innovations (JSM-3)

The signal observed at each node in this model is assumed to be composed of an arbitrary common component z_c and a sparse innovation component z_i. This differs from the JSM-1 in the fact that the assumption regarding the common component being sparse is no longer a requirement. Such a model can be adopted in scenarios where it is either impossible or complicated to obtain a sparse representation of the common signal in any basis. Therefore it is suitable for situations where the intersignal correlations are dominant compared with the almost nonexistent intrasignal correlations.

10.4.2 DCS reconstruction algorithms

The joint recovery could be performed via ℓ_0-minimization, but as discussed earlier in this chapter, it is more reasonable to relax it and recover the signal ensemble via the ℓ_1-minimization. Without loss of validity of the standard compressed sensing reconstruction methods, many algorithms were proposed for DCS, just as for CS itself. Their properties differ based on the nature of the signal set, application, and requirements of the system, mainly the computational complexity and the tolerated reconstruction error. Additionally, various reconstruction strategies were proposed for each JSM model to meet the desired specifications. For example, OSGA solves the DCS data ensemble modeled using the JSM-1. Moreover, Trivial Pursuit (TP) is a greedy algorithm designed for the JSM-2. It demands a large signal set to perform well. Furthermore, DCS-Simultaneous Orthogonal Matching Pursuit (DCS-SOMP) [227] is a DCS-based Simultaneous Orthogonal Matching Pursuit (SOMP) (which is a variant of OMP). This strategy requires a small number of measurements that is proportional to the sparsity k for a moderate number of signals. Sarvotham et al. [224] showed that reconstructing one signal of the set could be achieved using $k + 1$ measurements as the number of signals tends to infinity. More algorithms are available in the literature, but the choice of discussing TP and DCS-SOMP was made as they were some of the first reconstruction strategies for DCS.

10.5 Compressed sensing for communications

The challenging tasks of the 5G mobile environments are evolving toward becoming extremely heterogeneous and complex. The recent success of the CS technique has made it a potential solution to help manage this inevitable data deluge. As previously addressed, the CS reconstruction algorithms have the fascinating ability of being stable, flexible, and scalable. As a result, this disruptive technique has established its implication in various practical applications that involve noise and sparse or compressible signals. It has a conspicuous impact on several applications, including, but not limited to, sampling theory, medical imaging (MRI, ECG, EMM, etc.), radar, sparse approximation, error correction, and matrix completion. Recently, it has proliferated into communications systems and upgraded its profile from being a signal processing tool to a solid component in communications, essentially bringing a new perspective to the way we process data. It reduces the overhead for channel estimation and feedback in massive Multiple Input Multiple Output (MIMO) systems [228]. Currently, it is a hot research topic for interference cancellation, symbol detection, spectrum sensing, and so on. Being in an era of ephemeral data, the choice and decision whether to use or lose the data collected has to be rapidly made. No system is able to permanently handle all circulating data. Compressed sensing allows us to avoid such an extreme decision. Even in harsh network conditions, an approximation of the data could be made. Following the topic of this book, we shed light on the implications of CS in Wireless Sensor Networks (WSNs) as they became a crucial component of the upcoming 5G standard.

10.5.1 Compressed sensing for WSN

The dawn of IoT includes a huge number of smart devices that are dedicated to the Industry 4.0, wireless communications, or home appliances, to name a few. They rely on massive numbers of sensors deployed in smartphones, robots, or smart gadgets (haptic jackets and gloves) to enable the extension of humans' observations of physical environments. Such a disruptive proliferation of sensors in all IoT devices has led to generating massive amounts of data, which surpass the threshold expected by all data forecasting systems. This complicates the procedures of processing and transmitting by such power constrained devices. This data avalanche is urging to find solutions, first how to acquire and process this amount of data efficiently and effectively and then how to store it. On the other hand, the fascinating observation about real-life signals is that they can be well approximated by sparse signals. Employing compressed sensing could foster the data collection and dissemination, especially in very large and lossy networks.

10.5.2 Kronecker compressed sensing

The main compressed sensing works proposed focused on problems that involve solemnly 1-D or 2-D signals, such as sensor readings and images, respectively. Kronecker Compressed Sensing (KCS) [229] was introduced to deal with the problems that have multidimensional signals, as many compressed sensing candidate applications involve higher-dimensional signals, including coordinates, spectral, and other dimensions. Such a class of compressed sensing can also be adopted in scenarios where spatial and temporal dimensions are involved, for example, environmental sensors [230], camera and microphone arrays, and so on. Due to the algebraic properties preserved by the Kronecker product, involving it in the compressed sensing problem is not limited to measurement matrices multiplication, but it is also valid to use it for sparsity bases.

10.5.2.1 Kronecker product sparsifying bases

The Kronecker product of a set of sparsifying bases for each of the d-sections of a multidimensional signal results in having one single sparsifying basis for the entire signal. Specifically, we denote $X \in \mathbb{R}^{n_1 \times \cdots \times n_d} \equiv \mathbb{R}^{\prod_{d=1}^{D} n_d}$, and we assume that every d-section is either sparse or compressible in the basis denoted by Ψ_d. The sparsifying basis resulting from the Kronecker products is expressed as $\bar{\Psi} = \Psi_1 \otimes \cdots \otimes \Psi_D$. Thus the compressed sensing problem could be reformulated as

$$vec(X) = \bar{\Psi}\Theta, \qquad (10.28)$$

where $vec(X)$ is the vector-reshaped representation of X, and Θ is the coefficient vector for the signal ensemble.

Previous empirical results show significant low compression error for Kronecker product sparsifying bases compared with the wavelet transforms (space and frequency) performed on the real-world hyperspectral datacube [229]. However the compression errors of the different approaches are approximately the same when the number of coefficients increases. Finally, some recent works have focused on designing Kronecker dictionary learning, which provides fast operators while maintaining a significant degree of flexibility [231,232].

10.5.2.2 Kronecker product measurement matrices

This refers to the scenario where measurement matrices are designed using the Kronecker products. This is equivalent to having independent measurement processes on portions of the multidimensional signal. The overall measurement matrix is defined as $\bar{\Phi} = \Phi_1 \otimes \cdots \otimes \Phi_D$. The Kronecker compressed sensing is thus formulated as

$$Y = \bar{\Phi}X. \qquad (10.29)$$

Most importantly when dealing with compressed sensing problems is the measurement matrices obeying the main properties for an effective reconstruction. It has been proved that the mutual coherence across Kronecker products is conserved [233, 234]. Additionally, KCS preserves the guarantees of a variety of standard CS reconstruction algorithms. As for numerical results, it was shown by Duarte et al. [229] that KCS performs considerably better than the independent type of recovery. Nevertheless, the main issue of the Kronecker CS is the computational complexity due to the exploding size of the Kronecker product matrices, that is, it is proportional to the product of the dimension of the signal and the data partitions in one dimension. This hinders its adoption in many real-life applications that have considerably higher dimensions.

PART

5

Building the testbed

Outline

After explaining the main theory parts for the book, we now dive into the practice part by presenting our ComNetsEmu environment, which is composed out of Mininet and Docker. The latter two are introduced in one chapter each, before we combine those two tools into the ComNetsEmu.

PART 5

Building the testbed

Mininet: an instant virtual network on your computer

11

Zuo Xiang[a], **Patrick Seeling**[b]

[a]*Technische Universität Dresden, Dresden, Germany*
[b]*Central Michigan University, Mount Pleasant, MI, United States*

Simplicity is prerequisite for reliability.
Edsger W. Dijkstra

11.1 Introduction

In networked system designs and evaluations, rigorous and sufficient experimentation is required to ensure that theories work within their contexts. In recent years, the importance of providing not only results of experimentation, but also of enabling reproducibility has increased. The Association for Computing Machinery (ACM), for example, defines three main levels of reproducibility: i) repeatability (where one team obtains the same results with the same setup), ii) replicability (where another team obtains the same results with the same setup), and iii) reproducibility (where a different team obtains the same results with a different setup); see, for example, [235, 236] for more details. Modern experimentation should, subsequently, demonstrate the behavior of the system under consideration with dynamic network parameters and provide reproducible evaluation results.

Simulation and emulation are two commonly utilized approaches for networked system experiments. Simulators generate results typically with platform-independent modeling code; discrete event simulators are among the most popular approaches to simulation. Offering significant benefits for reproducibility, simulations typically require major time and resource commitments to develop well-suited models of real-world implementations of hardware and software components that constitute a networked system. Network emulators, in contrast, utilize software (e.g., OS kernels or network applications) on a computer that is part of a real networked system to perform experiments in continuous time [237]. Therefore network emulation is the preferred approach to performing practical experiments for networked systems.

Mininet is a lightweight network emulation orchestration system for the rapid prototyping of a complete networked environment. It utilizes GNU is Not Unix with Linux added (GNU/Linux) OS-level virtualization technologies to create a realistic, virtualized network running hosts, switches, routers, and network applications

Computing in Communication Networks. https://doi.org/10.1016/B978-0-12-820488-7.00025-6

on a single physical machine [238]. Mininet supports using the OpenFlow protocol for SDN network emulation. Because of its simplicity and reproducibility, Mininet is used in *Reproducing Network Research* teaching experiments at Stanford University [239]. A Mininet network with two hosts connected directly to a switch is illustrated in Fig. 11.1.

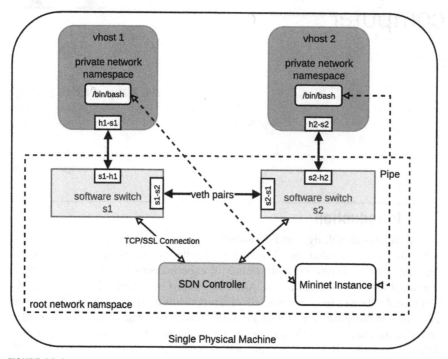

FIGURE 11.1

Mininet overview.

Mininet utilizes the container mechanism provided by the GNU/Linux kernel to emulate nodes in the network. By default all Mininet hosts are regular processes that share the same OS kernel, process IDs, user names, and file systems. Each Mininet host has an independent network stack and corresponding network resources, including network interfaces, Address Resolution Protocol (ARP) caches, and routing tables. Each host additionally features a virtual interface that can be connected to a virtual (software) switch (e.g., Open vSwitch [240]) via a virtual link with configurable parameters (e.g., bandwidth, latency, or loss rate). The virtual device and link are emulated via GNU/Linux Virtual Ethernet Device (veth) [241]. Compared to physical test beds and VM-based heavy-weight emulators, Mininet utilizes these lightweight technologies to enable scaling to relative large topologies (i.e., to over hundreds of nodes [238]).

Using Mininet for networked system emulation and experiments provides the following main benefits:

- Lightweight emulation significantly accelerates the run-debug-evaluate loop. Emulations can be executed on a single laptop with sufficient scalability.
- All components in the network are fully customizable with straightforward and friendly Python APIs.
- Mininet supports SDN concepts and switches in the emulated network and can be programmed with OpenFlow protocol.
- Emulation and evaluation scripts for Mininet are conveniently shared for reproducible experiments.
- Mininet is an open-source project with ongoing active development, detailed documentation, and available tutorials.

Besides an elegant design of Mininet, it also has some limitations:

- All components share the resources of the underlying physical machine. Resources need to be scheduled carefully among components.
- A shared GNU/Linux kernel is used by all hosts. Different OS kernels or the same OS kernel with different versions cannot be compared in the same emulation.
- All Mininet hosts reside in different network namespaces. All other namespaces, including Process ID (PID) and file system, are shared. This isolation is not sufficient for some NFV applications.
- Because of the shared file system used by all Mininet hosts, required software packages for the emulated hosts should be installed and configured in the hosting OS. Software dependency conflicts and easily reproducible single host configurations are difficult to handle in the current Mininet implementation.
- Default Mininet host type does not support deployment and management of containerized applications, for example, using Docker technology. The emulation of cloud-native applications with mobility is also not supported by the regular Mininet implementation.

After this high-level introduction to Mininet, we now shift to its common implementation workflow.

11.2 Mininet workflow

Typically, the customization and emulation of a networking system on Mininet require the following steps:

11.2.1 Create a network topology

In Mininet, parameterized network topologies can be created with its Python API [242]. Network nodes and links can be added and configured by overriding the `build()` method of the `mininet.topo.Topo` class. The following code snippet 11.1 demonstrates a simple topology that consists of N hosts connected to a single switch [242]. For a detailed description of the parameters for each class and corresponding methods, we refer the reader to the official documentation [243].

```
#!/usr/bin/python3
from mininet.topo import Topo
from mininet.node import CPULimitedHost
from mininet.net import Mininet
from mininet.link import TCLink

class SingleSwitchTopo(Topo):
    """ N hosts connected to a single switch """
    def build(self, n):
        switch = self.addSwitch("s1")
        for h in range(n):
            host = self.addHost(
                "h%s" %(h+1),
                ip="10.0.0.%s" % (h+1), cpu=0.5/n)
            self.addLink(switch, host,
                bw=10, delay="50ms", loss=3,
                max_queue_size=1000, use_htb=True)

    def perfTest():
        topo = SingleSwitchTopo(n=3)
        net = Mininet(topo=topo,
            host=CPULimitedHost, link=TCLink)
        net.start()
        net.pingAll()
        print("Test the bandwidth between h1 and h3")
        h1, h3 = net.get("h1", "h3")
        net.iperf((h1, h3))
        net.stop()
```

Listing 11.1: Simple topology with a single switch.

In addition to the configuration of the types of nodes and their connectivities, performance limitations can be designed by employing special node and link classes. In code snippet 11.1, the maximal allowed CPU bandwidth of each host is limited to $50\%/n$ of the global system resources. Each link between host and switch is bidirectional with delay and loss rate parameters emulated by the Linux NetEM [244] utility. A veth pair is created for each link with NetEM attached to each virtual interface. The performance characteristics are applied to all packets outgoing from the attached interface to provide a bidirectional performance-parameterized link connection. In this example the delay is set to 50 ms (the units can be ms, µs, and s), and the loss rate (with an independent random loss probability [244]) is given as percentage. In addition to delay and packet losses, the bandwidth of the link is also limited to 10 Mbit/s with a maximum queue size of 1000 packets using the Hierarchical Token Bucket (HTB) rate limiter [242].

11.2.2 **Interact with a network**

After a network is successfully started, arbitrary commands can be executed on any node in the topology. Each host in Mininet is fundamentally an interactive bash shell process, running in its own network namespace. Executable commands can be sent to the standard input of the shell with the cmd() method of each node instance. This method waits for the output of the command and returns this output in string format. Additional methods to communicate with nodes are provided in Mininet's Python API; see [243].

Mininet also has a built-in mininet.cli.CLI class to provide a Command Line Interface (CLI) for running interactive commands during the emulation. The CLI can be invoked by calling the CLI() method on a running Mininet instance, CLI(net). Useful options that are included in the CLI include, among others:

- Python scripts can be executed with py command.
- The status of a created link and switch can be configured with link and switch commands.
- Basic performance tests can be performed, including bandwidth tests using Iperf and latency tests with ping.
- Terminal windows of each node (Xterm by default) can be created to execute commands interactively.

11.2.3 **Programmable network with SDN**

Switches in Mininet networks can be programmed using the OpenFlow protocol. We refer the interested reader to Chapter 6 for a more in-depth discussion of OpenFlow. However, to utilize OpenFlow, an SDN controller needs to be configured in the Mininet object. By default, the built-in Stanford reference controller is chosen when Mininet is installed. As shown in the example Listing 11.2, an existing SDN controller can be added into the network with the help of the RemoteController class. The RemoteController should be used as a class constructor here.

```
from mininet.net import Mininet
from mininet.topo import SingleSwitchTopo
from mininet.node import RemoteController
from functools import partial

net = Mininet(topo=SingleSwitchTopo,
    controller=partial(RemoteController,ip='127.0.0.1', port=6633)
)
```

Listing 11.2: Usage of the remote SDN controller.

To automatically start and stop the controller program in an emulation, a subclass of mininet.node.Controller should be created with start() and stop() methods overridden.

11.3 Demystifying Mininet

Mininet utilizes a set of built-in isolation and virtualization features directly provided by the Linux kernel. This section provides an introduction of the main technologies utilized by Mininet to support lightweight and high-fidelity emulation.

11.3.1 Resource management and isolation

We now briefly review how Mininet manages resources and achieves their isolation.

11.3.1.1 Linux NS

Network Slicing (NS) wraps a particular global operating system resource to provide OS-level resource isolation [245]. The processes within one namespace have their own isolated global resource. Each new process created via the fork() system call is in the same sets of NS of its parent [246]. NS is one fundamental kernel feature for the implementation of containers.

In the Linux kernel (v4.15) used by the ComNetsEmu test bed, the following NS types are supported [245]: i) Cgroup, the Control Groups (Cgroups) root directory, ii) IPC, the System Inter-Process Communication (IPC) and Portable Operating System Interface (POSIX) message queues, iii) Network, for Network devices, stacks, ports, and other network resources, iv) Mount for mount points, v) PID for Process IDs, vi) User for user and group IDs, and vii) UTS for hostname and domain name.

One method to check the namespaces of one process is reading the symlink (symbolic link) located in its NS pseudofilesystem(/proc/[pid]/ns/). Processes that show the same symlink are in the same NS. Several system calls, including clone, unshare, and setns, can be used to manage the NS of an individual process.

The mount and network NSs are used by Mininet to create the default isolated host. The startShell(self, mnopts=None) method of the host instance starts a new shell (Bourne-Again SHell (BASH)) process with the mount namespace of the caller and a new network namespace. NS-related utility functions are implemented in mnexe.c. Therefore individual Mininet hosts have their own network resources, but all share the same file system with the host OS.

11.3.1.2 Linux Cgroups

Linux Cgroups are a kernel feature enabling resources (such as CPU, memory, or block Input/Output (I/O) bandwidth) management and monitoring of processes organized into hierarchical groups [247]. Cgroups have two main components: i) a method to group processes hierarchically and ii) resource controllers to control and monitor processes in a Cgroups.

The management of Cgroups can be performed via a pseudofilesystem interface (/sys/fs/cgroup). (See Listing 11.3.) Therefore all Cgroups operations can be performed using files in the filesystem. Cgroups is heavily used in many applications, including containers and systemd [248]. Cgroups currently have two versions (v1 and v2), and both exist in the Linux kernel (v4.15). Version 2 solves several implementation issues of the older version and is intended to replace v1 in the future. When systemd [248] is used in the GNU/Linux distribution, all Cgroups v1 resource controllers are automatically mounted after boot. Currently mounted Cgroups systems and corresponding controllers can be evaluated by running the following command in a shell:

```
# mount | grep cgroup
cgroup on /sys/fs/cgroup/pids type cgroup (pids)
cgroup on /sys/fs/cgroup/cpuset type cgroup (cpuset)
...
```

Listing 11.3: Display mounted file systems related to cgroup.

Cgroups version 1 have following resource controllers [247]:

- cpuset: Pin a Cgroup to one CPU or a subset of available CPUs.
- cpuacct: Expose the CPU usage of a Cgroup.
- cpu: Management of CPU cycles provisioned for a Cgroup. Proportional-weight division mode and bandwidth control mode are supported:
  - Proportional-weight division: shares is used to specify the proportion of CPU provisioned to a Cgroup. This limitation works only heavy competition for CPU happens.
  - Bandwidth control: the CPU maximal quota of a Cgroup is limited by the quotient of cpu.cfs_quota_us and cpu.cfs_period_us. This limitation works even when there is no competition for CPU.
- memory: Limit the memory usage of a Cgroup.
- devices: Manage the list of devices that can be accessed by a Cgroup.
- pid: Limit the number of processes in a Cgroup.
- freezer: Suspend and resume processes in a Cgroup. This feature is used in container migration.
- net_cls: Tag outgoing packets sent by members in a Cgroup.
- net_prio: Control the priority of output network traffic of members in a Cgroup.
- blkio: Limit I/O operations on block devices with similar policies used by cpu controller.
- perf_event: Perform perf event [249] monitoring for a Cgroup.
- hugetlb: Limit the huge page resources available for members of a Cgroup.

Compared to Cgroups version 1 (v1), the Cgroups version 2 (v2) introduced the following main improvements:

- v2 uses single hierarchy for all controllers.
- The consistent inheritance rule of all controllers are applied with detailed documentation.
- The blkio controller is replaced by a more general I/O controller.
- The freezer and hugetlb controllers are currently not supported (at Linux kernel v5.0).

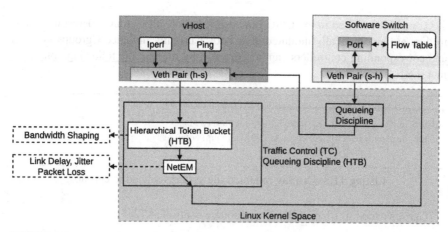

FIGURE 11.2

Configurable data plane.

Mininet has a `CPULimitedHost` host type, which uses CPU-related Cgroups to limit the fraction of a CPU available to this host.

11.3.2 Configurable data plane

Software tools including veth, Traffic Control (tc), and OVS are utilized by Mininet to provide a fully programmable data plane to support SDN-based emulations. (See Fig. 11.2.)

11.3.2.1 Linux virtual ethernet pairs (veth pairs)

In Mininet, veth pairs are used to connect isolated hosts to a software switch. The Linux kernel provides veth mainly to create tunnels between network namespaces. A veth pair is a full-duplex link with two separate interfaces in the same or different NSs. Each interface of a veth pair behaves like a regular Ethernet interface and can be configured with `ethool`. Packets transmitted from one interface are directly forwarded to the other interface in the same pair [241].

11.3.2.2 Linux traffic control

Mininet uses Linux kernel traffic control to configure the bandwidth, delay, loss rate, maximal queue size, and other characteristics of a virtual link. Traditional elements of traffic control include shaping, scheduling, classifying, or dropping. With shaping, packets are delayed in a queue to meet a configured output rate. Scheduling can be used to prioritize packets in a queue. Whereas classifying is the mechanism to separate traffic into different queues, dropping is used to discard a entire packet. Shaping and scheduling are performed on egress traffic, whereas classifying and dropping are applied on both ingress and egress traffics [250]. In the following, we focus on managing and manipulating the traffic.

In the Linux kernel, Queueing Discipline (QDISC), classes, and filters are the main components for traffic scheduling, shaping, and classifying. When a packet is sent to an interface by the Linux kernel, it is initially enqueued to the configured qdisc for scheduling. Multiple qdiscs can be applied to packets before they reach the network adapter driver [251]. For egress traffic, qdiscs need to be attached to the *root* qdisc. Ingress traffic is controlled by qdiscs attached to the *ingress* qdisc. The QDISCs provide two usable types, classful and classless qdiscs:

Classful QDISCs   This type allows adding classes and provides a handle to attach filters. It is useful to apply different treatments to various kinds of traffic. Classes in a classful qdisc can have a single or multiple child classes. A packet entering a classful qdisc needs to be classified into any of the classes within using filters. These classes build a tree structure, as illustrated in Fig. 11.3. Leaf classes are used as the terminal classes, which cannot have a child class. Two representative classful qdiscs are:

- HTB is a hierarchical qdisc with an arbitrary number of token buckets. These tokens buckets can be nested with different classifying mechanisms. The most common usage of HTB is for bandwidth shaping. It can be used to simulate multiple slower links with one physical link [250].
- Hierarchical Fair Service Curve (HFCS) is a qdisc that provides precise bandwidth and also delay allocation for all leaf classes. Detailed information of HFCS is provided, for example, in [252].

Classless QDISCs   This type can be either used as the primary qdisc or attached inside a leaf class of a classful qdisc. One typical example of this type is *fifo_fast* (First In, First Out) without any special traffic modification. Other common classless qdiscs include:

- *fq_codel* (Fair Queueing Controlled Delay) is the default classless qdisc since `systemd` 217. *fq_codel* aims at providing a fair share of bandwidth to all the flows in the queue.
- The Network Emulator (NetEM) is also a classless qdisc that allows us to add configurable delay, packet loss to traffic outgoing from a network interface. Currently, it supports independent random loss, 4-state Markov loss, and Gilbert–Elliot loss models [244].
- Token Bucket Filter (TBF) is a simple traffic shaper using tokens and buckets. It can be used to slow down the speed of the transmitted traffic to a given rate.

Another important component of tc is filtering. Filters can be used to classify packets into classes when a classful qdisc is attached to the root qdisc [251]. Examples of filters include i) bpf, which filters packets using Extended Berkeley Packet Filter (eBPF), ii) flow, which filters packets based on the flow metadata, and iii) route, which filters packets using routing tables.

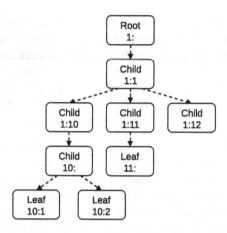

FIGURE 11.3

Classful Queueing Disciplines Hierarchy [250].

In Mininet, traffic control can be used by adding virtual links of type `TCLink`. The `TCLink` object constructs its veth pairs with class `TCIntf`, which can add qdiscs on both virtual interfaces. If BandWidth (BW), delay, jitter, loss, and maximal queue size are configured in the `addLink()` method, then a pipelined classful qdisc is created. By default, Mininet creates an HTB qdisc with a single HTB class for bandwidth shaping. A NetEM qdisc is attached as the leaf class to emulate delay, jitter, random loss, and maximal queue size of a link.

11.3.2.3 Virtual switch

By default, Mininet employs OVS, which is an open-source multilayer virtual switch with production quality characteristics [240]. Besides featuring built-in support for many standard interfaces and protocols, OVS is reprogrammable through the Open-Flow protocol. OVS is one of the most-used SDN switches, and OVS of version higher than 2.8 supports OpenFlow versions 1.0 to 1.4.

The architecture of OVS is illustrated in Fig. 11.4. The basic OVS implementation features three separate major components:

datapath-kernel-module is a high performance kernel module for packet I/O including forwarding, modification, sampling, and dropping. The behavior of the datapath module is determined by the instructions (called actions) added by the `ovs-vswitchd`. If the datapath has no any rules for a packet, then this packet is forwarded to the `ovs-vswitchd`. The actions received from the daemon are commonly cached by the datapath for the similar group of packets.

`ovs-vswitchd` is a daemon running in userspace that manages the behavior of the datapath. OpenFlow flow tables are received by this daemon to add new actions in the datapath.

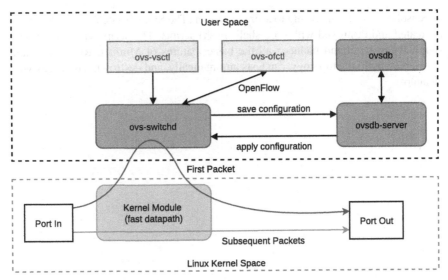

FIGURE 11.4

Open vSwitch architecture [253].

ovsdb-server is a DataBase (DB) server used by OVS for switch configuration. The `ovs-vswitchd` gets its configuration from this database. The OVSDB management protocol can be used to access this database [254]. Typical operations defined in this protocol include creation, modification, or removal of Open-Flow datapaths running in an OVS instance.

OVS provides multiple administration and management command-line tools. The most frequently used ones include:

ovs-vsctl is used to configure and query the `ovs-vswitchd` daemon with OVSDB management protocol. Typical usages of this command create a new bridge and manage its ports.

ovs-ofctl is used to manage and monitor OpenFlow switches (including OVS). This command provides a lightweight approach to manage flow tables without using SDN controllers. Typical usages of this command add a new flow entry and monitor the flow status of all created ports on a bridge.

Mininet provides a node type named `OVSSwitch` to manage OVS.

11.4 Create a tiny topology from scratch

A practical example of using all technologies introduced in this section can be found in the ComNetsEmu emulator repository [255]. In this example a basic topology with three isolated hosts connecting to a single switch (The SingleSwitchTopo in Mininet) is created and destroyed with pure shell (bash) scripts. The script was written with simplicity in mind and includes all the basic features of Mininet. Run `run.sh` and `clean.sh` to build the topology, run tests automatically, and destroy the topology with cleanups.

Docker: containerize your application

12

Alexander Kropp, Roberto Torre

Technische Universität Dresden, Dresden, Germany

Container love, the years have passed, the rainbow shows he fell in love, not far from there, decation grows container love...

Phillip Boa and the Voodooclub

12.1 Introduction to Docker

Docker is a platform for developers with the main purpose of developing, deploying, and executing virtualized applications in sandbox environments called containers [81]. A container is an instance of an image (a standalone executable package that runs an application) paired with its state. An image includes the components necessary to run the application: i) the application itself, ii) the libraries needed, iii) environment variables, and iv)configuration files. A container in execution can thus be interpreted as a modified version of an original image that received changes in the programs, libraries, or other content. The underlying technology behind Docker, runC [256], is a sandbox environment that abstracts the underlying host without the need for a complete revision of the application. runC is a production-level technology mainly designed for security and isolation from the host. Fig. 12.1 displays a schematic of container virtualization.

The main characteristics that make containers a leader in virtualization technologies are [81]:

Flexibility: Containerization technology is able to virtualize and instantiate lightweight applications that utilize only a small number of libraries up to heavy applications with images reaching GB in terms of size.

Lightweight: The world is moving to a microservice-based model, where multiple but very small services coexist in the same host. Containers excel in lightweight application provisioning, because they share the host kernel. This reduces their size and, as a result, the booting and processing times.

Interchangeable: The flexibility that Docker provides enables developers and system administrators to deploy program updates and upgrades on the fly. Differently put, the application itself is running while its new version is deployed. Afterwards, the newer version takes over the tasks of the older one.

Computing in Communication Networks. https://doi.org/10.1016/B978-0-12-820488-7.00026-8

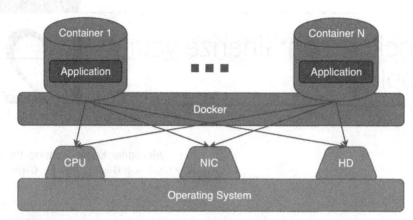

FIGURE 12.1

Virtualization with containers.

Portable: `runC` provides a sandbox environment that allows applications (container images) to be built locally, uploaded to the cloud, downloaded, deployed, and executed anywhere. The only constraint is that the kernel of the host where the container was built and the kernel of the host where the application is to be deployed are the same, to allow for a successful deployment.

Scalable: The number of replicas deployed can be easily increased (with a single click). Increases and decreases of instances can be automated, so the number of replicas can dynamically adjust to the current requirements of users.

Stackable: In case multiple services share dependencies and can be orchestrated together, Docker enables the possibility to create a *stack*. This feature groups services together, thus saving time, resources, and container size. This stacking can be performed on-the-fly, that is, while the application runs.

12.2 Containers vs virtual machines

The main competitors of containers for function virtualization are VMs. However, they also complement themselves by targeting different use cases of functionality for virtualization. Fig. 12.2 illustrates how virtualization is achieved in VMs. In contrast to container-based virtualization, a VM is a virtual environment where a full operating system runs. In turn, VMs provide more isolation than containers because the guest systems do not share the host kernel. A Virtual Machine Monitor or hypervisor controls the instantiation and removal of VMs and assigns resources to them. The hypervisor also provides access and control of the physical hardware (as virtualized hardware) to each guest system. However, VMs normally provide a virtualized environment with more resources than the application needs. This increases the required size, resulting in increased booting and processing times. On the other side, contain-

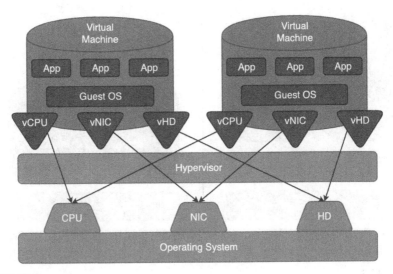

FIGURE 12.2

Virtualization with VMs.

ers share the same kernel, and the hypervisor program establishes different policies to separate containers from the host machine.

A representative size comparison between VMs and containers is provided in Fig. 12.3. In the case of virtualization using VMs a host operating system is employed directly on top of the host hardware. Then a virtualization technology implementation, such as VMware [257], VirtualBox [258], or KVM [259], provides the hypervisor that is able to run isolated virtual environments. Finally, the individual VMs run on top of the hypervisor layer. Each VM requires its own operating system, with its set of binaries and libraries. On top of this stack, the application runs. Virtualization using containers, on the other hand, employs a program as hypervisor program, such as provided by Docker [81] or LXC [260], that runs inside the host OS. Containers execute natively on the host OS using the host kernel, and they run as a single process inside the host OS, thus using system memory as any other executable, which reduces the required resources.

To summarize, Table 12.1 provides an overview of the most important features, differences, and similarities between VMs and containers.

12.3 Management, orchestration and external tools

Docker is a well-rounded tool to run applications in containers (and in Section 12.4, we will provide a set of dedicated examples supporting this claim). However, there are different tools that either assist or make use of Docker in a different plane. In this

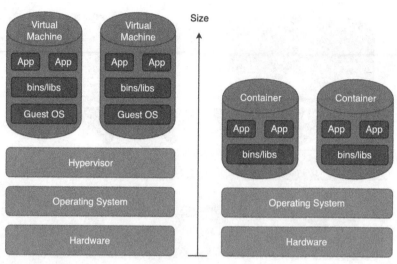

FIGURE 12.3

Size comparison between VMs and containers.

Table 12.1 Comparison between containers and virtual machines.

Feature	Virtual Machines	Containers
Security	Each VM runs its own OS	Kernel shared
Diversity	Multiple OS	Host and Guest OS must be the same
Booting time	Slow	Fast, around ~ ms
Size	Large, difficulties to copy and push	Lightweight
Migration	Long migration time, heavy copy, low downtime	short migration time, light copy, high downtime (precopy not supported by Docker)
Resources	Different flavors, shared resources. Heavy.	Ability to provide the number of cores when booting. Normally more containers than VMs can coexist in the host
Isolation	VM processes cannot see each other	Container processes cannot either

section, we briefly introduce these tools, highlight their importance and benefits, and describe their relation to Docker.

12.3.1 Kubernetes

Kubernetes [261] is an open-source software designed to manage and orchestrate containerized applications in a cluster. It interacts with Docker by launching, updating, moving, scaling, and removing Docker containers from the network. Kubernetes comprehends a number of virtualized environments that contain different levels of abstraction. A *Pod* is the smallest level of abstraction Kubernetes offers, and it lies on top of a Docker container. Normally, each *Pod* has a single Docker container. However, there is the possibility that one *Pod* contains multiple containers. Each *Pod* is mortal, that is, it is born, and when it dies, it is not resurrected. A *Service* in Kubernetes is an abstraction level defined by a set of *Pods* and policies. Whereas the *Pods* die, the *Service* can stay alive. *Namespaces* is the last level of abstraction Kubernetes provides and uses to isolate different resources inside the cluster. This is ideal when multiple working groups coexist in the same cluster and a number of resources need to be optimized.

The infrastructure to organize a Kubernetes cluster is combination of a Kubernetes master and the Kubernetes worker nodes. The Kubernetes master is a collection of processes responsible for maintaining the cluster and can be replicated for extra resilience and resource availability. The client interfaces communicate with the Kubernetes master, who translates the command and forwards it to the worker nodes.

The main features that Kubernetes offers are the following [261]:

Service discovery and load balancing: Kubernetes runs an internal network with a DNS server that provides IP addresses to the containers. An address translation protocol runs in the master node, in case the containers want to communicate with external networks. The master node performs load balancing across the worker nodes, running new Docker containers in the least loaded nodes.

Storage orchestration: Kubernetes provides a platform to mount most storage systems, whether they are located locally or remotely.

Automated rollouts and rollbacks: Running updates in systems has always been an issue, since these had to be shut down. Kubernetes allows to roll out changes with the application running. Moreover, it monitors the application state and checks for system failures. If the new version fails, then it can automatically roll back to the previous version.

Batch execution: Kubernetes enables the replacement of containers in the event of failures; it manages batch files for a safe deployment of the applications.

Automatic bin packing: Kubernetes monitors the network and reads the application requirements. This helps the master node instantiate containers in the optimal location, taking critical applications with higher priority into account.

Self-healing: Kubernetes runs a self-check to ensure a proper functionality of the running containers and restarts failed containers. If a self-check fails, then it can either restart the container in the same node (in case there was an application issue) or in a different node (if the node was disconnected from the network).

Secret and configuration management: Kubernetes provides a secure environment with the use of the *secret* object. This object contains passwords, tokens, keys, and so on that need to be kept secure. This reduces the risk of accidentally exposing sensitive data.

Horizontal scaling: Kubernetes can run multiple replicas of an application by modifying a single parameter. Moreover, this number can either increase or decrease with a simple command, adjusting the number of resources consumed on the fly. This adjustment can also be performed automatically, depending on the CPU usage.

12.3.2 Docker Swarm

A *Docker Swarm* allows the orchestration of multihost *Docker* applications. Compared to *Kubernetes*, a *Docker Swarm* offers a reduced feature set. It is more lightweight, since it is integrated into *Docker* itself. The features that *Docker Swarm* is offering as well are the following [81]:

1. Service discovery and load balancing,
2. Automated rollouts and rollbacks,
3. Batch execution,
4. Self-healing, and
5. Horizontal scaling.

A *Docker Swarm* consists of several nodes designated to be either a manager node or a worker node. A worker node has almost no rights in a swarm; it can only execute containers. A manager node is responsible for orchestration of the swarm through two main tasks it performs, namely manager nodes i) can create, update, or remove stacks from a swarm and ii) can manage other nodes of the swarm, such as setting labels of a node (which is important for placement restrictions), changing the role or availability (active, pause, drain) of a node, or simply removing a node from the swarm.

12.4 Getting started with Docker

This section provides a hands-on introduction to Docker. After an explanation of the most important commands, we show how to create a Docker image using a blueprint, the so-called *Dockerfile* and, subsequently, how to compose complex Docker services. We assume that the reader is familiar with the general Linux command line and either has access to a Linux distribution or follows the examples in Linux synonymously in different operating environments.

12.4.1 Basic commands

We initially note that most Docker commands require root privileges. The command `docker version` outputs basic information about the installed docker version. The current state of docker (information about containers, images, and drivers) and general system information of the host are shown via `docker info`.

12.4.1.1 Docker images

Docker images can either be created (as we will describe with an example in Section 12.4.2) or pulled from a Docker registry. A Docker registry is a storage solution for Docker images; the most popular Docker registries are Docker Hub, Google Container Registry, and Amazon Elastic Container Registry. However, it is also possible to individually host a Docker registry. Within a Docker registry, a Docker repository contains all images with the same name but with different tags (tags are used to identify different versions of an image). Docker repositories from Docker Hub can be searched using `docker search <name>`. There are also options to filter and format the output. We give an example showing only the three highest-rated *Hello World* images.

```
$ sudo docker search --limit 3  --format "table {{.Name}}\t{{.StarCount}}\t{{.
    IsOfficial}}" hello-world
NAME                                          STARS      OFFICIAL
hello-world                                   1009       [OK]
tutum/hello-world                             62
ansibleplaybookbundle/hello-world-apb         0
```

The first image in the list is the official *Hello World* image from Docker Hub.

Once identified, images can be pulled from the Docker Hub and stored locally with `docker pull <image>`. Afterwards, the locally stored images can be listed via `docker image ls` or just `docker images`. If the option `-q, --quit` is appended to the command, then only the IDs are displayed. The output of `docker image ls` can be formatted equivalently to `docker search`.

```
$ sudo docker pull hello-world
Using default tag: latest
latest: Pulling from library/hello-world
1b930d010525: Pull complete
Digest: sha256:6540
    fc08ee6e6b7b63468dc3317e3303aae178cb8a45ed3123180328bcc1d20f
Status: Downloaded newer image for hello-world:latest
docker.io/library/hello-world:latest
$ sudo docker images
REPOSITORY    TAG      IMAGE ID       CREATED        SIZE
hello-world   latest   fce289e99eb9   7 months ago   1.84kB
```

In addition, it is also possible to load images from a *tar* file or save existing images as a *tar* file. Images can be deleted with `docker rmi <image>` or `docker image rm <image>`. The following example shows how an image can be saved, deleted, and finally restored from the created *tar* file.

```
$ sudo docker image save hello-world > hello-world.tar
$ sudo docker rmi hello-world
Untagged: hello-world:latest
Untagged: hello-world@sha256:6540
    fc08ee6e6b7b63468dc3317e3303aae178cb8a45ed3123180328bcc1d20f
```

```
Deleted: sha256:
    fce289e99eb9bca977dae136fbe2a82b6b7d4c372474c9235adc1741675f587e
Deleted: sha256:
    af0b15c8625bb1938f1d7b17081031f649fd14e6b233688eea3c5483994a66a3
$ sudo docker images
REPOSITORY   TAG      IMAGE ID      CREATED       SIZE
# docker load < hello-world.tar
af0b15c8625b: Loading layer [============>]  3.584kB/3.584kB
Loaded image: hello-world:latest
# docker images
REPOSITORY   TAG      IMAGE ID .    CREATED       SIZE
hello-world  latest   fce289e99eb9  7 months ago  1.84kB
```

Local images can be uploaded to the Docker Hub, which requires an account. It is possible to login and logout to Docker Hub in the command line via `docker login` and `docker logout`. Part of this process is to enter the required Docker Hub credentials. As images need to be separated in their repository, they require a tag before they can be uploaded. We provide an example that demonstrates how to tag and upload an image to Docker Hub.

```
$ sudo docker login
$ sudo docker images
REPOSITORY   TAG      IMAGE ID      CREATED       SIZE
hello-world  latest   fce289e99eb9  7 months ago  1.84kB
$ sudo docker tag fce289e99eb9 <UserName>/hello-world:latest
$ sudo docker image ls
REPOSITORY              TAG      IMAGE ID      CREATED       SIZE
<UserName>/hello-world  latest   fce289e99eb9  7 months ago  1.84kB
hello-world             latest   fce289e99eb9  7 months ago  1.84kB
$ sudo docker push <UserName>/hello-world
$ sudo docker logout
```

The created repository is publicly available by default but can be set to private. It is also possible to delete repositories once they are no longer needed.

12.4.1.2 Docker containers

Containers can be created via `docker create <image>`. Running containers can be shown with the commands `docker container ls` or `docker ps`. All containers are listed with the option `-a`. The displayed output of the command can be limited to the IDs of the containers with the option `-q`, `--quiet`. The output of `docker container ls` can be formatted equivalent to `docker search` and `docker image ls`. The following example shows how a container named *MyFirstContainer* based on the *Hello World* image is created.

```
$ sudo docker create --name MyFirstContainer hello-world
$ sudo docker container ls -a --format "table {{.ID}}\t{{.Image}}\t{{.Names}}"
CONTAINER ID   IMAGE         NAMES
f09f8e9ae077   hello-world   MyFirstContainer
```

It may be useful to specify some additional options for some images, such as port forwarding for the *Nginx* web server. This creates an interface between the container and the outside world. In the following example, *Nginx*, which is inside the container running on port 80, is made available to the outside world on port 5001. This makes it possible to access the web server via the IP address of the host.

```
$ sudo docker create --name MyOwnWebserver -p 5001:80 nginx
```

There are a several commands for Docker containers, which can be referenced either via their IDs or their unique names. A container can be started, stopped, renamed, or removed. The following example shows how to execute the *Hello World* container and display its output in the command line.

```
$ sudo docker container start -i MyFirstContainer
```

Containers can also be created and started simultaneously with the command `docker run <image>`, with options similar to `docker create`. Following the *Nginx* example, it might also be useful to detach the container with the `-d` option when using the `run` command; otherwise, the output of the console will be permanently streamed to by *Nginx* messages. A container can be deleted via `docker rm <container>`.

Containers are dynamic objects that can be modified. Desired changes can be kept by committing, which will create a new image. The following example creates a new file inside a container based on the *Alpine* image, which contains a minimal Linux distribution. Afterwards, a new image is created based on the modified container.

```
$ sudo docker run --name MySecondContainer alpine touch /home/myfile
$ sudo docker commit MySecondContainer myalpine
```

Similar to images, it is also possible to directly save exported containers so that later they can be imported again.

```
$ sudo docker export mySecondContainer > myContainer.tar
$ sudo cat myContainer.tar | docker import - mycontaineralpine:latest
```

12.4.2 Building an image – Dockerfile

We continue from the overview of Docker commands with an explanation of how to
create a basic Docker image from scratch. In the following example, we show how to
create a Docker image for an application that is used to calculate Fibonacci numbers.

Initially, a new directory is required to hold the required files, for example, with
the following few commands.

```
$ mkdir MyFirstImage
$ cd MyFirstImage
$ touch Dockerfile fibo.py
```

The newly created directory has to contain a Dockerfile and other required data
for the image. Here we create the fibo.py Python program with the following example content to calculate the first 50 Fibonacci numbers.

```
print("This is my first Docker image.")
print("It prints the first 50 Fibonacci numbers.")
a = 0; b = 1
for i in range(1,51):
    print("Fibonacci %i: %i" % (i, b))
    a,b = b, a+b
```

The Dockerfile contains all information on how to build an image, with content
for our example as follows.

```
# A lightweight Python image as parent
From python:3.7-alpine
# Set the working directory to /dirInContainer
WORKDIR /dirInContainer
# Copy the python app fibo.py in the working directory
COPY fibo.py .
# Execute the python program fibo.py
CMD ["python", "fibo.py"]
```

In this example, a minimal Linux distribution with Python 3 is chosen as the parent image. Then a working directory within the image (respectively, container) is set,
and the fibo.py Python program is copied from the current directory into the working directory of the image. Finally, the Python program is executed. A Docker image
can subsequently be built via docker build .; we add options to specify a different
location for the *Dockerfile* (-f) and tag the image (-t).

```
$ sudo docker build -t fibonacci -f ./Dockerfile .
```

Once the image is created, it can be used to create containers just as in the prior
examples of pulled images.

```
$ sudo docker image ls
REPOSITORY   TAG      IMAGE ID      CREATED            SIZE
fibonacci    latest   d669d08043e9  About a minute ago  98.7MB
$ sudo docker run fibonacci
This is my first Docker image.
It prints the first 50 Fibonacci numbers.
Fibonacci 1: 1
Fibonacci 2: 1
Fibonacci 3: 2
Fibonacci 4: 3
Fibonacci 5: 5
...
```

Whereas this example highlights the utilization of Docker for the generation of a basic image and container from scratch, significantly more complex images can be created, only limited by the imagination of the maintainer.

12.4.3 Services and stacks

Docker also offers the ability to create complex applications, called *Stacks*. A *Stack* consists of services that are based on an image and defines the parameters for the execution of the image. There are two ways to create an application or service: i) directly with docker-compose, which allows us to create single-host applications and ii) by deploying the application in a *Docker Swarm* cluster, which is explained in Section 12.3.2. The advantage of *docker-compose* is that it supports creating images on the fly, which makes it better suited for development. *Docker Swarm*, on the other hand, requires already compiled images. A significant benefit of *Docker Swarm* is that it enables distributed deployment of applications on multiple hosts, making it better suited for larger and real-world deployments.

Whereas a Dockerfile configures individual images, so-called Compose files configure services, that is, complete applications. A Compose file is structured as a YAML file and contains information, for example, about the images which are used, network configurations, dependencies, placement options, and a deployment strategy. The dependency parameter ensures a correct startup order, if some services depend on each other. The placement parameter is very important for edge computing, since it allows us to exactly specify on which node or set of nodes a container should be deployed. For example, the provisioning strategy can include settings for resource allocation (maximum, minimum), the number of replicated containers, and the procedures to follow if an error occurs or if the application is updated. In addition, the Compose file also defines the underlying network and volumes for persistent storage.

The following example creates a simple *Stack* based on a newly created image called *myserver*. The Python program runs a web server on localhost on port 80. When a client opens the web page, the hostname of the server is returned. The contents of the underlying Python application and the Dockerfile are shown in the two following listings.

```python
from flask import Flask
import os, import socket
app = Flask(__name__)
@app.route("/")
def hello():
    html = "<h3>{hostname}</h3>"
    return html.format(hostname=socket.gethostname())

if __name__ == "__main__":
    app.run(host='0.0.0.0', port=80)
```

```dockerfile
# A lightweight Python image as parent
From python:3.7-alpine
# Set the working directory to /dirInContainer
WORKDIR /dirInContainer
# Copy the python app fibo.py in the working directory
COPY . .
# Install python package flask
RUN pip install flask
# Execute the python program server.py
CMD ["python", "server.py"]
```

Subsequently, a stack is defined with the file MyFirstStack.yml:

```yaml
version: "3.7"
services:
  web:
    #replace username/repo:tag with your name and image details
    image: myserver
    deploy:
      replicas: 5
      resources:
        limits:
          cpus: "0.1"
          memory: 50M
      restart_policy:
        condition: on-failure
    ports:
      - "5000:80"
    networks:
      - webnet
  visualizer:
    image: dockersamples/visualizer:stable
    ports:
      - "5001:8080"
```

```
    volumes:
      - "/var/run/docker.sock:/var/run/docker.sock"
    deploy:
      placement:
        constraints: [node.role == manager]
    networks:
      - webnet
networks:
  webnet:
```

This *Stack* consists of two services: i) a service called *web*, which is based on the image *myserver* and consists of five instances of that image, and ii) a service called *visualizer*, which is based on an official Docker image. The *web* service configures limited resources for each container, and in case of a failure the respective container is restarted. Additionally, port 80 of a container will be exposed to port 5000 of the outside world. The *visualizer* image is used to monitor a *Docker Swarm* cluster and the corresponding containers; it can be accessed via port 5001. Both services use *webnet*, which is a simple load-balancing network.

Before deploying the stack, a *Docker Swarm* cluster needs to be initialized. The stack can then be deployed as follows.

```
$ sudo docker swarm init
$ sudo docker stack deploy -c MyFirstStack.yml MyFirstStack
```

The option `-c` is used to point at the `Compose` file, and the last string in the command is the chosen name for the *Stack*. Existing *Stacks* can be listed via `docker stack ls` and removed via `docker stack rm`. The command `docker stack services <stack>` lists all associated services for a given *Stack*, and `docker service ps <service>` lists all associated containers for a given service. All containers inside a *Stack* can be shown with `docker stack ps <name>`. The command `docker service ls` shows all existing services, no matter if they belong to a certain *Stack* or not. *Stacks* can be updated on the fly by simply editing the `Compose` file and rerunning the deployment command.

12.4.4 Docker Swarm

Should the previous example be extended to include multiple nodes in a network using solely *Docker*, the built-in *Docker Swarm* can be employed as a lightweight approach. A new swarm can be initiated via `docker swarm init` by any host that runs *Docker*. The `--advertise-addr <IP>` option is useful when a host has multiple network interfaces to ensure that the swarm is accessible via the correct network interface. This command immediately returns the worker *join-token*, which is required by other hosts to join the swarm as a worker. In addition, the join tokens for workers and managers can be listed via `docker swarm join-token <role>`. The command `docker swarm join --token <token> <Swarm IP>:<Port>` is used to join a swarm from another host. Stacks can be deployed and managed as already shown in Section 12.4.3. All nodes of a swarm can be monitored by a manager. A manager can list and filter all nodes with `docker node ls` and inspect specific nodes in detail via `docker node inspect <node>`. Of course, a manager can also list all containers of a swarm with `docker node ps` or only all containers of a specific node with `docker node ps <node>`. A swarm can be left with `docker swarm leave`.

ComNetsEmu: a lightweight emulator

13

Zuo Xiang[a], Juan A. Cabrera G.[a], Sreekrishna Pandi[a], Patrick Seeling[b], Frank H.P. Fitzek[a]

[a]*Technische Universität Dresden, Dresden, Germany*
[b]*Central Michigan University, Mount Pleasant, MI, United States*

> *Working Code Trumps All Hype.*
> **Mahadev Satyanarayanan**

13.1 Introduction

As illustrated in Fig. 1.6, we will introduce a variety of emerging innovations and novelties into future communication networks due to the underlying softwarization of formerly dedicated and specialized network system components. This change opens tremendous possibilities to develop, implement, and evaluate novel ideas for 5G communication networks and beyond without the need for any specialized equipment. In order to not only theoretically study these innovations, but also provide a shared platform for development, testing, and reproducibility, a holistic SDN/NFV test platform is required. As concluded in Section 11.1, *communication network emulation* is the preferred approach to performing practical experiments for networked systems.

The ComNets Emulator (ComNetsEmu) is a holistic communication network emulator with support for softwarization technologies that can not only be deployed for research and development, but also for active and hands-on learning experiences for students, which we will describe in this chapter. In general, any emulator that could potentially be used needs to meet several main requirements for general deployment across different use case scenarios:

Simplicity: The emulator should be lightweight and able to run most experiments on a Commercial Off-The-Shelf (COTS) computer, for example, on a student's laptop. As a result, students can i) experience all the built-in application examples and ii) create new and innovative prototypes, all without having to manage many hardware devices. The test bed installation, reinstallation, snap-shooting, and other administrative operations should be automated and have appropriate version control. Users should be enabled to set up the test bed rapidly and share the common emulation environment across different host operating systems.

Reproducibility and shareability: The emulator should support reproducible experiments that are explained in Section 11.1. All the components required to perform the experimentation, including the runtime environment, source code, dependencies, and configurations, should be shareable through human-friendly templates or documents.

Mapping real-world implementations and deployments as accurately as possible: The emulator should support and use concepts and architectures deployed in the real world as much as possible, taking into account the lightweight and simplicity. The emulator should not only avoid oversimplified design and implementation of new innovative approaches, but also generate evaluation results that are comparable to the real-world systems.

State-of-the-Art (SoA) practical technologies without high complexity: To combine theory with actual prototyping and evaluation, emulators should support the use of trends and promising technologies from academia and industry to build experiments. In contrast to production-oriented platforms, such as OpenStack [262] or Kubernetes [261], the test bed should only provide the necessary and orthogonal functionalities for testing innovative concepts without introducing significant complexity due to production-level deployment and orchestration.

Extensibility: The emulator should have a modularized design and support extensions. Users should be enabled to easily extend the emulator based on their own requirements.

A simplified scenario of using NC in a multicloud deployment is illustrated in Fig. 13.1. In the following, we will employ this scenario to explain the outlined requirements in a practical example. In this example the mobile client can send raw data to the remote cloud for low-latency processing services, for example, for computationally intensive and power-hungry applications.

The original data is transmitted with RLNC (introduced in Chapter 9) to mitigate transmission channel losses. The data (encapsulated in coded packets) is initially transmitted to the edge cloud. For this example scenario, it is assumed that due to a lack of resources, the edge cloud must forward the packets to a centralized cloud for final processing. A recoding function is executed on the edge cloud to perform RLNC recoding on the received packets before forwarding them to the centralized cloud. As typically encountered for deployments in real-world data centers, applications are commonly containerized (e.g., packaged inside Docker containers) and managed by an orchestration system to run on a cluster of physical servers. When the client moves to another region, a handover can occur, and the data would subsequently need to be transmitted to a new edge cloud. The recoding function may also need to be redeployed or migrated to the new edge cloud to maintain the quality of the service when handover happens. The traffic redirection to a new edge cloud relies on the aggregation switch forwarding received packets to one of the alternative physical servers inside the cluster for recoding (assuming that each server can run the containerized recoding function).

One straightforward research problem in this scenario is that of the actual server selection. Which server should be chosen depends on many aspects, including the server network topology, the preparation and transmission delay of the links in the topology, and the available computational resources on each server. An SDN-based routing algorithm can be designed for this typical scenario and evaluated on the test bed.

The unmodified, standard Mininet emulator introduced in Chapter 11 is one competitive candidate for the implementation of a test bed for this scenario. However, this regular version of Mininet has the following characteristics with respect to the aforementioned desired requirements:

- Mininet supports simplicity, reproducibility, and shareability with its lightweight network node emulation and configurable data plane.
- As mentioned in Section 11.1, only the network `namespace` of hosts (lightweight containers) is isolated. All default hosts share the same file system, process IDs, and other resources that can be controlled by the `Cgroup`. Although this approach reduces the resource overhead of each host for the emulation, it has drawbacks for practical deployment emulations of fully containerized applications on the virtual host node:

  1. By default all application processes running on the same host share the same file system. Therefore dependencies required by the application must be installed on the host OS. Applications may require the same dependency in different versions; potential conflicts are difficult to avoid in this scenario. Additionally, since all dependencies must be installed on a host system that is typically bound to a specific OS distribution, sharing the settings and configurations of the complex software environment can quickly become unmanageable.
  2. Process ID isolation is a required feature to implement application container migration between different physical machines.
  3. The built-in `CPULimitedHost` node supports CPU resource management via `Cgroup`. To emulate a variety of different real-world applications, the emulator should provide more comprehensive and fine-grained resource management functionalities.

ComNetsEmu was developed as an extension of Mininet to overcome the outlined limitations of the original Mininet emulator. It extends and puts forward the concepts and work in the Containernet project [263,264]. It uses a different approach to extend the Mininet when compared to Containernet. Its main focus is using *sibling containers* to emulate network systems with computing. ComNetsEmu is used for the holistic emulation of all softwarization innovations introduced in this book.

The basic goal of ComNetsEmu is achieving practical emulation of emerging network applications while minimizing the system complexity. To resolve the limitation of the standard Mininet for emulating practical containerized applications, *Docker-*

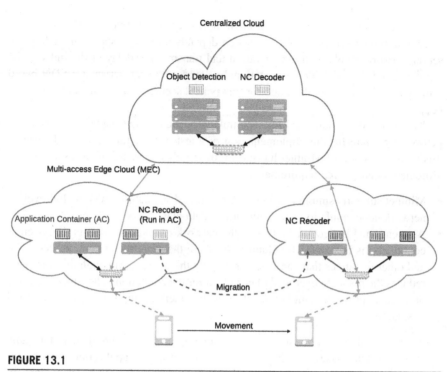

FIGURE 13.1

Typical deployment of a network coding application in a multi-cloud scenario.

in-Docker (sibling containers) is used by ComNetsEmu as a lightweight emulation of *nested virtualization*. The Docker host with deployed internal Docker containers is used to mimic an actual physical host that runs containerized applications. Since it is designed and developed for teaching purposes, its components and dependencies prefer using Free and Open-Source Software (FOSS) to enable future expansions and incur no additional costs. ComNetsEmu follows the guidelines proposed in the Mininet hackathon [265] to extend the original Mininet implementation via composition. In addition to its main software modules (developed with `Python 3`), ComNetsEmu also provides:

1. Handy scripts and receipts to create, recreate, and destroy the test environment VM with minimal overhead.
2. A collection of basic examples to demonstrate core functionalities of the emulator.
3. A collection of application examples to demonstrate all innovations and technologies introduced in this book.
4. Detailed documentation for the usage of the emulator and all application examples.

In turn, the whole emulator can run smoothly inside a VM on a COTS laptop.

13.2 ComNetsEmu in a nutshell

This section provides an introduction of the main functionalities provided by the ComNetsEmu and its implementation details. ComNetsEmu employs Vagrant [266] to manage the test/development VM environment with all dependencies automatically installed. *Sibling Containers* are utilized to support lightweight emulation of application containers deployed on a cluster of physical machines. The Docker platform (Engine-Community edition) introduced in Chapter 12 is chosen as the management and orchestration platform for all containers in the emulator.

ComNetsEmu is published on a public Git repository [255] to facilitate easy distribution and continuous development. This repository contains the following main components:

1. The `comnetsemu` Python module (developed with Python 3.6), which extends the original Mininet with user-friendly APIs.
2. An `examples` folder with programs for core functionalities of the `comnetsemu` module.
3. An `app` directory containing representative examples of innovations and technologies introduced in this book, including network coding for transport, network coding for distributed storage, machine learning for routing, object detection, network slicing, compressed sensing, Software-defined Radio, and so on.
4. A `util` directory for handy scripts for environment management.
5. A `test_containers` folder for Dockerfiles and dependencies required for the sample programs in `examples`.

13.2.1 Test environment management

For providing a cross-platform and convenient environment to build, develop, and manage the test bed, ComNetsEmu provides a preconfigured VM environment managed by `Vagrant` in an easy-to-use workflow. Vagrant is a free and open-source tool to manage and share portable virtual machine environments [266]. Vagrant supports GNU/Linux, macOS, and Windows as hosting OS. By default, Vagrant can use Virtualbox, Hyper-V, Docker containers, VMware, and others as the VM hypervisors. The configuration and provisioning steps of the ComNetsEmu VM is described in the `Vagrantfile` located in the root directory of its Git repository [255]. Virtualbox is used as the default hypervisor, since it is free and open-source [258]. As Mininet requires the Linux kernel, a GNU/Linux distribution should be installed as the guest OS. The Ubuntu server edition [267] is chosen as the base to build the emulation environment. If Git, Vagrant (v2.2.5 and beyond), and Virtualbox (v6.0 and beyond) are already installed on your host OS, then the emulation environment can be easily created with shell commands listed in 13.2.1:

```
$ git clone Remote_URL_OF_COMNETSEMU ./comnetsemu
$ cd ./comnetsemu && vagrant up comnetsemu
```

When the VM is created for the first time, it requires some time to download the base VM image and install all software dependencies via the built-in `Bash` installer script. After the build and provision of the emulation VM is completed, the commands listed in 2 can be utilized to use the VM:

```
# Check the state of the VM managed by the Vagrant
$ vagrant status
# SSH into the running Vagrant VM
$ vagrant ssh comnetsemu
# shuts down the running machine Vagrant is managing
$ vagrant halt comnetsemu
# Stop the running the VM and destroy all its resources. This command can be
    used to reset to clean state if any issues happen. The VM can be re-
    created with up command.
$ vagrant destroy comnetsemu
```

By default the directory containing the `Vagrantfile` on the host OS is synchronized with the folder `/home/vagrant/comnetsemu` on the guest VM. The Python module is installed with the `development` mode. Development of the Python module can be performed on the host OS within the user's customized workflow, and the tests can be executed inside the development VM without data copying and software reinstallation, making for a convenient workflow.

ComNetsEmu and its built-in example programs require following minimal dependencies:

- Mininet (v2.3.0d6 and beyond) with its minimal dependencies: Since the Mininet latest version (with Python 3 support) has no official binary release and the ComNetsEmu wants to use the latest stable release as soon as possible, the Mininet Python module is installed from source code. As minimal dependencies, Open vSwitch, Stanford OpenFlow1.0 reference controller, and Wireshark with OpenFlow support are also installed.
- Ryu SDN controller (v4.32 and beyond) [268]
- Docker Engine-Community (19.03.0 and beyond)
- Docker SDK Python (v3.7.2 and beyond) [269]
- Wireguard (0.0.20190702-wg1 bionic and beyond)

All these requirements are installed automatically with the installer located in `./util/install.sh` with -a option. Each component can additionally be installed separately with proper options if desired. The help information can be printed with -h option of the installer.

13.2.2 Application container management

As discussed in Section 13.1, the goal of ComNetsEmu is providing a lightweight platform to emulate containerized application deployed on clusters of physical machines. One typical approach to enable this emulation is using VM-based nested virtualization. With nested virtualization, the hypervisor can run inside the guest OS to manage multiple VMs inside each VM [270]. This approach has the following difficulties, which are in the way of using the ComNetsEmu emulator:

- Nested virtualization requires support from both, underlying hardware and software hypervisor. For example, Virtualbox 6.0 (the default hypervisor used by ComNetsEmu) does not support nested virtualization on all Intel processors. This limits the usage of ComNetsEmu as a general teaching test bed for COTS laptops.
- VM-based nested virtualization can introduce significant performance overheads, which can limit the scale and level of fidelity within network emulation that can be performed on the emulator. The emulator should support emulation of typical network containing hundreds of connected nodes on a single laptop.

Due to these limitations, ComNetsEmu addresses the lightweight emulation implementation with a *Docker-in-Docker* approach. This approach is inspired by the design of Pod [271] in the de facto standard container orchestration platform Kubernetes. A Pod is a single container or a group of containers with shared storage and shared networking stack that can be used as the smallest deployment unit on Kubernetes. Containers in the same Pod are in the same network `namespace`. Therefore they share the same IP address and port range [271]. They can communicate with each other via the `localhost` interface, and they communicate with containers outside the Pod using the shared Pod interface. Docker supports one networking mode named `container` mode [272]. It allows a newly created container to run its network stack on top of another already created container. Subsequently, the network resources of this container (called *internal* container) are isolated from other containers (called *external* container). Furthermore, Docker also supports customized `cgroup` configuration via the `cgroup-parent` option. This option allows the user to assign a specific `cgroup` to run the container. Due to the hierarchical architecture of the `cgroup`, introduced in Section 11.3, this option can be used to limit all resources managed by the `cgroup` of the internal container to the constraints of the external container. With this network isolation and resource limitation, the nested container approach is able to emulate the containerized application orchestration in a significantly lightweight approach, especially when compared to the full VM-based nested virtualization. This approach is different from the real *Docker-In-Docker* approach implemented in the repository [273] as follows. The external Docker container does not run its own isolated Docker daemon to manage internal containers. All internal and external containers share the same Docker daemon running on the OS with sufficient compute and network resource isolation. The external Docker container with one or multiple deployed internal Docker containers is used to mimic an actual physical host that runs containerized applications. To support the *Docker-In-Docker* approach for SDN/NFV applications, ComNetsEmu extends the regular Mininet mainly with a new node type called `DockerHost` and an application container manager.

Fig. 13.2 illustrates the abstracted architecture of ComNetsEmu for the example scenario we previously presented in Fig. 13.1. Following this example, each physical server located in the edge cloud is emulated by the external `DockerHost`

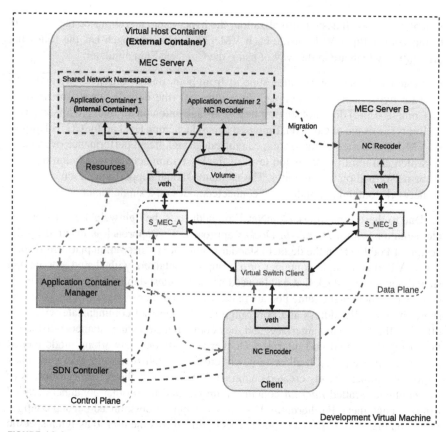

FIGURE 13.2

Fundamental architecture of ComNetsEmu for the network coding with mobile edge clouds.

instance. This host type behaves exactly like the built-in `CPULimitedHost`. It is connected to the configurable data plane managed by Mininet and provides an interactive shell to execute commands. Containerized applications are managed by an additional `application container manager` and deployed as internal containers on the `DockerHost` instances. For example, with this approach, the migration of the containerized application is equal to migration of the internal container to another external container.

13.3 Examples for getting started

In this section, we guide the reader through the process of using ComNetsEmu with the help of two examples. The first one emulates the deployment of a simple function in a computing node. The second example emulates multiple network nodes that are limited in terms of computing resources. To do this, we exploit the functionality of Docker that limits the CPU resources assigned to each container. This is useful if the reader wants to emulate multiple network nodes that are heterogeneous in computing capabilities, for example, to study the best placement of a network function based on link parameters and computing delays.

13.3.1 Echo server

The files referenced here are located in the `examples/echo_server` folder. The first step to deploying a function is developing the function itself. This function can be as complex as the required application, but for teaching purposes, we assume that the function consists of an echo server, constituted by only one Python file called `server.py`. This script is a TCP echo server that waits for connections from clients at the TCP socket 65000, receives data, and transmits it back to the client.

Once the function is developed, we can containerize it as described in Chapter 12. To do this, we need the `Dockerfile` shown in Listing 13.1 as the recipe for the container. The best option would be to employ a minimal Docker image capable of running Python as the base image. A quick search suggests using `python:3.6-alpine3.9`. This is an image that features Python version 3.6 installed on an Alpine image version 3.9. Then we need to copy the Python script to the Docker image.

```
FROM python:3.6-alpine3.9
COPY ./server.py /home/server.py
CMD python /home/server.py
```

Listing 13.1: Dockerfile for the image of the echo server.

When the Dockerfile is ready, we can create the image with a simple command found in the file `build_docker_image.sh` and shown in Listing 13.2. We chose `echo_server` as the tag for our image.

```
$ docker build -t echo_server --file ./Dockerfile .
```

Listing 13.2: Building the Docker image from the Dockerfile.

Once the function is containerized, we can emulate a network topology with multiple hosts. Subsequently, we can deploy the function in any emulated host as a Docker container. By the end of the file we would have had emulated the topology shown in Fig. 13.3. In our topology, we have two emulated hosts `h1` and `h2`. These hosts are each connected to switches `s1` and `s2` correspondingly, and the switches are connected to each other. Each link has a capacity of 10 MBps and a latency of 10 ms.

Next, we employ the ComNetsEmu to emulate the topology shown in Listing 13.3. The following listings are read from the file `topology.py`. In Listing 13.3

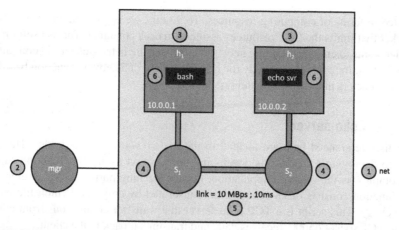

FIGURE 13.3

Emulated topology for the echo server example.

the first step is to create *net*. We create this with a *Containernet* object. In the arguments for its creation, we specify the type of the controller we are going to use and, more importantly, the type of link. In this case, it is a TCLink, which is a type of link that allows us to set limit the bandwidth and latency. We also create the *mgr* object and add a controller to *net*.

```
net = Containernet(controller=Controller, link=TCLink, xterms=False)
mgr = VNFManager(net)

net.addController("c0")
```

Listing 13.3: Instantiating net and mgr.

The next step is creating the hosts. In Listing 13.4, we show that the arguments needed are i) the Docker image to use, ii) the IP address that each host will have in the network, and iii) the arguments for the Docker containers that emulate the network hosts. It is important to note that for the network hosts, we use dev_test as the Docker image. This is the image that allows us to run Docker-in-Docker. We will deploy our network functions as Docker containers inside these host containers, thus the need of Docker-in-Docker.

```
h1 = net.addDockerHost("h1", dimage="dev_test", ip="10.0.0.1", docker_args={"
    hostname": "h1"})
h2 = net.addDockerHost("h2", dimage="dev_test", ip="10.0.0.2", docker_args={"
    hostname": "h2"})
```

Listing 13.4: Creating the emulated hosts.

Following the creation of the hosts, we create the switches and the communication links, as shown in Listing 13.5. The argument needed to create the switches is simply the desired name. The arguments needed to create the links are the end-points of the link and the link parameters, such as bandwidth in MBps and delay.

```
switch1 = net.addSwitch("S1")
switch2 = net.addSwitch("S2")
net.addLink(switch1, h1, bw=10, delay="10ms")
net.addLink(switch1, switch2, bw=10, delay="10ms")
net.addLink(switch2, h2, bw=10, delay="10ms")
```

Listing 13.5: Creating the emulated hosts.

The last step is creating the network services. This is shown in Listing 13.6. The arguments to create these services are, in respective order, i) the name of the service, ii) the name of the host where to execute them, iii) the name of the docker image to use, iv) the command to run inside the docker image, and v) the arguments for the Docker commands to instantiate the service. We select host h2 to run the echo server. We denote it srv1, and we use the Docker image echo_server (note that this is the Docker image created as in Listing 13.2). Next, a bash terminal is needed in the network to transmit TCP packets to the server. For this, we create srv2 with any Docker image able to run bash. We chose dev_test.

```
srv1 = mgr.addContainer("srv1", "h1", "echo_server", "python /home/server.py",
    docker_args={})
srv2 = mgr.addContainer("srv2", "h2", "dev_test", "bash", docker_args={})
```

Listing 13.6: Creating the services in the hosts.

Once the reader runs the script, access to a bash terminal (srv2) running as a Docker container inside h1 becomes available. From this terminal, we can use any tool we want to send a TCP packet to the echo server at port 65000. For example, we can use the tool nc.

```
$ echo "Hello ComNetsEmu" | nc 10.0.0.2 65000
```

We should see the transmitted data printed in the console.

13.3.2 Docker-in-Docker for resource limitation

After going step by step through the example of creating and deploying the echo server, the reader is ready to review the next example. This file is located in examples/dockerindocker.py, and it shows the user-friendly APIs to limit the CPU resources of the hosts and the services deployed. It also deploys a CPU-stress application to show the reader the computing limitations of the different hosts.

With this initial overview of ComNetsEmu, in this book, we now transition to making the complex topics of *Computing in Communication Networks* intuitively accessible for everyone by employing ComNetsEmu in practical hands-on examples.

Examples

PART

6

Outline

Based on the ComNets Emulator, we describe how some of the aforementioned concepts can be realized with the help of several examples and how to include the identified innovations.

Realizing network slicing

14

Fabrizio Granelli[a], Truong Giang Nguyen[b], Huanzhuo Wu[b]

[a]*University of Trento, Trento, Italy*
[b]*Technische Universität Dresden, Dresden, Germany*

Different paths, different business...
Confucius

Network Slicing, as described in Chapter 3, is a technique to allocate different partitions (or slices) of the same physical network and computation infrastructure to satisfy different sets of QoS. In this chapter, we provide a practical implementation of NS with the help of implemented software-defined networking. We initially start with a brief discussion of the basic concepts of NS combined with an example using Mininet. Subsequently, we demonstrate three NS scenarios based on the ComNetsEmu,[1] described in Chapter 13, to enable practical implementation of NS utilizing Ryu APIs. Additionally, we introduce three methods to validate the functionalities of slices: i) connectivity (e.g., slice topology verification), ii) bandwidth (e.g., slice capacity verification), and iii) data flow (e.g., service slicing verification). The goal of this chapter is to show that different requirements are able to be fulfilled on a shared physical infrastructure by using NS.

14.1 Network slicing in Mininet

14.1.1 Introduction

In this section, we review basic concepts on generating topologies using the Mininet network emulator. We proceed from simple to complex topologies.

The easiest and fastest way to generate regular topologies in Mininet is using the built-in topologies. Those are accessible by using the `-topo` parameter while launching the software. There are three types of built-in topologies available in Mininet:

1. *Single*: One switch is connected to a number of hosts.
2. *Tree*: A tree of a given depth and a given number of children per node are generated.

---

[1] This project is currently hosted on ComNets Gitlab; see https://git.comnets.net/public-repo/comnetsemu.

Computing in Communication Networks. https://doi.org/10.1016/B978-0-12-820488-7.00029-3

3. *Linear*: Several switches are connected one after another, with one host per switch.

These three examples of usage of the built-in network topologies can be implemented using Mininet as follows:

1. `sudo mn --topo single,4` will generate a topology composed of one switch connected to 4 hosts.
2. `sudo mn --topo tree,depth=3,fanout=2` will generate a tree topology of depth 3, with two children per node.
3. `sudo mn --topo linear,3` will generate a topology consisting of 3 switches, one after the other, with one host per switch.

14.1.2 Link capacity slicing

This section provides a first and simple example of slicing. Slicing the capacity of a link refers to the allocation of different capacities (or bandwidth) for different partitions sharing the same link. For this case, we will partition the output bandwidth of a switch by using virtual OpenFlow buffers and Linux queue management functions. The reader should recall that slicing requires two functionalities:

1. the capability to manage and assign the resources into slices (i.e., *resource slicing*) and
2. the capability to associate traffic flows to slices (i.e., *flow mapping*).

These concepts will be implemented through the following steps:

1. *resource slicing*: OpenFlow commands will create two virtual queues at the output of one port of the switch and assign those different maximum capacities using *Linux HTB – Hierarchy Token Bucket*[2];
2. *flow mapping*: OpenFlow flow tables will be employed to classify traffic into two different flows and to map them to the two queues.

Initially, we generate a simple topology by using the following command:

```
$ sudo mn --mac --switch ovsk --topo single,3
```

Fig. 14.1 shows the selected topology, a built-in Mininet star topology with one switch and three hosts. Within this topology, we want to partition the output bandwidth of port `eth3` of the switch and allocate 100 Mbits/s to connections between host `h1` and host `h3`. Furthermore, we allocate 200 Mbits/s to the connections between hosts `h2` and `h3`.

To achieve this goal, we need to build two virtual output queues to the `eth3` interface of the switch. In particular, we will build virtual queue `10` with maximum

---

[2] https://linux.die.net/man/8/tc-htb.

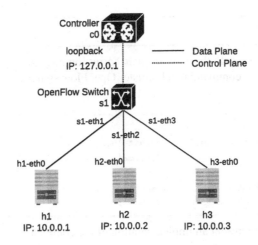

FIGURE 14.1

Topology used for link capacity slicing.

rate of 100 Mbits/s and virtual queue 20 with maximum rate of 200 Mbits/s with the following command (ovs-vsctl is used to configure the ovs-vswitchd configuration database):

```
$ sudo ovs-vsctl set port s1-eth3 qos=@newqos -- \
--id=@newqos create QoS type=linux-htb \
other-config:max-rate=10000000000 \
queues:10=@1q \
queues:20=@2q -- \
--id=@1q create queue other-config:min-rate=50000000 \
other-config:max-rate=100000000 -- \
--id=@2q create queue other-config:min-rate=50000000 \
other-config:max-rate=200000000
```

In practice, the commands requests the OVS to define two virtual queues in output to port eth3. The queues will be managed by the Linux Hierarchy Token Bucket service discipline.

For an in-depth description, please refer to the *Open vSwitch Cheat Sheet*.[3] We can verify the tc queues we generated with:

```
$ tc class list dev s1-eth3
```

and the service discipline with:

---

[3] http://therandomsecurityguy.com/openvswitch-cheat-sheet/.

```
$ tc qdisc show
```

Next, we need to allocate the flow to their corresponding queues. This is done by using the `ovs-ofctl` command to administer OpenFlow switches as in the following two command lines:

```
$ sudo ovs-ofctl add-flow s1 \
ip,priority=65500,nw_src=10.0.0.1,nw_dst=10.0.0.3,\
idle_timeout=0,actions=set_queue:10,normal
sudo ovs-ofctl add-flow s1 \
ip,priority=65500,nw_src=10.0.0.2,nw_dst=10.0.0.3,\
idle_timeout=0,actions=set_queue:20,normal
```

These commands represent an implementation of the command:

```
$ ovs-ofctl add-flow <bridge> \
<match-field>actions=set_queue:<queue>,normal
```

which ensures that the packets matching the <match-field> expression will be output to the specified queue. The addition of the *normal* action means that, in addition, packets will be processed as device normal OSI Layer 2 (L2)/OSI Layer 3 (L3).

We can check that flows are installed with the following OpenFlow administration command:

```
$ sudo ovs-ofctl dump-flows s1
```

The readers are recommended to refer to *ovs-ofctl Common Commands*[4] for further reference.

Finally, we can test and validate this example by running the following commands:

```
mininet> iperf h1 h3
*** Iperf: testing TCP bandwidth between h1 and h3
*** Results: ['95.6 Mbits/sec', '104 Mbits/sec']
mininet> iperf h2 h3
*** Iperf: testing TCP bandwidth between h2 and h3
*** Results: ['191 Mbits/sec', '203 Mbits/sec']
```

To show the programmability of SDN and control over the sliced resources, it is possible to modify the bandwidth allocated to the two flows by running the following commands:

---

[4] https://docs.pica8.com/pages/viewpage.action?pageId=3083175.

```
$ sudo ovs-vsctl clear port s1-eth3 qos
$ sudo ovs-vsctl set port s1-eth3 qos=@newqos -- \
--id=@newqos create QoS type=linux-htb \
other-config:max-rate=10000000000 \
queues:10=@1q \
queues:20=@2q -- \
--id=@1q create queue other-config:min-rate=50000000 \
other-config:max-rate=10000000 -- \
--id=@2q create queue other-config:min-rate=50000000 \
other-config:max-rate=20000000
```

In this case, the corresponding bandwidth allocation will be one-tenth of the previous value, that is, 10 Mbits/s and 20 Mbits/s, respectively.

14.2 Network slicing in ComNetsEmu

In this section, we apply the overall approach described in Section 14.1 to implement NS using the Ryu API in a more complex topology. Instead of the regular topology in Section 14.1, we employ the multihop one illustrated in Fig. 14.2. The topology consists of four hosts and four switches, which are connected by virtual Ethernet cables that allocate 10 Mbits/s and 1 Mbits/s of bandwidth, respectively; see Listing 14.1 for the corresponding code example.

```
# Create host
for i in range(4):
    self.addHost('h%d' % (i+1), **host_config)

# Create switch
for i in range(4):
    sconfig = {'dpid': "%016x" % (i+1)}
    self.addSwitch('s%d' % (i+1), **sconfig)

# Define links
http_link_config = dict(bw = 1)
video_link_config = dict(bw = 10)

# Add switch links
self.addLink('s1', 's2', **video_link_config)
self.addLink('s2', 's4', **video_link_config)
self.addLink('s1', 's3', **http_link_config)
self.addLink('s3', 's4', **http_link_config)

# Add host links
self.addLink('h1', 's1', **host_link_config)
```

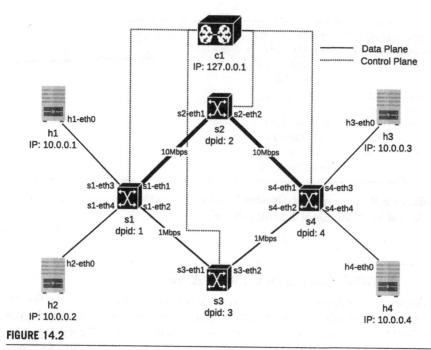

FIGURE 14.2

Network example for NS.

```
self.addLink('h2', 's1', **host_link_config)
self.addLink('h3', 's4', **host_link_config)
self.addLink('h4', 's4', **host_link_config)
```

Listing 14.1: Creating the network.

This network topology is loaded by Mininet and the following actions are automatically performed. The switches are set as OVS supported, and the MAC address of each host is set according to its IP address, for example, the MAC of host `h1` is `00:00:00:00:00:01` when its IP address is `10.0.0.1`. Additionally, all-pair ARP entries are used to remove the need of handling broadcast. In the end the SDN switches are configured to connect to a remote SDN controller `c1` with IP `127.0.0.1` on port 6633. The code provided in Listing 14.2 can be utilized to load the network.

```
net = Mininet(topo=topo, switch=OVSKernelSwitch, build=False, autoSetMacs=True
    , autoStaticArp=True, link=TCLink)
controller = RemoteController( 'c1', ip='127.0.0.1', port=6633 ) net.
    addController(controller)
```

Listing 14.2: Loading the network.

As described in Section 14.1.2, the following functionalities are required to implement NS:

1. the capability to manage and allocate the resources into slices and
2. the capability to associate traffic flows to slices.

In the following two subsections, we will proceed step by step using the Ryu APIs instead of the command line to implement two NS scenarios. In addition, the following three manners to validate network slices will be presented:

1. *ping* to verify connectivity,
2. *iperf* to verify bandwidth, and
3. *ovs-ofctl* to verify flows.

The implementation examples can be found in the folder *./app/network_slicing/* of the ComNetsEmu distribution accompanying this book.

14.2.1 Example 1: topology slicing

The first NS scenario to be implemented is the slicing of network topology resources. The topology of the network should be isolated into two layers:

- Upper slice: h1 -> s1 -> s2 -> s4 -> h3, 10 Mbits/s
- Lower slice: h2 -> s1 -> s3 -> s4 -> h4, 1 Mbits/s

Fig. 14.3 shows that each slice has its own view of the network nodes and the quality of the connection. In such a case, s1, s2, and s4, including their connection bandwidth, are available for h1 and h3 from the upper slice point of view, that is, the rest of the network resources are not visible to these network nodes. In a similar manner, network nodes h2, h4, s1, s3, and s4 can only be connected to each other in the lower slice. However, it is impossible to send data over other slices, for example, from h1 to h4, and impossible to use network resources in the upper slice, for example, 10 Mbits/s bandwidth.

14.2.1.1 Implementation

1. Managing and assigning resources into slices.
   To achieve the goal, multiple flows should be defined for each switch on each output port. In this case the ports will be connected in the same slice, and the ports in different slices will be isolated.
   For instance, at s1, we define a flow on s1-eth1 connected to s1-eth3 and another flow at s1-eth3 connected to s1-eth1 in the upper layer. For the same reason, two more flows connecting s1-eth2 and s1-eth4 in the lower layer are defined as s1. The same principle is used to define other flows on all ports of s2, s3, and s4. An example of this is provided in Listing 14.3.

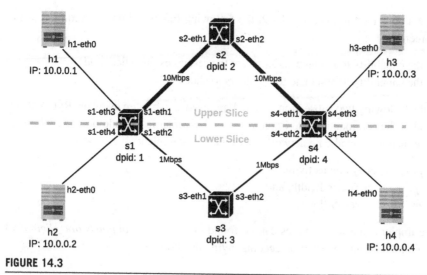

FIGURE 14.3

Topology slicing example.

```
def __init__(self, *args, **kwargs):
    super(TrafficSlicing, self).__init__(*args, **kwargs)

    # out_port = slice_to_port[dpid][in_port]
    self.slice_to_port = {
        1: {1:3, 3:1, 2:4, 4:2},
        4: {1:3, 3:1, 2:4, 4:2},
        2: {1:2, 2:1},
        3: {1:2, 2:1}
    }
```

Listing 14.3: Resource to slice.

2. Associating traffic flows to slices.

According to the *match* rule in the *flow table*, traffic packets can be classified into the correct flows and can perform their *actions* to fulfill the required slicing defined in the previous step.

In this example, the controller first detects a new incoming packet by the source switch datapath.id and the input port in_port. These form a *match* to ensure that the packet is associated with the correct flow. Based on the defined slicing, an output port out_port can be looked up for the packet. Forwarding the packet to the selected port is the *action* of this packet. The tuple of [datapah, match, actions] as *flow entry* is sent along with the packet to the *flow table* on the source switch. The example implementation code is provided in Listing 14.4.

```
def _packet_in_handler(self, ev):
    msg = ev.msg
    datapath = msg.datapath
    ofproto = datapath.ofproto
    in_port = msg.match['in_port']
    dpid = datapath.id

    out_port = self.slice_to_port[dpid][in_port]
    actions = [datapath.ofproto_parser.OFPActionOutput(out_port)]
    match = datapath.ofproto_parser.OFPMatch(
        in_port=in_port
    )

    self.add_flow(datapath, 1, match, actions)
    self._send_package(msg, datapath, in_port, actions)
```

Listing 14.4: Flows to slice.

14.2.1.2 Validation

1. Using `ping` to validate connectivity.
   As described at the beginning of Section 14.2.1, h1 can only be connected to h3, and h2 to h4. Therefore we use the command `ping` to test their connectivity:

```
mininet> pingall
*** Ping: testing ping reachability
h1 -> X h3 X
h2 -> X X h4
h3 -> h1 X X
h4 -> X h2 X
*** Results: 66% dropped (4/12 received)
```

As shown above, the connectivity of the upper and lower slices is as expected.
2. Using `iperf` to validate bandwidth.
   h2 and h4 should not have permissions to use the 10 Mbits/s bandwidth, because they are in the lower slice. `iperf` can be used to validate the bandwidth available in both slices. In turn, TCP traffic streams are generated between h1 and h3 and between h2 and h4.

```
mininet> iperf h1 h3
*** Iperf: testing TCPandwidth between h1 and h3
*** Results: ['9.50 Mbits/s', '12.4 Mbits/sec']

mininet> iperf h2 h4
*** Iperf: testing TCP bandwidth between h2 and h4
```

```
*** Results: ['958 Kbits/s', '1.76 Mbits/s']
```

These results show that the maximum available upload bandwidth between h1 and h3 is 9.50 Mbits/s, and the download bandwidth is 12.4 Mbits/s, which can only be provided in the upper slice. The maximum upload and download bandwidths between h2 and h4 are 958 kbits/s and 1.76 Mbits/s, which can be achieved at the lower slice.

3. Using dump-flows to check flow entry.

Another way to check how SDN forwards incoming packets is looking up the data flow at each switch. For example, on s1 a packet comes in at s1-eth1. This means that the packet comes from s2, which belongs to the upper slice. It should go to h1 via s1-eth3, so the s1 takes action to send this packet out on s1-eth3, resulting in incoming s1-eth3 to outgoing s1-eth1. Likewise, packets on port s1-eth2 and s1-eth4 are in the lower slice, so they are sent out on port s1-eth4 and s1-eth2.

```
mininet> sh ovs-ofctl dump-flows s1
 cookie=0x0, duration=3214.300s, table=0, n_packets=44, n_bytes=3380,
     priority=1,in_port="s1-eth1" actions=output:"s1-eth3"
 cookie=0x0, duration=3214.203s, table=0, n_packets=17, n_bytes=1382,
     priority=1,in_port="s1-eth3" actions=output:"s1-eth1"
 cookie=0x0, duration=3214.170s, table=0, n_packets=44, n_bytes=3384,
     priority=1,in_port="s1-eth2" actions=output:"s1-eth4"
 cookie=0x0, duration=3213.705s, table=0, n_packets=17, n_bytes=1382,
     priority=1,in_port="s1-eth4" actions=output:"s1-eth2"
 cookie=0x0, duration=3214.451s, table=0, n_packets=4, n_bytes=348,
     priority=0 actions=CONTROLLER:65535
```

14.2.2 Example 2: service slicing

The second NS scenario is the slicing of NetServ. As illustrated in Fig. 14.4, two types of slicing are defined:

- A video slice with 10 Mbits/s bandwidth for video traffic, consisting of UDP packets with destination port 9999 and
- a nonvideo slice with 1 Mbits/s bandwidth for other traffic.

14.2.2.1 Implementation

As with Example 1 in Section 14.2.1.1, we need two steps to define the slices and to associate traffic flows to slices.

1. Managing and allocating resources into slices.

On the end switches s1 and s4, two flows are defined to the destination hosts, respectively. At s1, a flow on out_port=3 goes to the destination MAC address of h1, to showcase one example. Additionally, bandwidth needs to be assigned

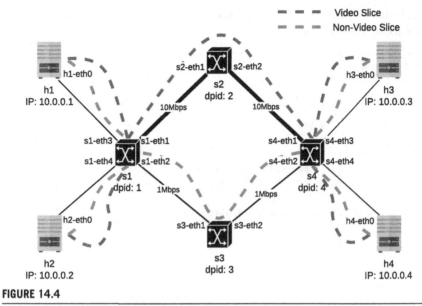

FIGURE 14.4

Service slicing example.

to both services. At `s1` and `s4` the video slice (slice number 1) is assigned the `out_port_1`, which has the 10 Mbits/s bandwidth. Likewise, the nonvideo slice (slice number 2) obtains `out_port_2` with 1 Mbits/s bandwidth. The UDP port number is defined as 9999 to identify packets in the next step.

```python
def __init__(self, *args, **kwargs):
    super(TrafficSlicing, self).__init__(*args, **kwargs)

    self.end_swtiches = [1 ,4]

    #outport = self.mac_to_port[dpid][dst_mac_address]
    self.mac_to_port = {1: {'00:00:00:00:00:01': 3, '00:00:00:00:00:02':
        4}, 4: {'00:00:00:00:00:03': 3, '00:00:00:00:00:04': 4}}

    #outport = self.slice_ports[dpid][slicenumber]
    self.slice_ports = {1 : {1 : 1, 2 : 2}, 4 : {1 : 1 , 2 : 2}}
    self.slice_TCport = 9999
```

Listing 14.5: Resource to slice.

2. Associating traffic flows to slices.

To assign traffic packets to the flows, their *match* rules should be defined. Next, the classified traffic packets are handled with *actions*. Therefore we will first define some *match* rules to classify the incoming packets to distinguish between video and other traffics. The implementation of this example is shown in Listing 14.6 in the form of pseudocode.

```
def _packet_in_handler(self, ev):
    save datapath of packet
    get in_port of packet

    get dpid of switch

    if switch is end switch:
        if packet from middle switch:
            send packet to the out_port of its destination

        elif UDP packet and destination port 9999:
            slice_numer = 1
            look up out_port
            actions
            match

        elif UPD packet and destination port not 9999:
            slice_numer = 2
            look up out_port
            actions
            match

        elif TCP packet:
            slice_numer = 2
            look up out_port
            actions
            match

        elif ICMP packet:
            slice_numer = 2
            look up out_port
            actions
            match

    if switch is middle switch:
        flood port
        actions
        match

    add_flow
    send_package
```

Listing 14.6: Flows to slice.

At end switches s1 and s4, the packets from s2 and s3 are forwarded to the out_port, connected to their destination. Each packet to s2 and s3 is verified whether it is a UDP packet and its destination port is 9999, that is, video traffic. The video traffic is classified to the video slice; in the example codes the slice number is 1. The other packets, for example, UDP with other destination port, TCP, and Internet Control Message Protocol (ICMP), belong to the other traffic category. This other traffic is considered as nonvideo slice, slice number 2 in the example codes. All these verifications build *matches*. The out_port can be looked up from the flows defined in the previous step. Each switch should handle the incoming packets as *actions*. On the switches s2 and s3 in the middle, packets are flooded to the out port, because the traffic is already assigned to corresponding slices.

14.2.2.2 Validation

1. Using ping to validate connectivity.
   Since ping creates ICMP packets that belong to the nonvideo slice, it can be determined how all hosts are reachable.

```
mininet> pingall
*** Ping: testing ping reachability
h1 -> h2 h3 h4
h2 -> h1 h3 h4
h3 -> h1 h2 h4
h4 -> h1 h2 h3
*** Results: 0% dropped (12/12 received)
```

2. Using iperf to validate bandwidth.
   From h1 to h3 we generate a video service that is a 10 Mbits/s UDP traffic with destination port 9999. This service should use the video slice.
   Log into h1 and h3 in a new terminal:

```
mininet> xterm h1 h3
```

Start listening to UDP packets at port 9999 on the h3 as reveiver:

```
$ iperf -s -u -p 9999 -b 10M
```

Start sending UPD packets with destination port 9999 on the sender h1 to h3:

```
$ iperf -c 10.0.0.3 -u -p 9999 -b 10M -t 10 -i 1
```

The results in Figs. 14.5 and 14.6 show that h1 keeps sending video traffic and h3 is able to receive the video traffic with 8.97 Mbits/s bandwidth, which means that the video traffic uses the video slice we created.

```
-------------------------------------------------------------
Client connecting to 10.0.0.3, UDP port 9999
Sending 1470 byte datagrams, IPG target: 1121.52 us (kalman adjust)
UDP buffer size:  208 KByte (default)
-------------------------------------------------------------
[  5] local 10.0.0.1 port 51201 connected with 10.0.0.3 port 9999
[ ID] Interval        Transfer     Bandwidth
[  5]  0.0- 1.0 sec  1.25 MBytes  10.5 Mbits/sec
[  5]  1.0- 2.0 sec  1.25 MBytes  10.5 Mbits/sec
[  5]  2.0- 3.0 sec  1.25 MBytes  10.5 Mbits/sec
[  5]  3.0- 4.0 sec  1.25 MBytes  10.5 Mbits/sec
[  5]  4.0- 5.0 sec  1.25 MBytes  10.5 Mbits/sec
[  5]  5.0- 6.0 sec  1.25 MBytes  10.5 Mbits/sec
[  5]  6.0- 7.0 sec  1.23 MBytes  10.3 Mbits/sec
[  5]  7.0- 8.0 sec  1.28 MBytes  10.7 Mbits/sec
[  5]  8.0- 9.0 sec  1.25 MBytes  10.5 Mbits/sec
[  5]  9.0-10.0 sec  1.25 MBytes  10.5 Mbits/sec
[  5]  0.0-10.0 sec  12.5 MBytes  10.5 Mbits/sec
[  5] Sent 8918 datagrams
[  5] Server Report:
[  5]  0.0-11.2 sec  11.9 MBytes  8.97 Mbits/sec   0.000 ms  406/ 8918 (0%)
[  5] 0.00-11.16 sec  30 datagrams received out-of-order
```

FIGURE 14.5

Video slice on sender.

```
-------------------------------------------------------------
Server listening on UDP port 9999
Receiving 1470 byte datagrams
UDP buffer size:  208 KByte (default)
-------------------------------------------------------------
[  5] local 10.0.0.3 port 9999 connected with 10.0.0.1 port 51201
[ ID] Interval        Transfer     Bandwidth        Jitter   Lost/Total Datagrams
[  5]  0.0-11.2 sec  11.9 MBytes  8.97 Mbits/sec   0.366 ms  406/ 8918 (4.6%)
[  5] 0.00-11.16 sec  30 datagrams received out-of-order
```

FIGURE 14.6

Video slice on receiver.

3. Using `dump-flows` to check flow entry.

   On switch s1, a UDP packet comes in the port s1_eth3, and its destination address is 00:00:00:00:00:03 with destination port 9999. This means that this packet should be in the video slice. s1 looks up its *flow table*, and the corresponding *actions* are output as s1_eth1. This output port maps the packet to the flow with 10 Mbits/s bandwidth.

```
mininet> sh ovs-ofctl dump-flows s1
 cookie=0x0, duration=716.049s, table=0, n_packets=8922, n_bytes=13490064,
     priority=2,udp,in_port="s1-eth3",dl_dst=00:00:00:00:00:03,tp_dst=9999
       actions=output:"s1-eth1"
 cookie=0x0, duration=704.853s, table=0, n_packets=3, n_bytes=4536,
     priority=1,dl_dst=00:00:00:00:00:01 actions=output:"s1-eth3"
 cookie=0x0, duration=704.849s, table=0, n_packets=3, n_bytes=1770,
     priority=1,icmp,in_port="s1-eth3",dl_src=00:00:00:00:00:01, dl_dst
     =00:00:00:00:00:03 actions=output:"s1-eth2"
 cookie=0x0, duration=818.963s, table=0, n_packets=75, n_bytes=10576,
     priority=0 actions=CONTROLLER:65535
```

In practical scenarios, service slicing is used in a demonstrator of industrial conditional monitoring [274,275].

14.2.3 Example 3: SDN proxy-based slicing

The previous examples provide an effective solution for implementing NS in scenarios where a single controller is expected to manage the entire slicing process. Nevertheless, there are other scenarios where multiple tenants require full control within their own slices. In several cases, this implies the possibility to run different SDN controllers on the same infrastructure.

In those scenarios the simplest solution is to introduce an SDN proxy device capable of generating virtual networks as slices of the existing infrastructure and to enforce isolation among the slices. An example of an SDN proxy is FlowVisor. In this related example the slices illustrated in Fig. 14.3 will be implemented with FlowVisor.

14.2.3.1 Implementation

To include FlowVisor as a component of ComNetsEmu, it is necessary to generate a proper Docker container by configuring the corresponding Dockerfile.

The following script shows how to build the Dockerfile for building a container for FlowVisor. This represents a general example on how to extend the functionalities of the ComNetsEmu by adding new functionalities through additional containers. The container is built by using a CentOS Linux image, adding the FlowVisor package, and finally configuring it. Moreover, the scripts for running the example are uploaded in the container to make them available at run time.

```
#
# About: Image for FlowVisor: A transparent proxy between OpenFlow switches
        and multiple OpenFlow controllers
# Ref  : https://github.com/fernnf/vsdnemul
#

FROM centos:6.10
RUN yum update -y && yum install wget sudo nano -y
WORKDIR /root
RUN wget http://updates.onlab.us/GPG-KEY-ONLAB
RUN rpm --import GPG-KEY-ONLAB
RUN echo -e "[onlab] \nname=ON.Lab Software Releases \nbaseurl=http://updates.
    onlab.us/rpm/stable \nenabled=1 \ngpgcheck=1" >> /etc/yum.repos.d/onlab.
    repo
RUN yum update -y
RUN yum install flowvisor -y

# Run FlowVisor configuration
RUN fvconfig generate /etc/flowvisor/config.json flowvisor flowvisor
```

```
RUN sed -i 's/"run_topology_server": false/"run_topology_server": true/' /etc/
    flowvisor/config.json
RUN fvconfig load /etc/flowvisor/config.json
RUN sed -i -e "s/\/sbin\/flowvisor /\/sbin\/flowvisor -l /ig" /etc/init.d/
    flowvisor
ENV TERM=vt100
ENV HOME /root
ENV BUILD_NUMBER docker
RUN fvconfig load /etc/flowvisor/config.json && \
    chown -R flowvisor:flowvisor /usr/share/db/flowvisor/

# Add Tini --- A tiny but valid init for containers
ENV TINI_VERSION v0.18.0
ADD https://github.com/krallin/tini/releases/download/${TINI_VERSION}/tini /
    tini
RUN chmod +x /tini
ENTRYPOINT ["/tini", "-g", "--"]

# Add scripts for demos
COPY flowvisor_script* ./
COPY fvpassword ./

CMD ["bash"]
```

Once the container is ready, it is possible to run it via

```
$ sudo docker run -it --rm --network host flowvisor:latest /bin/bash
```

However, it is best to first run the Mininet environment by running the topology (see above) and then to force all OpenFlow switches to use only version 1.0 of OpenFlow using the command:

```
$ sudo ovs-vsctl set bridge s1 protocols=OpenFlow10
$ sudo ovs-vsctl set bridge s2 protocols=OpenFlow10
$ sudo ovs-vsctl set bridge s3 protocols=OpenFlow10
$ sudo ovs-vsctl set bridge s4 protocols=OpenFlow10
```

Once the Mininet environment is ready and the FlowVisor container is running, we can run FlowVisor using the commands:

```
FlowVisor_docker> sudo -u flowvisor flowvisor > fv.log 2>&1 &
FlowVisor_docker> cat fv.log
```

The file fv.log should contain the log of FlowVisor startup and signal any anomaly. Then we can define the slices on the Mininet topology. In this example, we will generate the upper slice presented before, but this time we will use FlowVisor. This is achieved by the following commands:

```
FlowVisor_docker> fvctl -f fvpassword add-slice upper tcp:localhost:10001
    admin@upperslice
FlowVisor_docker> fvctl -f fvpassword add-flowspace dpid1-port1 1 1 in_port=1
    upper=7
FlowVisor_docker> fvctl -f fvpassword add-flowspace dpid1-port3 1 1 in_port=3
    upper=7
FlowVisor_docker> fvctl -f fvpassword add-flowspace dpid2 2 1 any upper=7
FlowVisor_docker> fvctl -f fvpassword add-flowspace dpid4-port1 4 1 in_port=1
    upper=7
FlowVisor_docker> fvctl -f fvpassword add-flowspace dpid4-port3 4 1 in_port=3
    upper=7
```

The first line builds the upper slice and assigns its controller. In this case, we assume that the SDN controller for the upper slice will be available on the localhost at port 10001. Therefore we will run it on our Virtual Machine. The parameter -f fvpassword is used to avoid entering FlowVisor password but using the one stored in the corresponding file.

The following lines define a topology slice by associating with the upper slice ports 1 and 3 of switch s1, ports 1 and 3 of switch s4, and all ports of switch s2. This is achieved by adding the corresponding flowspaces.

In order for the switches within the network slice to correctly operate, we need to start an SDN controller listening on port 10001. The simplest way to achieve it is to run the Mininet controller embedded in the VM:

```
$ controller -v ptcp:10001 &
```

14.2.3.2 Validation

1. Using ping to validate connectivity:
   In this scenario, ping is extremely useful, as we built a network slice using topology slicing. Moreover, since only one slice is available, we can run the pingall command and check connectivity on the upper slice:

```
mininet> pingall
*** Ping: testing ping reachability
h1 -> X h3 X
h2 -> X X X
h3 -> h1 X X
h4 -> X X X
```

2. Using `iperf` to validate bandwidth.

   Since h1 to h3 belong to the same slice, we can test the corresponding bandwidth with

   ```
   mininet> iperf h1 h3
   ```

   Of course, it is possible to perform bandwidth testing between nodes belonging to different network slices.

3. Using `dump-flows` to check flow entries.

   To validate in detail how the switch handles packets of different services, it is possible to run the OpenFlow `ovs-ofctl` command on them. As an example, for switch s1, we run the command

   ```
   mininet> sh ovs-ofctl dump-flows s1
   ```

Realizing mobile edge clouds

15

Zuo Xiang[a], Carl Collmann[a], Patrick Seeling[b]

[a]*Technische Universität Dresden, Dresden, Germany*
[b]*Central Michigan University, Mount Pleasant, MI, United States*

Any program is only as good as it is useful.

Linus Torvalds

15.1 Introduction

In this chapter, we provide a practical hands-on example with ComNetsEmu demonstrating how to improve the latency performance of a migrated application in the MEC with the help of SDN and NFV technologies. We refer the interested reader to a comprehensive introduction of the Mobile Edge Cloud (MEC) in Chapter 4. Similarly, in Chapter 13, we introduce the ComNetsEmu environment. The source codes for the examples in this chapter are available online [276].

According to the multiaccess edge computing architecture, latency-sensitive applications can be deployed on MEC application servers to perform related processing tasks closer to a mobile client. Due to the mobile nature of the edge client, handovers between BSs can happen when clients move across different locations. A typical handover and application migration scenario with the MEC paradigm is illustrated in Fig. 15.1.

In this simplified scenario, we identify three main components:

Some type of mobile client: A typical example for a mobile client is a smartphone. Mobile clients can communicate over the network to utilize specific services provided by a remote cloud.

Base Station (BS): Typically, the BS is assumed to forward data between the mobile clients and the core network.

Mobile Edge Cloud (MEC): A cluster of MEC application servers. Depending on the design of the cellular network system, this cluster can be deployed directly on a specific type of BS or on the RAN controller.

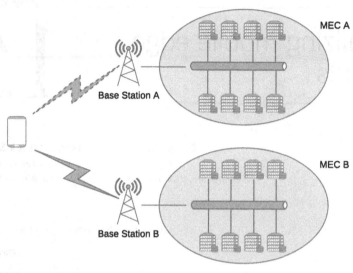

FIGURE 15.1

Typical handover and migration scenario with mobile clients switching between different base stations.

As illustrated in Fig. 15.1, the connection of the mobile client can be switched from BS A to BS B. To provide a continuous high-level of QoS, the application deployed on BS A is required to be seamlessly migrated to BS B to avoid negative service impacts. Before a more detailed elaboration of the system design and implemented mechanisms, the following assumptions are made to limit the difficulty of this emulation example:

1. All servers are able to deploy the application even if they are already experiencing a heavy workload. There is no denial of service for the deployment and also no performance guarantee. The constraints of computational resources for each server are assumed to be identical. Therefore the data processing delay of the application deployed on a heavy-loaded server should be higher than a light-loaded server.
2. At any given moment, there is only one instance of the service application supposed to be deployed on only one of the servers to ensure overall resource efficiency.
3. The application to be deployed is stateless. State synchronization and migration problems are out of scope and not addressed in this example.
4. This example does not simulate a specific application. Instead, the application is simulated by i) the client sending UDP datagrams with random payload to the edge cloud and ii) the application performing *dummy* data processing with random delays.

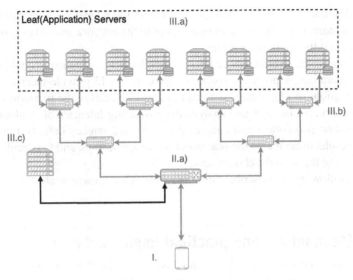

FIGURE 15.2

Simplified MEC cluster with servers connected in a tree topology.

As illustrated in Fig. 15.1, multiple physical application servers are deployed in a data center to distribute the service workload. These machines are connected via network devices with a specific topology. A representative topology designed for this scenario is the fat tree network [277] illustrated in Fig. 15.2.

In the example here a simple binary tree topology is used connecting multiple components in *ComNetsEmu* to emulate the scenario in Fig. 15.2:

1. The virtual switch (II a) is used to emulate the gateway of a BS. It is the access point for all traffics from and to the mobile edge client. When the switch receives the first packet of a specific flow from a newly connected client, a proper path should be chosen to forward this packet and any following packets in this flow to the destination server with the required service already deployed.
2. Aggregation virtual switches (II b) are used to build the tree topology and forward traffic with configured rules.
3. Application (leaf) servers (III a) are emulated with external Docker hosts on which application containers can be dynamically deployed to handle traffic forwarded by the network.

In the example for this chapter, the gateway is assumed to receive a new flow from a mobile client that moved into the region handled by this MEC. Subsequently, there initially arises the problem of *which physical server* should be chosen to deploy the migrated application. Additionally, a decision of *how to forward the traffic* to this server needs to be made as well. A dedicated metric needs to be employed to decide between different available candidate servers – here we employ the service latency for the mobile client as the metric to focus on. The end-to-end service delay contains two main parts, namely (i) the transport delay spent in the network and (ii) the processing delay spent on the application server.

To minimize the overall service latency, both latency components should be taken into consideration. A proposed solution approach should provide mechanisms and implementations of functionalities to (i) monitor and analyze the transport latencies in the network, (ii) monitor and analyze the processing latencies of available application servers, (iii) collect and analyze the monitored latency data, (iv) utilize the analysis results to decide on the placement of the application, and (v) decide on how to forward the traffic to the chosen server.

In the following, we describe a mechanism to achieve these goals.

15.2 Mechanisms and practical implementation

A comparison between approaches without and with SDN/NFV technologies is performed as part of the example of this chapter to highlight the significance of network softwarization technologies for performance enhancements.

15.2.1 Without SDN/NFV technologies

Without SDN/NFV technologies, the behavior of network devices requires a definition with a static set of protocols and rules. In the scenario of this chapter, network devices can utilize the traditional server discovery mechanism to choose the destination server. The approach implemented in this example is to utilizing ARP to monitor the transmission latency characteristics for each server as follows. When the gateway switch receives the first packet of a flow from the mobile client, it broadcasts an ARP request based on the destination IP address encapsulated in the packet. To simplify the IP Address Management (IPAM) in this example, the broadcast IPv4 address [278] is used as the destination address for all packets sent from the client. Following this approach, all application servers respond to the ARP request. As the ARP protocol is a simple link layer protocol implemented in the OS kernel, generating the ARP response does not require significant computational resources. Subsequently, the server should send an ARP response in a relatively short time, even if it is under heavy computational workload. The delay between this ARP request and a server response can be employed as a reasonable estimation of the transmission delay. The gateway switch is configured to follow a straightforward static rule: It forwards all traffic of a flow to the server that has the minimal ARP response delay.

15.2.2 With SDN/NFV technologies

The overview of the enhanced mechanism enabled by SDN/NFV technologies is illustrated in Fig. 15.3. Here only the half of the servers (servers 1–4) constituting the

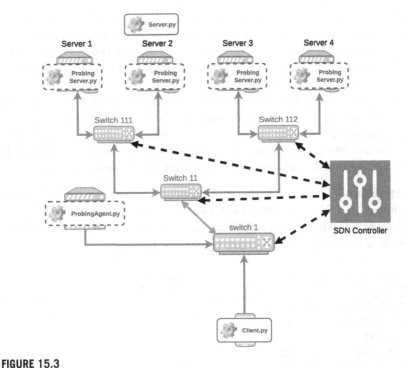

FIGURE 15.3

Mechanisms with SDN/NFV technologies.

binary tree with the depth of two are presented for clarity. The main approach of the SDN/NFV enhanced solution is including the processing delay of all servers into the selection mechanism. In addition to the ARP monitoring introduced before, an active probing is orchestrated by the SDN controller to evaluate the processing delay as follows:

1. In addition to the four application servers, a dedicated server is used to deploy the probing agent network function. The probing agent can inject packets into the network based on the type of traffic received from the agent.
2. Injected probing packets are received by individual servers and processed by the probing network function deployed on all application servers under consideration. The probing server function should perform the same operation on probing packets as the service application to be placed.
3. All probing packets are monitored by the SDN controller. When probing packets from the probing agent arrive at the controller, the time when they are being sent out to their destined locations is stored in a data structure. After these packets have been processed by the servers and rearrive at the controller, their delay can be measured with use of the previously saved timestamp. This delay is an esti-

mation of the *total* service latency of each server. The server with the minimal service delay is chosen as the best server, and the service application is deployed on this server. As an example in Fig. 15.3, server 2 is chosen to deploy the service application program (in the code example provided later in this chapter, this refers to server.py).

4. To demonstrate the dynamic and adaptive orchestration provided by the network softwarization, the active probing monitoring is performed during the whole service session. The controller checks after each round of probing whether there is a server with lower latency than that currently running the service application. In such a case the service application will be migrated to the new best server.

The pseudocode for the monitoring mechanism of the SDN controller is described in Algorithm 15.1. The algorithm is implemented in the handler of the PacketIn message, which is triggered once the first packet of a flow arrives at a switch that does not have matching forwarding rules for this flow [279].

Algorithm 15.1: Pseudocode for the algorithm of SDN controller.

 1: **if** packet type ARP **then**
 2:    update link latency
 3: **end if**
 4: **if** packet type UDP **then**
 5:    **if** probing packet from probing agent **then**
 6:       flood packet out
 7:       flood ARP packet
 8:    **end if**
 9:    **if** probing packet from server **then**
10:       update service latency
11:       **if** MESSAGE_COUNT > 80 **then**
12:          find optimal host
13:          **if** NEW_OPTIMAL_HOST != OLD_OPTIMAL_HOST **then**
14:             remove old Flows
15:             add new Flow
16:             send Server change message
17:          **end if**
18:       **end if**
19:    **end if**
20: **else**
21:    send packet out
22: **end if**

We now shift to the implementation of these scenarios with the ComNets Emulator (ComNetsEmu).

Table 15.1 Link delays in the emulated topology.

Link	Delay in ms
switch 1 – client	200
s1 – probe agent	50
switch 1 – switch 11	10
switch 11 – switch 111	10
switch 11 – switch 112	10
switch 111 – server 1	100
switch 111 – server 2	150
switch 112 – server 3	200
switch 112 - server 4	250

15.3 ComNetsEmu experimentation

The overall experimentation steps we describe in the following employ *ComNetsEmu* with details of its usage discussed earlier in this book. The SDN/NFV enhanced approach on the topology illustrated in Fig. 15.3 is emulated on the ComNetsEmu with link delays listed in the Table 15.1.

We note that there is no bandwidth limitation for any links. Each server node is configured to only use the first CPU core and maximal 25% of the CPU time. Several `Python` scripts are located in the application example directory [276]. The scripts to build the static topology with parameter configuration can be found in `only_forwarding.py`. The scripts for a more involved scenario include:

- `controller_probing.py`: The controller application (using the Ryu SDN framework) with implemented probing and server selection algorithms.
- `client.py` and `server.py`: Programs for the mobile client and service application to be deployed. UDP sockets are used to send and receive data packets.
- `probe_client.py` and `probe_server.py`: Programs to generate and response probing traffic.
- `only_forwarding.py`: The script to build the topology and handle the deployment of the service application.

Shell commands listed in List 15.1 can be executed in the *ComNetsEmu* test environment to run the example and generate latency measurement results. Note that two shell windows are required to interactively run the program. After running the emulation scripts, the SDN controller generates a `log.txt` file with latency measurement results.

```
# Shell Window 1: Enter the mec example directory
$ cd ./app/mec/

# Shell Window 1: Before running, a dedicated docker image needs to be created
    with:
```

```
$ sudo docker build -t mec_test -f ./Dockerfile.mec_test .

# Shell Window 1: Start the controller program with log file and TCP listen
    port (For OpenFlow) specified.
$ ryu-manager --verbose --log-file log.txt --ofp-tcp-listen-port 6633 ./
    controller.py

# Shell Window 2: Launch the emulation program
$ sudo python3 ./edgelcoud/edgecloud.py
```

Once the experimentation phase is completed, the generated results can be evaluated as exemplarily described further.

15.4 Emulation results

The results we present in the following are obtained for emulation performed on a PC with an Intel i5–4590 CPU @3.30 GHz CPU and 8 GB RAM.

15.4.1 Latency measurement results on SDN controller

The latency measurement results of active probing based monitoring are illustrated in Fig. 15.4, and numeric values are presented in Tables 15.2 and 15.3 for link and service latencies, respectivly.

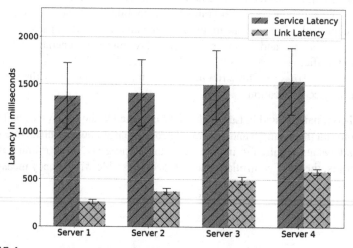

FIGURE 15.4

Link and service latency of four servers.

Table 15.2 Measured link latency for four servers.

server	link latency (ms)	# packets	adjusted link latency (ms)	# packets
I	261.91 ± 53.98	21672	259.04 ± 24.00	15796
II	377.35 ± 75.41	21215	371.33 ± 33.62	17021
III	496.06 ± 78.40	20461	489.23 ± 39.48	15425
IV	594.47 ± 65.34	21079	583.68 ± 30.20	14035

Table 15.3 Measured service latency for four servers.

server	service latency (msec)	# packets	adjusted service latency (msec)	# packets
I	1360.32 ± 616.12	15624	1373.54 ± 350.13	9358
II	1401.05 ± 607.99	15598	1409.89 ± 348.63	9030
III	1487.74 ± 637.34	15604	1498.21 ± 363.04	9456
IV	1532.38 ± 607.06	15574	1537.55 ± 350.95	9115

Fig. 15.4 illustrates the averages obtained and the standard deviations as error bars from each of the servers for two types of delays for the recorded traffic. We note that the calculated values are created with additional postprocessing, initially rejecting outlier latency values for link and service latencies that do not exceed borders of 1000 ms and 2500 ms respectively, then reiterating over the packets. The link and service delay without and with this postprocessing step are listed detailed in Tables 15.2 and 15.3. From these tables we can observe that removing packets based on standard deviation have no major impact on the mean value, which indicates symmetry of the latency distribution. Furthermore, we note that packets with a latency higher than one standard deviation from the expected mean value lead to removal of about 60 to 75% of the packets, especially in case of the service latency. This is close to what a normal distributed packet latency would yield (i.e., 68%), suggesting that the packet latency is approximately normally distributed. Subsequently, this leads to the conclusion that the observed variation in latency originates from a source that is uncorrelated for each of the examined paths and can be described with a normal distribution.

To derive whether the chosen probing range was sufficient, it is possible to observe the Standard Deviation (STD) over time, whereas the STD is calculated based on the current latency values for the observed packets at the given time. If at a certain point the STD becomes stable and does not change significantly anymore, then a stable state is achieved. This refers to the amount of packets evaluated being sufficiently large and supporting the assumption that the performed measurement is valid. Fig. 15.5 illustrates the respective link and service latency STDs for servers 1 and 4, respectively. We observe that the STD in each case becomes stable after around 10000 packets, suggesting validity of the performed measurement. Closer inspection reveals that there appears a higher amount of probing packets for the link latency than for the service latency, especially considering the Algorithm 15.1 implemented on the SDN controller. The reason for this disparity is twofold:

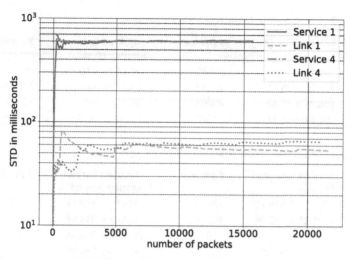

FIGURE 15.5

Standard deviation (STD) of link and service latency.

1. The controller is active even before the individual containers are instantiated, and during this starting phase of the network, numerous ARP packets are sent between all servers to fill their ARP tables.
2. During the emulation, each of the servers can independently send out ARP requests, independently from the actually performed experiment.

   Jointly, these two intricacies explain the observed disparity.

15.4.2 Latency measurement at client side

The monitoring of the latency from the perspective of the client is performed by running Wireshark on the host system and evaluating the timestamps of captured packets as postprocessing. Since all Docker containers run on the same underlying OS kernel and physical machine, no time synchronization is required for the nodes in the network. Fig. 15.6 shows the Cumulative Distribution Function (CDF) of the latency from the mobile client to each server. From these CDFs we can directly obtain that the client is typically best served by MEC server 1, which is a direct outcome of the service and link latencies we presented from the SDN controller's viewpoint before.

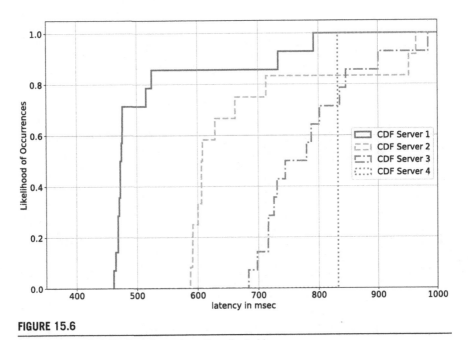

FIGURE 15.6

Cumulative probability for latency from the client side.

Machine learning for routing

16

Justus Rischke, Peter Sossalla

Technische Universität Dresden, Dresden, Germany

> This is your machine learning system? *Yup! You pour the data into this big pile of linear algebra, then collect the answers on the other side.* What if the answers are wrong? *Just stir the pile until they start looking right.*
>
> **xkcd comic #1838**

16.1 Introduction

A challenge for many network service providers is the rising and varying network traffic demand. A particular dilemma is that traffic peaks during the day can overload routes while state-of-the-art routing protocols do not consider demand or capacity. With SDN, a central software controller can shift traffic according to the current situation, as proposed by Schlinker et al. [126]. However, the questions of how and which routes to change remain. Ideally, routing protocols would optimize themselves, but currently there is no such protocol. In this chapter, we demonstrate how Reinforcement Learning (RL) can optimize routing without any additional engineering effort. The advantage of RL, a subdomain of ML, is that it does not require an underlying model of the network. The RL agent implemented in the SDN controller learns a beneficial routing configuration by interacting with the environment, that is, the network. In our example, we aim to minimize latency. This has two reasons: i) we want to reduce congestion, which causes an increased latency and reduces throughput, and ii) future applications, such as the *Tactile Internet*, require low latency. The problem of how to apply RL on tasks such as routing remains. In the next section, we explain a possible solution, but also we want to emphasize that this is still an open research topic and there are many open questions.

16.2 Fitting reinforcement learning to routing

The reinforcement learning agent, which may operate at the SDN controller, observes the environment (i.e., the network) by measuring latency at discrete time steps $t = 0, 1, 2, \ldots$. The observation consists of the environment *state* $S_t \in \mathcal{S}$ and a *reward* $R_t \in \mathcal{R} \subset \mathbb{R}$ at time step t.

This entire process is shown in Fig. 16.1.

Computing in Communication Networks. https://doi.org/10.1016/B978-0-12-820488-7.00031-1

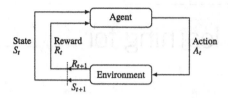

FIGURE 16.1

Interaction between the agent and environment, according to Sutton [167].

16.2.1 Designing state and action space

The communication from a given sender (source host) s_f to a given receiver (destination host) d_f is referred to as flow f, where $\mathcal{F}$ is the set of all flows. A path P_{s_f,d_f} from the set of all possible paths $P \in \mathcal{P}_{s_f,d_f} = \{P_{s_f,d_f,1}, P_{s_f,d_f,2}, \ldots\}$ connects source host s_f to destination host d_f, whereby the set $\mathcal{P}_{s_f,d_f}$ can be determined by a graph search algorithm, such as Depth-First Search (DFS). Consider the following topology depicted by Fig. 16.2 with four switches Sw and four host nodes H.

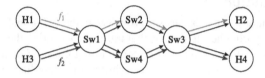

FIGURE 16.2

Example topology.

There are two flows f_1 and f_2, connecting *H1* to *H2* and *H3* to *H4*, respectively. In addition, there are two possible paths $\mathcal{P}_{f_1,f_2}$ for both flows, Sw1 → Sw2 → Sw3 and Sw1 → Sw4 → Sw3.

Formally, a state S from the set of states $S \in \mathcal{S} = \{S^1, S^2, \ldots\}$ is defined by the equation

$$S = \begin{cases} f_{s_1,d_1}: & P_{s_1,d_1,t}, \\ \vdots & \\ f_{s_i,d_i}: & P_{s_i,d_i,t}. \end{cases} \tag{16.1}$$

A state consists of pairs consisting of a flow f and the current path P_f. Transferred to our example, Fig. 16.3 depicts all possible states as a table.

An action A changes the state, that is, according to our previous definition of the states, it changes path P of flow f. This is expressed by

$$A_t = \{f_{s_1,d_1}: P_{s_1,d_1,t} \Rightarrow P_{s_1,d_1,t+1}\}. \tag{16.2}$$

In addition, the arrows in Fig. 16.3 show such an action. Consider the example of being in the lower left state in Fig. 16.3. There are three possible actions that

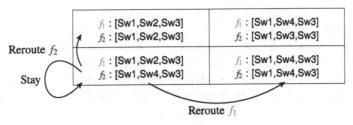

FIGURE 16.3

States and actions with respect to Fig. 16.2.

can be performed: i) reroute f_1, that is, change its path from $Sw1 \rightarrow Sw2 \rightarrow Sw3$ to $Sw1 \rightarrow Sw4 \rightarrow Sw3$, ii) reroute f_2 in the same fashion, or iii) stay in the current state, for example, when the routing configuration seems to be beneficial. The learning agent keeps a table with state–action pairs to determine which action to perform in a certain state. The agent estimates the quality of an action in a certain state by employing an algorithm called *Q-Learning* (Q as in quality), as explained in greater detail in Chapter 8, Section 8.4.2. This *Q-table* is shown by the equation

$$
\text{Q-Table} = \begin{cases} S^1 : \begin{cases} A^1 : Q(S^1, A^1), \\ \vdots \\ A^j : Q(S^1, A^j), \end{cases} \\ \vdots \\ S^i : \begin{cases} A^1 : Q(S^i, A^1), \\ \vdots \\ A^j : Q(S^i, A^j). \end{cases} \end{cases}
\tag{16.3}
$$

In the provided example, the Q-table is implemented as a nested dictionary with the states as outer keys and the corresponding actions as inner ones. The Q-values $Q(S, A)$ in this data structure are the values of the inner dictionary. The Q-value is calculated by the Q-learning algorithm, which takes the reward R into account.

16.2.2 Reward

As mentioned before, the RL agent considers the average latency per flow as objective to be optimized. In turn, the reward is defined accordingly as

$$
R_t = - \sqrt{\frac{\sum_{\forall f \in \mathcal{F}} L_f^2}{|\mathcal{F}|}} = - \sqrt{\frac{L_{f,1}^2 + L_{f,2}^2}{2}}.
\tag{16.4}
$$

This reward is calculated using the root mean square of the individual latencies per flow. Subsequently, outliers are weighted higher. In addition, since a higher latency is less desirable, a minus sign is required, because the reinforcement learning agent strives to maximize its reward. The latency per flow is determined by the SDN controller. For this, the controller sends probing packets between the switches to measure the latency, as proposed by Phemius et al. [280]. These interswitch latencies are then used to calculate the actual flow latency as

$$L_{f,1} = L_{Sw1,Sw2} + L_{Sw2,Sw3}, \tag{16.5}$$

$$L_{f,2} = L_{Sw1,Sw4} + L_{Sw4,Sw3}. \tag{16.6}$$

Here the composition of the flow latencies depends on the currently chosen path. This is necessary since an action would result in a changed average flow latency, which is then considered by the reward. In that way, the loop of state-action-reward is adapted by current paths-rerouting-measuring.

As explained before, the Q-table is used to determine which action to choose in a certain state. The RL agent establishes the corresponding Q-values through the process of *Exploration*, which is explained in the next section.

16.2.3 Exploration

Before the RL agent finds a beneficial routing configuration, it first needs to evaluate different state–action combinations (exploration). Once the agent finds a presumable optimum, it should stay in that state and not reroute the traffic anymore (exploitation). However, it must also be ensured that changes in the network are still detected and the routing is adapted accordingly. This problem is commonly known as *Exploration vs. Exploitation*. There are three well-known algorithms for selecting an action:

ϵ**-greedy:** ϵ-greedy is this simplest one, always choosing the action with the highest Q-value. It is therefore called *greedy*. However, this would not allow exploration. Therefore the agent chooses a random action with probability ϵ as follows:

$$A_t \doteq \begin{cases} a \text{ with probability } \epsilon, \\ \underset{a \in \mathcal{A}}{\text{argmax}} \ Q(S, A) \text{ otherwise.} \end{cases} \tag{16.7}$$

As a result, all possible state–action combinations will be tried. Unfortunately, this will be done infinitely, that is, even if a latency minimum is found, the routing will be sometimes randomly changed.

Softmax: The *Softmax* strategy converts the Q-values into the probabilities

$$p(A|S) = \frac{\exp[-1/(Q(S,A) \cdot \tau)]}{\sum_{b \in \mathcal{A}} \exp[-1/(Q(S,b) \cdot \tau)]} \tag{16.8}$$

for each action of the states and samples over the results, that is, actions with higher Q-values are preferably selected, which is a good trade-off between exploration and exploitation. The *Softmax* Eq. (16.8) in our example is slightly modified from the original as explained by Sutton et al. [167], since the scenario has negative rewards. Both ϵ-*greedy* and *Softmax* have the disadvantage that they do not decrease the exploration over time, also called *annealing*.

Upper Confidence Bound (UCB): The UCB approach in Eq. (16.9) already includes *annealing* by adding the bonus $b+$ defined in Eq. (16.10). UCB is counting the times an action has been chosen in a state $N(S, A)$ vs. the number of times the state has been visited $N(S)$. Thereby, actions which have been performed more often and which are resulting in better rewards are preferably chosen.

$$A_t \doteq \underset{a \in A}{\operatorname{argmax}} \left(Q_t(S, A) + cb^+ \right), \tag{16.9}$$

$$b^+ = \sqrt{\ln N(S)/N(S, A)} \tag{16.10}$$

The example provided in the following allows us to switch between different exploration algorithms and their corresponding parameters ϵ, τ, and c, which control the exploration. The parameters can be changed in the configuration file (./controller/config.py). This enables repetition of the following example and comparison of the outcomes of the learning processes.

16.3 Example

The following example demonstrates how routing is adapted by the SDN controller with reinforcement learning. For that, precise latency measurements are required. Therefore *libvirt* has to be used as a virtual machine provider instead of *Virtualbox*. See the explanation provided in the README.md of *ComNetsEmu*.

16.3.1 Setup

Fig. 16.4 showcases the topology with two possible paths used in the hands-on example of this chapter. In this scenario, three flows exist with $f_{H1,H4}$ requiring 2.75 Mbit/s, and $f_{H2,H5}$ and $f_{H3,H6}$ require 1.75 Mbit/s each. The optimal routing configuration, in terms of the minimum average flow latency, can be easily determined here, namely $f_{H1,H4}$ via route Sw1 → Sw2 → Sw3, and $f_{H2,H5}$, $f_{H3,H6}$ via route Sw1 → Sw4 → Sw3. Any other routing of these three flows would result in a congested path with an increased latency.

The example scenario creates the load with the *Iperf* tool (see Chapter 27), which creates a stream of UDP packets.

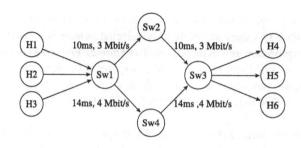

FIGURE 16.4

Example topology.

16.3.2 Running the example

This example is located in `comnetsemu/app/machine_learning_for_routing` and requires two terminals. In addition, the example needs additional libraries for statistical calculations. These can be installed with the command in Listing 16.1.

```
$ sudo ./install_dependencies.sh
```

Listing 16.1: Terminal 1: Installation of missing dependencies.

The example can then be started with the command in Listing 16.2, which consists of a *Mininet* script and three additional threads starting *Iperf*.

```
$ sudo python3 ./example_scenario.py
```

Listing 16.2: Terminal 1: Starting the *Mininet* topology with load scenario.

In the code repository, there is an additional script (`./tinker_example.py`) to enable rapid experimentation variation. This script does not automatically start *Iperf* and allows different communication sessions to be started manually.

After the example scenario is executed, the SDN controller needs to be started with the command in Listing 16.3 in the second terminal.

```
$ ryu-manager ./controller/remote_controller.py
```

Listing 16.3: Terminal 2: Starting the SDN controller.

After some initial delay, the controller will start to learn, which can be observed in its log output similar to Listing 16.4.

```
...
State: {('10.0.0.1','10.0.0.4'): [1, 4, 3],
        ('10.0.0.2','10.0.0.5'): [1, 4, 3],
```

```
                ('10.0.0.3','10.0.0.6'): [1, 4, 3]}

Action: [('10.0.0.1','10.0.0.4'), [1, 2, 3]]

Next State: {('10.0.0.1','10.0.0.4'): [1, 2, 3],
             ('10.0.0.2','10.0.0.5'): [1, 4, 3],
             ('10.0.0.3','10.0.0.6'): [1, 4, 3]}

PrevReward: -100.11005401611328
Average Latency: 28.06532382965088 Reward: -28.31982606468689
...
```

Listing 16.4: Terminal 2: Example log output of the controller (formatted).

As can be seen from the logged information, the RL agent in the controller tries different routing configurations, for example, it routes all three flows over path Sw1 → Sw4 → Sw3. However, this exceeds the capacity of the path and results in congestion with increased latency and a low reward. After some time, the agent will have tried more beneficial states and will converge to them, which results in a higher reward.

16.3.3 Discussion

The controller logs its training process and the log data can be plotted with the provided scripts. Fig. 16.5 shows the result of multiple runs of the learning process, where the shaded areas represent the 5 and 95 percentiles, respectively. At first, the routing configuration is randomly initialized, resulting in an increased latency. In the training phase, we can observe several peaks, both high and low. This settles after a while, depending on the chosen exploration strategy.

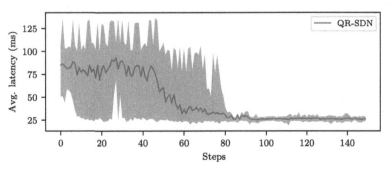

FIGURE 16.5

Learning process of RL agent with *Softmax* ($\tau = 0.00005$) as exploration strategy.

16.3.4 Changing parameters

The provided SDN controller with RL-aided routing can be configured with the dedicated configuration file. This allows changing the exploration strategy (exploration_mode) and the corresponding parameter shown in Listing 16.5.

```
$ cat controller/config.py
...
# for eps-greedy
epsilon = 0.05
# for Softmax
temperature = 0.00005  # tau
# for UCB
exploration_degree = 30  # c
...
exploration_mode = ExplorationMode.SOFTMAX
...
```

Listing 16.5: Parameters of exploration algorithms.

Changing, for example, the default exploration strategy from *Softmax* to ϵ-*greedy* causes quite observable latency peaks in the exploitation phase. Finally, we would like to encourage the reader to set various parameters and examine their influence on the learning process.

Machine learning for flow compression

17

Máté Tömösközi

Technische Universität Dresden, Dresden, Germany

The nice thing about standards is that you have so many to choose from.
Andrew S. Tanenbaum

17.1 Introduction

As the Internet grows and increases in complexity catalyzed by the adoption of numerous next generation wireless network applications, such as the IoT and Vechicle-to-Everything (V2X), billions of devices will need to communicate in the same interconnected digital environment. To facilitate this, one has to provide sufficient metadata to tackle addressing, routing, synchronization, and error recovery concerns, among other things. However, this requires an encapsulation overhead that can even exceed the size of the logical payload.

The IP has been widely adopted as the main network layer protocol. Typically, IP packets contain a protocol encapsulation overhead from higher layers of the network protocol stack as well, which are prepended to the logical payload (i.e., the binary information that needs to be transmitted). Generally, header compression implementations are integrated between the link and Internet layer of the Internet Protocol suit. Fig. 17.1 illustrates this via a simplified example of H.264/AVC video transmission with Robust Header Compression (RoHC). A common protocol combination in this context is Real Time Protocol (RTP), UDP, and IP (commonly denoted as RTP/UDP/IP), accounting for 40 bytes with Internet Protocol Version 4 (IPv4) and 60 bytes with Internet Protocol Version 6 (IPv6). Compression of this overhead can yield significant savings, especially with smaller payloads.

All of the above is widely known, but network developers cannot assume that in our rapidly advancing age, we have the perfectly fitting programming library (function) ready for every current and future application. One of the main limiting factors for the employment of header compression solutions for 5G networks is the constricting nature of these compression designs. Header compression is one of the technologies enabling packet-switched computer networks to operate with higher efficiency, even if the underlying physical link is limited. Since its inception in the 1980s, it was meant to improve the QoS for dial-up Telnet connections and has evolved into a complicated multifaceted compression library. Robust Header Compression (RoHC), as one example, has been integrated into third and fourth generation cellular networks,

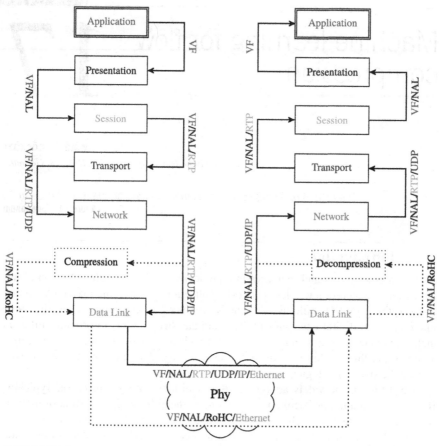

FIGURE 17.1

The conventional location of header compressions inside the protocol stack of the ISO/OSI model with a simplified example where the transmission of H.264/AVC video frames (VF) are compressed by Robust Header Compression (RoHC).

among others. Beyond the promised benefit of decreased bandwidth usage, header compression has shown that it is capable of improving the QoE for already existing services, such as real-time audio and video calls alike [281], and is currently a still developing hot topic. IPv6 packet delivery in Low Power Wireless Personal Area Networks (6LoWPAN) [282] and Static Context Header Compression (SCHC) [283], for example, have undertaken the challenge of enabling IPv6 connections and general header compression concepts on resource-constrained low-power devices. Moreover, shorter packets result in shorter network interface activity as well, which reduces the drain on battery, as predicted by [284].

Although header compression already covers the IP protocol stack, the introduction of new compression profiles would limit its applicability in the future. Finalizing an appropriate RFC can take years and would deter the compression of future protocols – for example, QUIC [285] – from employing compression. With recent state-of-the-art networks, many new protocols and applications will emerge in the coming decade as well. Additionally, compression from the application layer headers' point of view has not been exploited as of yet. However, this can be performed with the methods discussed in this chapter, as a machine learning-enabled compression scheme can theoretically approximate the compression characteristics of any packet-based communication flow. Moreover, the accompanying implementation can interpret such characteristics and execute their compression.

One of the first header compression schemes was Compressed TCP (CTCP) by Van Jacobson [286], and it exclusively concerned itself with the compression of the TCP protocol. CTCP combines TCP and IP headers together for better results and lower complexity and extensively utilizes *delta coding* of the header fields. The main benefit of CTCP is the high compression ratio (for its time). Unfortunately, it is very susceptible to losses on the channel, as it was designed to operate only in wired networks. Consequently, it lacks sufficient countermeasures to avoid or correct decompressor desynchronization, which normally results in the discarding of successfully received compressed packets. Moreover, it lacks any internal error detection scheme and relies on the protection mechanism of other protocols. The first version of RoHC was introduced in [287] and incorporated the concept of extensibility. It had various compression profiles defined in separate Requests-For-Comments (RFCs) designed after its original publication. RoHCv1 itself was incorporated into the 3GPP-UMTS and WiMAX networks. However, the newest header compression standard, version 2 of *Robust Header Compression*, defined in [288], opts for simplicity in design over extensibility. It aims to be more robust under similar network conditions while increasing *compression gain*. Consequently, it gained widespread adoption in the LTE networks next to Robust Header Compression version 1 (RoHCv1).

Although both versions of RoHC achieve around 80–90% in gain (see, e.g., [289]), a lack of knowledge about the structure of the packet data – including any headers – normally necessitates the omission of header compression altogether. The various compression standards rely on fixed structures inside the headers. As this can only be determined at *design time* of the compression algorithm, we would need to construct separate compression schemes to accommodate compression of a particular type of stream. In the world of massive heterogeneity and standardization of the Internet, this is, of course, a quite impregnable hurdle. However, with the advancement of the increasing data processing capabilities of modern hardware, to achieve this goal, we can utilize such emerging concepts as NFV, Artificial Intelligence (AI), and ML.

In this chapter, we employ an *oracle* to determine the structure of packet flows and to enable their future compression. This concept and the accompanying compression scheme is called *Oracle-Structured Stream Compression* or *O2SC* [290]. An *oracle*, or an *oracle machine* in computational theory terms, is an abstract entity that can make decisions of higher complexity. It acts as a black box, which, given a certain input, produces the corresponding output in a single operation. In our case, we delegate the *design time* determination of the packet structure to an *oracle* entity, and we construct a compression scheme that, given an arbitrary set of packets, can query the aforementioned *oracle* and interpret its solution to determine how to compress the given type of stream. This, in turn, enables the compression of packet flows that would otherwise not be compressible at all (basically, any non-UDP/IP, TCP/IP stream) or only partially by current standards (various extensions and/or header combinations do not have standardized compression profiles).

In the following section, we briefly introduce the concept of the *Oracle-Structured Stream Compression*, and in Section 17.3, we detail the usage of the corresponding Python library. After that, in Section 17.4, we give examples for the compression of network headers and real-life packet streams. Finally, Section 17.5 concludes this chapter with a look at the interactive Docker-based environment in the context of the *ComNetsEmu*.

Please note that, the original concept of *O2SC* was initially published in [290] and this chapter closely follows the principles established in that article. The following Section 17.2 therefore reflects a concise version of the discussion in the aforementioned publication.

17.2 The compression oracle

To facilitate the compression of any set of consecutive network packets, we define two entities that operate on opposite sides of a (direct) channel, the compressor and the decompressor. The main goal of the compressor is guaranteeing that the decompressor is synchronized with it and is aware of every change in the compressed flow while making sure that only the bare minimum of data is transmitted. The decompressor, in turn, has to verify each incoming packet and determine whether it can safely update its context without corrupting it for the future recovery and decompression of messages.

The main metric of evaluation throughout this chapter is the *compression gain* or *savings*, which can be expressed as

$$S = 1 - \frac{\|T_{co}\|}{\|T_{uc}\|},\tag{17.1}$$

where $\|T_{co}\|$ and $\|T_{uc}\|$ are the total transmitted bytes during compression and without compression. Our initial evaluations of RoHC employ exactly this formula and express the amount of compression that can be achieved for a given packet. For further details, we refer the interested reader to [289].

Since we utilize the *oracle machine* concept for the determination of the *design time* compression structure, we need to rigorously construct its input and output. First, we specify the available knowledge of the *oracle* as

$$\mathbf{P}_n = \{p_1, \ldots, p_n\},\tag{17.2}$$

where $p_i \in \mathbb{F}_{2^8}^m$, $n, i \in \mathbb{N}$, $1 \leq i \leq n$, $\mathbb{F}_{2^8}^m$ is a *finite field* representing a packet of length m bytes, and $\mathbf{P}_n$, in turn, is a set of n available uncompressed messages (i.e., a history of previous packets). Subsequently, we define the *oracle* as a function

$$f: \quad \mathbb{F}_{2^8}^m \times \mathbf{P}_n \to \mathbb{F}_{2^2}^m. \qquad (17.3)$$

The corresponding output of the function is $\mathbb{F}_{2^2}^m$, which is of the same length as the message and represents a certain compression pattern for a given packet from $\mathbb{F}_{2^8}^m$. The output pattern consists of a series of 2-bit flags signaling whether a given field is *static*, *sequential*, or *random*. These in turn are defined as follows:

Static: These fields have constant or very rarely changing values between packets. Their transmission should generally be performed once per lifetime of the flow. Examples include IP source and destination addresses, UDP source and destination port numbers, and RTP payload types.

Sequential: Fields of this type change in a well-defined manner, which refers to the delta between consecutive packets remaining relatively constant throughout the transmission. Fields of this class can be IPv4 identification, RTP timestamp, RTP sequence number, and so on.

Random: Lastly, the fields of this group do not exhibit any specific pattern and change in an erratic way. They include various checksums or an already compressed logical payload. Note that some could be deduced in the knowledge of the underlying algorithms, for example, CRCs and checksums. However, the current *oracle* design does not consider this.

For the determination of the compression *oracle*, we employ machine learning approaches, specifically, classification models. Theoretically, any classification algorithm could be employed; however, some criteria must always be kept in mind during the evaluation process. Primarily, the prediction should be as fast as possible to avoid delays during the transmission. Secondly, the training of the model should be efficient, both in relation to the time required and the amount of data needed. The former is important if the model is constructed online (parallel to and during a live transmission), and the latter in the cases where only a very small number of packets (i.e., one to ten) are available at the start of the transmission. Lastly, the model should be accurate, since incorrect predictions decrease the efficiency of the compression quite dramatically.

Some of the examples in this chapter use the basic *logistic regression* (see Section 8.2.3 for more detail) as the *oracle*, which in general provides a fairly accurate and fast estimation of the packet structure.

17.3 The O2SC library

The most up-to-date version of the *O2SC* compression is written in Python and can be accessed via the URL found in Listing 17.1. However, we must keep in mind that this is an experimental implementation for the purposes of academic research with updates published on a frequent basis. This also means that there are certain assumptions and restrictions on its usage, which can change over time and are detailed in the accompanying README of the repository.

```
$ git clone https://tomoskozi.visualstudio.com/DefaultCollection/o2sc/_git/
    o2sc/
```

Listing 17.1: Accessing the *O2SC Python* library with git.

In order for the library to function properly, the system PATH should give access to the following packages at least: numpy, pandas, matplotlib, scipy, scapy, scikit-learn, and imbalanced-learn. Since the above list of dependencies may change in the future, the simplest way of making sure that all of the prerequisites are available on the system is to install them via the pip tool and the requirements.txt located in the project root directory, as seen in Listing 17.2.

```
$ pip install -r requirements.txt
```

Listing 17.2: Installation of the *O2SC* Python library dependencies with pip.

For a quicker setup, pip can be also used to install the library directly from the repository, as shown in Listing 17.3.

```
$ pip install git+https://tomoskozi.visualstudio.com/DefaultCollection/o2sc/
    _git/o2sc/
```

Listing 17.3: Installation the *O2SC* Python library with pip.

In the following, we demonstrate the basic features of the library with some simple code examples. The interested reader is invited to experiment further to obtain a hands-on understanding of the structures provided by the code. First of all, we assume an artificial packet sequence shown as a Python array in Listing 17.4.

```
packets = [
    [0xff, 0xff, random.randint(0x00, 0xff), 0x00, 0x00, 0xfd],
    [0xff, 0xff, random.randint(0x00, 0xff), 0x00, 0x00, 0xfe],
    [0xff, 0xff, random.randint(0x00, 0xff), 0x00, 0x00, 0xff],
    [0xff, 0xff, random.randint(0x00, 0xff), 0x00, 0x01, 0x01],
    [0xff, 0xff, random.randint(0x00, 0xff), 0x00, 0x01, 0x02],
    [0xff, 0xff, random.randint(0x00, 0xff), 0x00, 0x01, 0x03]]
```

Listing 17.4: Example of an arbitrary sequence of compressible packets containing *static (columns 1–2,4)*, *random (column 3)*, and *sequential (columns 5–6)* values.

This example only contains six packets for brevity and should be extended with more packets based on the same pattern. Following the definitions seen in Section 17.2, the first two bytes of the packet can be clearly identified as *static*, and the third as *random*, whereas the last two bytes are logically *sequential*, where the most significant bits can be also grouped as *static*.

17.3.1 Examples of predefined oracles

First, we create a basic *oracle* that exploits our observations and assigns every byte to one fixed predefined pattern. This approach is mostly interesting when we have a very thorough theoretical understanding of the given compressible stream and its dynamics. However, this knowledge can be acquired through, for example, a meticulous analysis of the compressible stream, which might not be easily obtained. However, if available, it can be applied to achieve little to none computational overhead for determining the patterns. This, in turn, can be utilized as an efficient strategy for aiding compression in resource-constrained environments, such as the IoT.

Each *oracle* has to derive from the *O2SC* Oracle class, which provides the train and predict methods for implementation. Since in this example there is no learning step, the train method is left empty, and the predict method returns always the same class in a character vector, which is seen in Listing 17.5.

```
class OneClassOracle(oracle.Oracle):
    def train(self, packets=None):
        pass

    def predict(self, packet=None):
        if packet == None:
            raise "No packet specified"
        return '003111'
```

Listing 17.5: One-class *oracle* for the packet stream shown in Listing 17.4.

The library distinguishes between *static* (0x0), *increasing* (0x1), *decreasing* (0x2), and *random* (0x3) fields. The presented pattern classifies the fourth byte (third index) as part of the *increasing* field. However, in reality, this could also be *random* or a simple *static* field, since at this stage, either of these could be true. For such ambiguous fields, choosing a patter could be difficult and is a balancing act in itself. If chosen incorrectly, in some cases the performance penalty can be quite high.

The model can be evaluated on the same packets via the Tester class. This class and the oracle.train() method are wrappers of all the functions related to the training of an *oracle* and the compression of a packet stream formulated as pandas DataFrames. For brevity, we omit a step-by-step explanation about how we transform DataFrames into scikit-learn compatible input. However, examples can be found in the repository. The get_gain() method of the tester returns the calculated *com-*

pression gain based on Eq. (17.1). Listing 17.6 presents how this evaluation can be formulated in Python.

```
# training the oracle on packets
oracle = OneClassOracle()
oracle.train(pd.DataFrame(packets))

# compression of packets via tester
tester = tester.Tester(oracle)

# calculate compression gain
gain = tester.get_gain(pd.DataFrame(packets))
print("Compression gain = " + str(gain))
```

Listing 17.6: Evaluation of the *oracle* from Listing 17.5.

Although this *oracle* can predict the compression pattern quite well, it fails to account for the overflow of the *sequential* field. A more complex *oracle* would provide a new class – flagging the fifth byte as *sequential* instead of *static* – upon encountering this behavior. To simulate how incremental learning can handle this, we now define two separate classes to tackle the overflowing of this multiple-byte-long *sequential* field. This results in two separate classes of patterns as shown in Listing 17.7.

```
class TwoClassOracle(oracle.Oracle):
    prevPacket = None
    def train(self, packets=None):
        pass

    def predict(self, packet=None):
        if packet == None:
            raise "No packet specified"
        pattern = ''
        if self.prevPacket == None or packet[4] == self.prevPacket[4]:
            pattern = '003001'
        else:
            pattern = '003011'
        self.prevPacket = packet
        return pattern
```

Listing 17.7: Two-class *oracle* for the packet stream shown in Listing 17.4.

17.3.2 Defining oracles using machine learning

To avoid the tedious manual definition of a hard-coded *oracle*, we can turn to the machine learning literature to create a classification model around the principles demonstrated before. To do this within the library, we can simply instantiate the `BasicModel` and pass it to the `Tester` as *oracle*. This model performs the preprocessing of the packets and the training of a basic *logistic regression*-based model. Listing 17.8 exemplifies how this is performed in code.

```
# training the oracle on packets
oracle = model.BasicModel()
oracle.train(pd.DataFrame(packets))

# compression of packets via tester
tester = tester.Tester(self.oracle)

# calculate compression gain
gain = self.tester.get_gain(pd.DataFrame(packets))
print("Compression gain = " + str(gain))
```

Listing 17.8: Logistic regression-based *oracle* for the stream shown in Listing 17.4.

A significant benefit of machine learning lies in the wide array of applicable models that can be used to solve a specific problem in many ways. We can easily adapt most of the classification methods from the `scikit-learn` library and use them in tandem with the `BasicModel` class, as demonstrated in Listing 17.9 with the `DecisionTreeClassifier` (see Section 8.2.5 for a discussion of the related theoretical background).

```
# instantiating the scikit-learn classifier
from sklearn.tree import DecisionTreeClassifier
clf = DecisionTreeClassifier()

# training the oracle on packets
oracle = model.BasicModel(clf)
oracle.train(pd.DataFrame(packets))

# compression of packets via tester
tester = tester.Tester(oracle)

# calculate compression gain
gain = tester.get_gain(pd.DataFrame(packets))
print("Compression gain = " + str(gain))
```

Listing 17.9: Decision tree based *oracle* for the packet stream show in Listing 17.4.

17.4 Examples

In this section, we apply the previously discussed methods to actual packet headers, where we simulate applied header compression for the reduction of encapsulation overhead with various real-life streams that were captured on the network interface. For pure header compression, we can utilize the following streams found in the pcaps directory of the project repository:

IP: In the case of IPv4 the header contains every mandatory field, resulting in 20 bytes of header information where the *identification* number always advances by exactly one for each consecutive packet. In case of version 6 of the Internet Protocol, only the base header is considered (i.e., no extension headers such as routing, hop-by-hop options, and so on are present)

UDP: The UDP packet header adds an extra eight bytes of overhead to the IP header. However, it only contains *static* and *random* fields (i.e., UDP *checksum*).

Easy RTP: With the addition of the RTP header, the structure of the packet becomes more complicated. For this evaluation, we only consider a constant size RTP header. This refers to the omission of optional and variable-sized fields (the Contributing Source (CSRC) identifiers and the extension header). In turn, the RTP adds extra 12 bytes of overhead.

Hard RTP: This stream is the same as the *easy RTP*, except that the *marker-bit* sets and unsets after randomly determined time intervals with a uniform distribution between the [5, 25] bounds. The same is true for the *packet type*, which randomly changes after [100, 200] packets and the *timestamp*, which varies with a delta between 150 and 300 every [150, 300] packets.

All these streams are contained in Packet Capture (PCAP) format and can be read in Python with the scapy library. We use scapy to load a stream and call the BasicModel's convert_pcap_to_df() method to transform the PCAP data into a processable format. Listing 17.10 shows how this can be done with the *Hard RTP* stream as a previous example. In case of these header streams, we omit the payload to measure the pure header compression.

```
# training the oracle on packets
oracle = model.BasicModel()

# loading the pcap
packets_train = scapy.rdpcap('pcaps/headers/hardRtpNoCsrcFixedPayl_train_100.
    pcap')
oracle.train(self.oracle.convert_pcap_to_df(packets_train))

# compression of packets via tester
tester = tester.Tester(self.oracle)
pkts_test = scapy.rdpcap('pcaps/headers/hardRtpNoCsrcFixedPayl_test_100_1000.
    pcap')
```

```
gain = self.tester.get_gain(pd.DataFrame(pkts_test), is_array=True)
print("Compression gain = " + str(gain))
```

Listing 17.10: Model building and compression using PCAP files.

Additional examples provided include real-world application network traffic captures. Most of the captured streams contain 14 bytes of Ethernet headers. For the creation of the models, we can take the first 100 packets of each stream as the training datasets and evaluate the compression with the following independent 900 packets.

Franka: This stream was captured on a direct link between a *Franka Emika* robotic arm[1] and its controlling computer. During the recording, the arm performed repeated circle drawing motions. The stream contains a single IPv4 header and a unique (and uninterpretable for Wireshark) payload of 235 bytes.

Asterisk: We connected the *Asterisk* VoIP server[2] to a fixed desktop client and an *Android* smartphone, both using the *ZoIPer VoIP* client software.[3] This configuration utilized the GSM 06.10 codec, which is the full-rate audio codec version, which results in 33 bytes of payload. The RTP *timestamp* and *sequence numbers* have a constant increase (delta) of 160 and one, respectively. The RTP *marker bit* is always zero, except in the first two packets.

Ekiga: This scenario represents a video conferencing session, which was originally generated using the *Ekiga* open source softphone software.[4] The packets contain RTP headers and relatively large payloads of 160 bytes in length.

Radio: A TCP acknowledgment stream of a digital radio station, which, in general, is efficient to compress, as no logical payload is present. This stream is the only one that contains no Ethernet header.

VLC: This scenario represents a high-fidelity audio transmission and was generated using the RTP streaming features of the popular VideoLan Client (VLC) open-source software.[5] This stream capture represents an RTP session again; however, the underlying network protocol in this example is IPv6, whereas the logical payload is 320 bytes.

An example of this can be seen in Listing 17.11.

```
# training the oracle on packets
oracle = model.BasicModel()

# loading the pcap
packets_train = scapy.rdpcap('pcaps/streams/franka_train_50.pcap')
```

[1] See https://www.franka.de/panda/ for details.
[2] See https://www.asterisk.org/ for details.
[3] See http://www.zoiper.com/ for details.
[4] See http://www.ekiga.org/ for details.
[5] See http://www.videolan.org/ for details.

```
oracle.train(self.oracle.convert_pcap_to_df(packets_train))

# compression of packets via tester
tester = tester.Tester(self.oracle)
pkts_test = scapy.rdpcap('pcaps/streams/franka_test_100_1100.pcap')
gain = self.tester.get_gain(pd.DataFrame(pkts_test), is_array=True)
print("Compression gain = " + str(gain))
```

Listing 17.11: Model training and compression based on captured real-life traffic.

In Fig. 17.2, we illustrate the achieved *compression gains* based on Eq. (17.1) when employing the experimental implementation of the *O2SC* compression. In case of the payloadless RTP streams, we can achieve at least 90% *gain*, whereas in comparison the *hard-to-compress* RTP flow ends up with less compression efficiency overall.

For the various captured streams – including any payload that is present – we observe that the highest *gain* is achieved for the *Radio* stream at about 83%, and the lowest is achieved with the *VLC* stream, which is below 20%. However, contrary to the normal header compression scenarios, O2SC has no any knowledge about the separation of network headers and payload, and it also considers the latter part for compression.

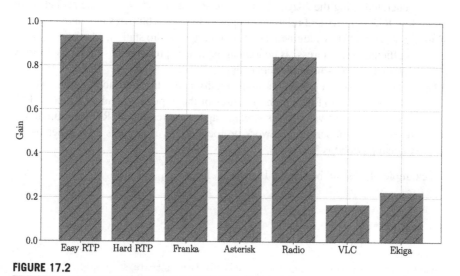

FIGURE 17.2

Compression gain using *O2SC* for various streams with logistic regression.

Illustrating the source of the observed *gains*, we show the uncompressed and compressed packet sizes in Fig. 17.3. We observe that in the case of RTP headers, we gain a minimum of 60 bytes. For the real-life flows, this resembles at least the amount of real headers that are present or, in some cases, even more (see the *Franka* stream).

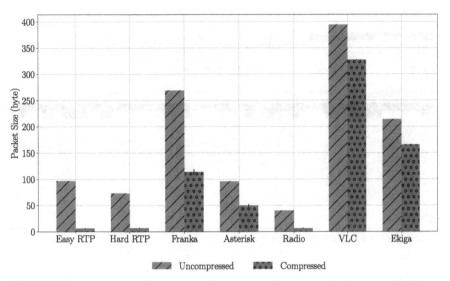

FIGURE 17.3

Average compressed and uncompressed packet sizes (and 95% confidence intervals) using *O2SC* for various streams with logistic regression.

17.5 The interactive environment

The repository also contains an interactive Docker environment, which can be integrated into the *ComNetsEmu* or be used stand-alone. To do this, we must first build the container with the `Dockerfile` found in the `docker` directory of the project root. Then at least one instance of the `o2sc` container can be run as a *compressor* (in dummy interface mode with the `-u` command line argument). The container is built and run with the following command:

```
$ sudo docker build -t o2sc .
$ sudo docker run -it o2sc -Du
```

Listing 17.12: Running the interactive Docker container in `bash` in dummy interface mode.

Since the container has to be executed in the interactive mode with the *compressor* staying in the foreground, the `-i` flag is always mandatory. Once running, the compressor provides a graphical interface for choosing the compression and oracle parameters, as well as the available compressible streams. The compressed stream can be sent, and a rudimentary bar chart will show the *compression gain* for each individual packet, and above it the overall compression statistics can be read as well. Fig. 17.4 shows one of the views of the Docker environment.

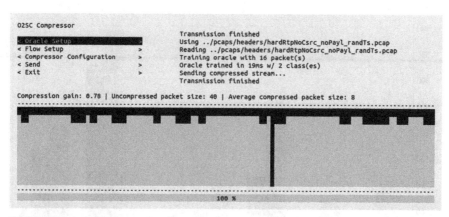

FIGURE 17.4

A screenshot of the interactive environment.

The *Oracle Setup* menu contains options for choosing the classification model (logistic regression, SVC, Bayes, etc.), the training sample size (16, 32, 64, etc.), naïve random oversampling, and online learning (different strategies). The *Flow Setup* contains the available compressible streams. These correspond to those presented in Section 17.4 and more. The *Compressor Configuration* option provides all the exposed header compression-related settings that are relevant to the scenario. The *Send* menu gives options for transmitting the chosen flow uncompressed or compressing it based on the oracle and compressor configuration.

The third option located in the *Send* menu is called *Listen for packets*. In the dummy interface mode, this would compress all the application streams multiplexed randomly together using online learning. However, this feature is more interesting when deployed in a network, such as that virtualized by the *ComNetsEmu* with the topology presented in Fig. 17.5.

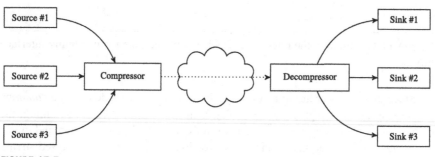

FIGURE 17.5

Simulated network topology of a multiflow compression setup with O2SC and the *ComNetsEmu*.

The *O2SC* repository provides a number of scripts to achieve this easily. Inside the *ComNetsEmu*, when the *O2SC* Docker image is ready, we can execute the *start_cne.py* script to setup a topology similar to the previous one. Once running, all the network nodes provide command line interfaces in separate windows. It is recommended to switch to `bash` right away. To start compression, just run the `node.py` with `python3` by specifying a *source* behavior on the `source_1`, `source_2`, and so on, nodes, and at least a *compressor* on the `compressor` node. The appropriate configuration of the script can be found out by querying its help with the `-h` flag. Once running, the compressor receives uncompressed transmissions with various application streams on the same UDP port, and after demultiplexing, it compresses and transmits them with an online learning *oracle*, which can be configured in the *Oracle Setup* menu.

Machine learning for congestion control

18

Christian Leonard Vielhaus
Technische Universität Dresden, Dresden, Germany

One must fight to get to the top, especially if one starts at the bottom...
Franz Kafka

18.1 Introduction

In the early Internet of the 1980s, ARQ protocols were widely employed. These protocols retransmit lost or damaged packets to ensure data integrity. At the same time, researchers observed poor Quality of Service (QoS) when data flows were routed over specific links within the network. The throughput dropped by three orders of magnitude, and a large percentage of all packets that traversed the links were retransmissions. *What happened?* The links in question were bottlenecks for a large number of flows they carried with bandwidths that were too small to process the incoming traffic. Overwhelmed by the load that took their route, the only choice was to drop packets and hope for the source nodes to slow down or give up. However, all sources that ran ARQ protocols did the opposite, they retransmitted the previously dropped packets without reducing their sending rate. An unwanted global synchronization among flows that shared the same bottleneck was established, a situation that is called congestive collapse [291].

Congestion control is a research area that deals with algorithms that prevent congestion from occurring. Sender-side algorithms are run directly at the source nodes to control the sending rate. They increase or reduce the rate, depending on congestion signals that are returned from the network. These algorithms are an ongoing research topic for more than three decades and play an important role for the stability and performance of networks [292]. We build this chapter upon Chapter 8.4 and demonstrate how RL can be used to learn a sender-side congestion control algorithm that adapts the sending rate.

18.2 Characterizing congestion

Congestion takes place when a node or link faces an incoming flow of data that exceeds its processing capabilities. The choking point that caps the throughput of a connection is referred to as its bottleneck and is the point of interest where congestion occurs. Fig. 18.1 shows the dumbbell topology that introduces a scenario in which a number of source nodes share the same bottleneck link and compete for the available bandwidth. It consists of k source nodes $S = \{s_1, \ldots, s_k\}$ that are connected with their corresponding destinations $D = \{d_1, \ldots, d_k\}$ via two gateways $G = \{g_1, g_2\}$, which function as store-and-forward routers. The nodes are connected by $2k + 1$ links $L = \{l_1, \ldots, l_{2k+1}\}$ with different capacities. A condition that ensures that link l_{k+1} is the bottleneck link for all sources is that its capacity $\text{cap}(l_i)$, measured in packets per second, is smaller than all other link capacities. It is assumed that each source s_i maintains exactly one flow with a sending rate $x_i(t)$ that is routed along the bottleneck to d_i. The condition

$$\sum_{i=1}^{k} x_i(t) \leq \text{cap}(l_{k+1}) \tag{18.1}$$

is the crucial network constraint imposed by the bottleneck link that limits the aggregated throughput that can be achieved by the k sources.

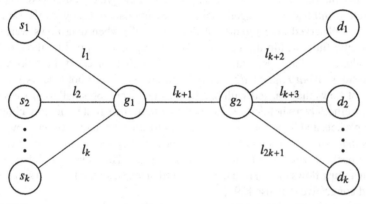

FIGURE 18.1

The dumbbell topology with a single bottleneck link l_{k+1}.

For the dumbbell topology, it is possible to locate the exact point inside the network that has the highest risk to become congested. Intuitively, this is the queue inside the output port of g_1 connected to the Network Interface Control (NIC) that injects packets into l_{k+1}. Fig. 18.2 shows the queueing system that models the output port.

The bundled packets of all k sources arrive with mean rate

$$\lambda = \sum_{i=1}^{k} x_i(t) \tag{18.2}$$

and are either enqueued or blocked. Throughout this chapter, we assume that a tail-drop queue is used, which operates with the First In First Out (FIFO) method and

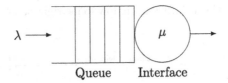

FIGURE 18.2

A queueing system that models an output port of a store-and-forward router.

blocks incoming packets only in case of overflow. The NIC pushes packets into l_{k+1} with rate $\mu = \text{cap}\,(l_{k+1})$, which is equivalent to the bottleneck capacity. As long as $\lambda > \mu$ and the queue still has buffer space, the queue grows over time, and each packet that is enqueued experiences a proportional queueing delay as a consequence. Thus there are two congestion signals that can be exploited to make conclusions about the congestion state of the network: packet drops and delay.

The challenge for congestion control algorithms is that both signals contain uncertainty. Packet losses can be caused by congestion, but also occur stochastically, especially in wireless networks. An increased packet delay can indicate a growth in queueing delay and thereby congestion. As packets often arrive in bursts, however, the added delay may only be temporal and vanish once the traffic peak is processed.

Most algorithms that are used today are loss-based and decrease their rates only if packets are dropped. These schemes interpret every loss as congestion and do not even evaluate Round-Trip Time (RTT) samples in any way. Both Linux and macOS use an algorithm called Cubic as their default setting as of today. Other algorithms are delay-based and adapt their sending rates based on RTT measurements. They react to a growth in queueing delay by reducing their sending rate and try to avoid building up standing queues that never drain entirely. A growth in queueing delay can be observed earlier than packet drops, so that these algorithms have the tendency to decrease their rates before loss-based variants do and are at risk to give up their fair share of the bandwidth.

The uncertainty of congestion signals and the necessity to compete against other flows make congestion control a complex, distributed problem. ML transforms this problem to a pattern recognition task that tries to correlate the most suitable action an RL agent can take given a history of congestion signals.

18.3 Congestion window

The first and foremost countermeasure to avoid congestion is to use ARQ with a sliding window scheme that limits the amount of unacknowledged packets in flight with a variable called the congestion window (cwnd) $w(t)$ [293]. Fig. 18.3 shows a schematic depiction of a sliding window. In this example, $w(t) = 6$, and the source is allowed to send out three additional packets in a burst with sequence numbers 7, 8, and 9 at time t. Afterwards, the source waits for a new Acknowledgement (ACK) of the fourth packet, which shifts the window to the right before being able transmit packet 10.

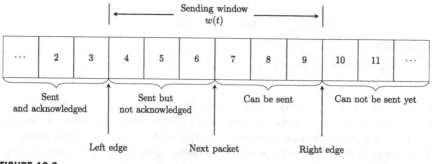

FIGURE 18.3

Sliding window scheme that limits the amount of unacknowledged packets in flight.

In case of congestion the RTT increases, and it takes longer until new ACKs arrive, so the source slows down. If packets are dropped, then its ACKs are not returned, and the window stalls. This is why the source is said to be ACK-clocked in a sliding window scheme, and it is an effective way to control the offered load. The average sending rate can be approximated if it assumed that all $w(t)$ packets are transmitted within one RTT $\tau(t)$ with

$$x(t) = \frac{w(t)}{\tau(t)}. \tag{18.3}$$

This means that instead of changing the sending rate directly, it is possible to control $w(t)$. But what is the target value? If there is no cross-traffic and only one transmission takes place in a network, the source can try to reach a sending rate that equals the bottleneck capacity c_l such that

$$w(t) = c_l \tau(t). \tag{18.4}$$

There is more than one solution to this equation, because a too large $w(t)$ creates a standing queue at the bottleneck, which in turn proportionally inflates $\tau(t)$. However, the smallest possible window that minimizes the RTT and still solves Eq. (18.4) is the optimal operating point, called the Kleinrock point of optimality w^* [294]. It equals the Bandwidth-Delay Product (BDP) of the path $w^* = c_l \tau_p$, where τ_p is the propagation delay of the path without queueing delay. Fig. 18.4 shows a qualitative graph of the sending rate and the RTT as a function of the window size [295].

In Fig. 18.4 the region $0 < w(t) < \text{BDP}$ does not use the entire bandwidth since $x(t) < c_l$. The sending rate can be increased until the Kleinrock operating point is reached with $w(t) = \text{BDP}$ and $x(t) = c_l$ without creating a standing queue at the bottleneck. Once this point is surpassed, additional queueing delay inflates the RTT, and the graph of the sending rate becomes flat. Once $w(t) > \text{BDP} + B_q$, where B_q is

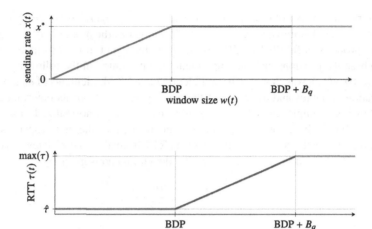

FIGURE 18.4

Different operating regions for the window size $w(t)$.

the size of the queue measured in packets, the standing queue is full on average and consequently has to block some arriving packets.

18.4 Designing the agent

A Reinforcement Learning (RL) agent obtains observations $o_n \in \mathcal{O}$ from its environment, with which it interacts with via actions $a_n \in \mathcal{A}$ it selects according to its policy π_n to maximize the expected cumulative reward given by Eq. (8.44). For congestion control, the agent controls the sending rate by adjusting $w(t)$. The feedback loop is illustrated in Fig. 8.15. When the agent updates $w(n)$ according to the selected action a_n at time step n, the observation and reward that is logically connected to this decision is extracted from ACKs that arrive one RTT later. Therefore the feedback loop of the agent has a delay of one RTT.

At the beginning of a time step, the agent chooses an action a_n and updates its cwnd with an action function $f_a(w(n), a_n)$. The source then transmits $w(n)$ packets in a sending phase that lasts approximately one RTT and records the first and last sequence numbers of packets transmitted during the sending phase of time step n. When ACKs arrive, it is now possible to identify the sending phase they refer to by inspecting the sequence numbers they acknowledge. The nth interval ends when an ACK arrives that acknowledges a packet of the next sending phase.

The feedback information is processed and summarized to an observation o_n, added to a history vector

$$\boldsymbol{h}_n = (o_n, o_{n-1}, \ldots, o_{n-\xi-1}) \tag{18.5}$$

and mapped to the reward $r_n = f_r(o_n; w)$. The triple (a_n, h_n, r_n) is all the agent needs to learn a policy that maximizes its reward. The larger ξ, the more likely it is that the Markov property is fulfilled for RL, as we show the agent more information of the past to base its decisions on. The implementation presented in the following section uses $\xi = 10$, which was shown to be sufficiently large [296]. Elements of o_n are called observation variables and define what the agent perceives from its environment. In the following example section an agent is tested that uses normalized observation variables, that is, these variables are largely independent of the link properties, such as the propagation delay. Let $\bar{\tau}(n)$ be the mean RTT from all gathered latency samples of step n. The first variable is called the RTT deviation and is given by

$$\eta_\tau(n) = \frac{\bar{\tau}(n) - \bar{\tau}(n-1)}{\bar{\tau}(n-1)}, \tag{18.6}$$

which reflects the relative change of the mean latency between two consecutive steps. The second variable, called the estimated queueing delay ratio, is given by

$$r_\tau(n) = \frac{\bar{\tau}(n) - \hat{\tau}}{\hat{\tau}}, \tag{18.7}$$

where $\hat{\tau} = \min_t \tau(t)$ is the smallest RTT sample observed during the entire connection. This variable is an estimator for the normalized queueing delay, and it is intended that the agent learns to avoid large values. The third variable is the loss rate of an observation interval given by

$$\varepsilon(n) = \frac{\text{packets sent} - \text{ACKs received}}{\text{packets sent}}. \tag{18.8}$$

Hence an observation $o = (\eta_\tau, r_\tau, \varepsilon)^T$ forms a mixed signal of the latency signal and packet loss rate but does not include any information about the throughput or window size. The latter is controlled by the agent by taking actions. To avoid sudden changes, the agent can only adjust $w(n)$ by increasing or decreasing it with a step size of 5% by choosing an action a_n from a discrete action space $\mathcal{A} = \{-5, 4, \ldots, 5\}$. In turn, a_n is translated to a $w(n)$ update by the action function

$$w(n+1) = f_a(w(n), a_n) = w(n)(1 + 0.05a_n), \tag{18.9}$$

which limits the window growth or decay to a factor of 1.25 per RTT. The agent chooses actions with the goal to maximize its expected return. The reward function $f_r(o_n; w)$ quantifies the objective to achieve the highest throughput with the lowest possible latency. Let

$$\Gamma(n) = \frac{a(n)}{\Delta t_n} \tag{18.10}$$

be the throughput measured with the number of received ACKs $a(n)$ and the elapsed time Δt_n between step $n - 1$ and n. With the mean RTT $\bar{\tau}(n)$ and the loss rate $\varepsilon(n)$, a weighted linear reward function can be defined as

$$f_r(o_n; \boldsymbol{w}) = w_1 \Gamma(n) + w_2 \bar{\tau} + w_3 \varepsilon. \tag{18.11}$$

For training, normalized weights were used with $\boldsymbol{w} = (4/c_l, -2/\tau_p, -2)^T$, where c_l is the bottleneck capacity, and τ_p is the propagation delay of the path, so that $r_n < 4$ for all rewards independent of the link properties.

A source node begins its transmission with a ramping phase called slow start, which is not controlled by the agent. In slow start, $w(t)$ is doubled every RTT until a packet loss is detected. Then $w(t)$ is halved, and the agent starts controlling all sending window adjustments.

18.5 **Example with ComNetsEmu**

A Python implementation of the RL congestion control algorithm that works on top of UDP can be run with the ComNetsEmu and is provided at the public repository inside the `app` directory as `app/machine_learning_for_congestion_control`. The first example can be run with

```
$ python3 dumbbell.py -a
```

and starts a transmission that lasts 30 seconds. It creates the dumbbell network of Fig. 18.1 for $k = 1$ with the following bidirectional link properties:

```
# bw takes bandwidths in Mb/s, stochastic loss probability = 0
net.addLink(source, g1, bw=50*8, delay='10ms', max_queue_size=1000)
net.addLink(g1, g2, bw=3*8, delay='50ms', max_queue_size=373)
net.addLink(g2, sink, bw=50*8, delay='10ms', max_queue_size=1000)
```

The queue size of the bottleneck matches the BDP of the path with a packet size of 1000 B with $B_q = 3$ MB/s $\cdot$ 140 ms/1000 B $= 420$. After the network is created, it runs `destination.py` at the destination and `source.py` at the source to start a transmission. The sending window of the source is managed in `source.py`, which creates an RL agent that needs the `keras-rl` module as a dependency [297]. Pretrained weights are provided in the file `weights.h5f` and loaded by default. An implementation of a destination that returns ACKs, when it receives data packets, is provided in `destination.py`. Since there are no competing flows in the first example, it is the task of the source to match cwnd to the BDP of the path. Fig. 18.5 illustrates the graph of the cwnd, and Fig. 18.6 illustrates the graph of the mean RTT.

After the initial slow start, cwnd is halved and further reduced by the agent until the RTT reaches values that are close to the propagation delay of the path. Another cycle of growth and reduction follows before $w(t)$ converges to a value of $w(5 \text{ s}) \approx 461$, which is larger than the BDP of the path. The agent accepts an error of

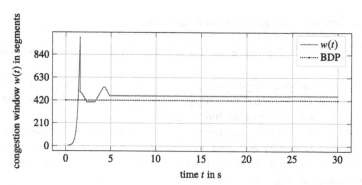

FIGURE 18.5

The congestion window $w(t)$ of one flow inside the dumbbell network with BDP ≈ 420 segments.

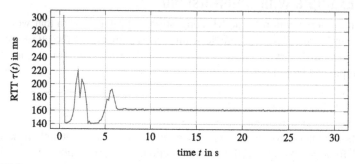

FIGURE 18.6

The RTT signal of the flow in Fig. 18.5. The path has a two-way propagation delay of $\tau_p = 140$ ms.

$e = w(5\,\text{s}) - \text{BDP} \approx 41$, which induces a standing queue at the bottleneck. This adds queueing delay of approximately 20 ms to all packets. This indicates that the agent learns to adjust cwnd until a queueing delay is observed with $w(t) > \text{BDP}$, which allows the conclusion that the sending rate reached the capacity and the path is fully utilized. The error and queueing delay can be reduced with further training.

The second example adds another flow to the dumbbell topology that is also controlled by the agent. When

```
$ python3 dumbbell.py -a -k 2
```

is executed, a dumbbell topology with $k = 2$ is created, and `destination.py` is run on both destinations, as well as `source.py` on both sources. Figs. 18.7 and 18.8 show the graphs of the cwnd and the mean RTT.

The script `source.py` of source s_1 is first run by `dumbbell.py`, which results in an earlier starting time of the flow and a higher value of $w_1(t)$ before slow start is

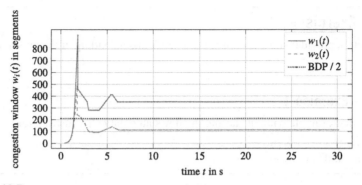

FIGURE 18.7

The congestion windows $w_i(t)$ of two competing flows inside the dumbbell network with BDP ≈ 420 segments.

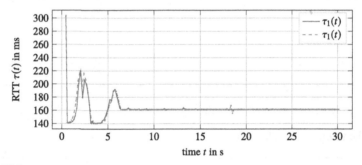

FIGURE 18.8

The RTT signals of two competing flows inside the dumbbell network with a propagation delay $\tau_p = 140$ ms.

exited. This initial advantage leads to an unfair equilibrium after convergence, for which $w_1(t)$ is larger than $w_2(t)$. The flows utilize the entire bandwidth but do not share the available resources equally. For this example, a fair allocation is given by $(w_1, w_2) = (\text{BDP}/2, \text{BDP}/2)$ with $\text{BDP}/2 = 210$ segments. Since both flows do not observe their own throughput and the course of the mean RTT of both sources is almost identical, source s_2 behaves as if it operated on a path with a smaller bottleneck capacity. When an agent is trained with competing flows, it learns a policy that is more aggressive to secure a larger share of the available bandwidth to maximize its reward.

In this chapter, we present a congestion control algorithm based on RL. Its performance has been demonstrated with two examples that can be run with ComNetsEmu. The Python scripts provide a basic implementation of the outlined algorithm on top of UDP, whose strengths and weaknesses have been discussed. The interested students are invited to work on open topics regarding the subject of congestion control, in particular, on solutions that involve ML.

18.6 Exercises

This section provides some exercises for a better understanding of the presented content.

18.6.1 Exercise 1

Run the first example without starting the flows automatically. Instead, run

```
$ python3 dumbbell.py
```

to gain access to a CLI of Mininet. Start two terminals at the source and destination hosts with

```
$ xterm source1
$ xterm sink1
```

and run

```
$ python3 destination.py
```

inside the terminal of sink1 and

```
$ python3 source.py -v
```

inside the terminal of source1. The verbose flag -v lets source.py print out information such as the observed throughput whenever a step ends. Use -h to display all available flags.

18.6.2 Exercise 2

Start a transmission with three sources by running

```
$ python3 dumbbell.py -a -k 3
```

and plot the results afterwards with

```
$ python3 plot_results.py
```

If Secure Shell (SSH) was used to connect to the VM, X11-forwarding must be activated.

18.6.3 Exercise 3

Train your own agent with

```
$ python3 dumbbell.py -t
```

for 50 episodes that last for 10 seconds. Newly generated weight files are created in intervals of 25 episodes, which capture the learning progress. To use your own weights afterwards, run

```
$ python3 dumbbell.py -a -w WEIGHTFILE
```

where WEIGHTFILE must be replaced with a legit file name.

Machine learning for object detection

19

Zuo Xiang[a], Renbing Zhang[a], Patrick Seeling[b]

[a]*Technische Universität Dresden, Dresden, Germany*
[b]*Central Michigan University, Mount Pleasant, MI, United States*

> *In fact, every time one combines and records facts in accordance with established logical processes, the creative aspect of thinking is concerned only with the selection of the data and the process to be employed and the manipulation thereafter is repetitive in nature and hence a fit matter to be relegated to the machine.*
> **Vannevar Bush**

19.1 Introduction

Daily news and reports about AI permeate research and professional boundaries, and by now the general public has become familiar with the term as part of daily life encounters. Among different applications proposed for AI, many have resulted in inspiring achievements in the domain of Computer Vision (CV). Within the environment of CV, one important and direct expression of AI is visual understanding. Visual understanding includes image analysis and machine vision tasks to enable higher-level decisions based on visual information. In turn, this approach has a close relationship to human eyes and the brain's subsequent information processing.

According to the Internet traffic forecast made by Cisco [298], 82% IP traffic will be video by the year 2022. The proportion of traffic that belongs to latency-sensitive video streaming applications can make up a significant portion of this traffic. Understanding the information of objects in video streaming is one popular research direction of CV, which is called real-time video analysis. As one implementation example, real-time object detection and analysis, such as for Google Lens, could be a future common service for everyone. Due to the high complexity of CV-related processing, connecting local processing with cloud services for more computationally expensive processing, for example, with MECs, provides a realistic application scenario.

In intelligent transport systems, pedestrian and car detection will be at the core of many AI applications. In connected autonomous driving an object detection service is helpful for decision-making, such as for braking and obstacle avoidance. In the driver view, for example, object detection services can help the car to protect Vulnerable Road Users (VRUs) such as pedestrians and bicycles, as illustrated in Fig. 19.1.

Computing in Communication Networks. https://doi.org/10.1016/B978-0-12-820488-7.00034-7

(A)   (B)

FIGURE 19.1

Object detection use cases including pedestrians and vehicles detection. (A) Pedestrian dataset [299] detection by YOLOv2; (B) Object detection on the street [300].

As human lives might be directly impacted in some real-time object detection or video analysis scenarios, providing an overall high-quality service is of the utmost importance and requires strong algorithmic and infrastructure support. Considering the fact that real-time video-related services require low-latency and high-level bandwidth, new optimizations for algorithms and networking are both urgently needed.

In recent years, several deep learning-based object detection methods have shown excellent results, including higher precision and higher recall.[1] They exceed most traditional feature extraction and classification-based methods and motivate more research based on deep learning. The development of CV is now focused on creating and improving deep learning-based methods such as CNN to obtain a more powerful ability to detect objects with higher precision in complicated scenes. Methods such as R-CNN [301], Faster R-CNN [302], or You Only Look Once (YOLO) [303] were proposed that achieve high precision while significantly improving the detection speed (also the inference speed). *A deep learning-based model can be abstracted as a function with input and output, namely feeding the image as input and obtaining the results as output.* This model can be trained and deployed for both research and industrial use cases. As hardware becomes more powerful and affordable, training tasks for deep neural network-based object detection methods are more convenient and efficient than before. Whereas the progress of improving the algorithms is at the core of AI-related research, for example, by increasing the precision of object detection, the deployment of resulting AI applications in the cloud is still an open topic.

---

[1] In information retrieval and classification tasks the precision is the fraction of correctly identified instances out of all identified ones, that is, True Positives/(True Positives + False Positives), whereas the recall or sensitivity is the fraction of correctly identified instances over all relevant instances, that is, True Positives/(False Negatives + True Positives).

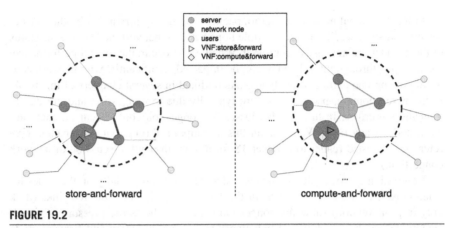

FIGURE 19.2

Structure of traditional and edge computing-based network.

For latency-sensitive object detection services, several performance-related aspects, such as resource usage, low-latency detection, and cost, should be carefully considered. The main challenges of deploying such services for more general use cases are minimizing the latency while retaining a relatively low-level cost and better management of resources, such as memory, CPU, and bandwidth. Based on the current network infrastructure, it is difficult to provide large-scale AI services. Cloud computing is a potentially good solution for large-scale computation and provides flexible resource management for computationally intensive tasks with a better user experience. They can also provide the environment for different kinds of services based on softwarization technologies. However, services that rely solely on a traditional network infrastructure connecting to the cloud-based resources are still facing significant challenges, especially for real-time use cases.

Network softwarization provides the opportunity to flexibly deploy applications such as for real-time video analysis, connected autonomous driving, and intelligent transport systems. Such services could be based on SDN, NFV, and SFC [304] related technologies with low latency to fulfill the requirements of real-time use cases. As the network becomes softwarized, Computing in the Network (COIN) and the MEC are promising avenues for efficiently utilizing the resources on the edge and offloading the workload from centralized servers. Computing in the communication network can significantly reduce the latency caused by current protocols and alleviate related issues, such as congestion between end-to-end nodes. Similarly, virtualization enables the deployment of computation at the edge and close to the user. Better management of VNFs will make it possible to deploy various types of services on any network nodes instead of placing all computational abilities in the center cloud. For example, the edge nodes can act as initial preprocessing units and reduce the latency pressure for the remaining network service chains. Fig. 19.2 shows the structure of the traditional store-and-forward network and the compute-and-forward strategy based on SDN and SFC.

In the traditional network scenario, packets are mostly forwarded by the rules in routers and switches. There are multiple hops between network nodes, and the status of those nodes are not easily monitored. When errors occur, the overall performance, for example, throughput or latency, can be impacted significantly, with network congestion being one particular challenge. As outlined in several chapters of this book, in the wired domain, packet losses are typically due to congestion and not due to transmission errors. In turn, packet losses by congestion could occur on each network node at any time. We maintain this assumption throughout this chapter. (We refer the interested reader to Chapter 18 for a more in-depth discussion of network congestion.)

Congestion can occur in any nodes inside a network, especially on the nodes located closest to the central server. In the left part of Fig. 19.2 the red links (dark gray in print version) show the congestion state, and the traffic pressure is visualized by the width of each line. The network nodes only use store-and-forward mode (diamond in yellow [white in print version] refers to store-and-forward mode being activated). As illustrated, the links closest to the server encounter high levels of traffic pressure, and the possibility of packet losses is higher than that of the links on the network edge. Considering the challenges mentioned before, compute-and-forward can be helpful to ease the pressure on those near-to-server links. Switching to a compute-and-forward networking approach, the workload of the server and the pressure on network links can be reduced, as unnecessary packet transmissions and retransmissions can be avoided through computation in edge nodes.

Fig. 19.3 illustrates an example of the ideas proposed in this chapter from a packet point of view.

In this figure the illustrated packet losses are assumed to originate from congestion, as an underlying wired network is considered. Returning to the store-and-forward example in Fig. 19.3A, when Alice continuously sends packets to Bob, network nodes (relays) handle the packets and forward them to the next hop without modifying the payloads. This strategy generally works well, but packet losses are possible when congestion occurs. Given the amount of data required, this is likely when transmitting video data. When packets are lost, retransmissions are necessary as video decoding without dedicated error handling mechanisms requires all data belonging to a video frame for decoding. Frequent retransmissions result in bandwidth waste and increase the end-to-end latency, which is highly undesirable in real-time scenarios, such as for the detection of VRUs.

If the nodes between Alice and Bob could perform computing tasks (e.g., preprocessing and information extracting), the total amount of output packets of the relays can be reduced, and the congestion problem can be alleviated. In Fig. 19.3B, relay $R1$ is capable of extracting useful information. In turn, the number of output packets is reduced by half when compared to the store-and-forward strategy. The packet losses caused by congestion on the receive queues of relay $R2$ will cease with the same conditions of store-and-forward strategy. In this case, compute-and-forward will bring benefits, such as reducing traffic on the links between end-to-end nodes, improving congestion and lowering the end-to-end latency.

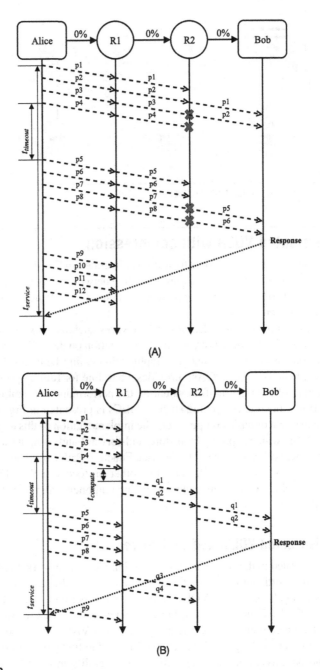

FIGURE 19.3

Alice and Bob example by store-and-forward and compute-and-forward.
(A) Store-and-forward; (B) Compute-and-forward.

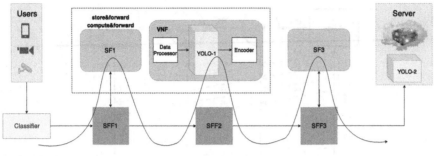

FIGURE 19.4

Overview of the distributed You Only Look Once (YOLO).

19.2 Distributed YOLO with compression

A network structure of compute-and-forward as illustrated in Fig. 19.2 is more likely to improve congestion and reduce the corresponding network latency. Before presenting the proposed distributed YOLO approach, we note two fundamental assumptions for this particular scenario:

i) commonly employed deep learning frameworks such as *Tensorflow* [305] can be deployed in VNFs and the server, and ii) the evaluation on the CPU of a regular PC without additional accelerators, such as Graphics Processing Units (GPUs) or Field Programmable Gate Arrays (FPGAs), provides a baseline for future optimizations.

The structure of the pipeline for evaluating the performance of deploying object detection services in edge computing such as MECs is presented in Fig. 19.4 with a detailed visualization of basic components. The implementation of this example is focused on the VNF, which supports both store-and-forward and compute-and-forward to adapt to the network state. The outer Service Function Path is not modified during computation, that is, the VNF will not affect other protocols or the SFC architecture. In the following, we introduce the two main components and their functions in greater detail.

19.2.1 Distributed YOLO: VNF and server

The VNF is activated in the compute-and-forward mode of a network node for reducing congestion by offloading part of the computation from the central server to the edge. In this example, we apply YOLOv2 for object detection tasks and deploy the model in both the VNF and the server. For convenience, the YOLOv2 model is split into two parts, whereby the first part is deployed in the VNF, and the second part is deployed on the server. In general, the complexity of the first part is lower than that of the second part, assuming that the computational ability in the VNF is less than that of the server. Therefore we split the model into two parts and make the first part the preprocessor for video frames. The detailed model split mechanism is related to the output size of each layer in the YOLOv2 model and will be introduced later.

The components in the VNF, *Data Processor*, *YOLO-1*, and *Encoder*, are packaged in a container, which includes all functions of object detection preprocessing, that is, data processing, information extracting, and feature maps compressing. The *Data Processor* collects the video packets from users and performs preprocessing tasks, such as video decoding, pixel value scaling, or image reshaping (the data format of different model inputs is slightly different). *YOLO-1* is the first part of the detection model (here YOLOv2), and the output of the layers in the CNN are called feature maps. Feature maps can be regarded as the extracted information of the original input image used for detection. The *Encoder* receives the output feature maps and encodes/compresses the data to reduce the bandwidth cost. Implementing an optimal encoder/compressor for feature maps is the core part of this example, for which we will evaluate two different encoding methods.

On the server the full functionality of YOLOv2 is deployed. The computational ability of the server is assumed to be more powerful than that of the network nodes. The server supports both store-and-forward and compute-and-forward strategies by following the instructions in the application header, which is top of the payload and is added by the VNFs. The data arriving at the server is decoded to raw images (in case of no preprocessing in-network) or to feature maps for further inference by YOLO-2. The results are directly sent back to the user as soon as the server-side processing is completed.

19.2.2 Model split

Typical components of CNNs for object detection are convolutional layers, pooling layers, fully connected layers, and batch normalization layers. The most common form of a CNN architecture in CV applications stacks several convolutional layers with proper activation functions, follows them with pooling layers, and repeats this pattern until the image has been merged spatially to a small size (Section 8.2.7). At some point, it is common to transition to fully connected layers. The last fully connected layer holds the output, such as the class scores [306]. Considering that edge nodes are commonly limited in available CPU and memory resources (physical or virtual), the total amount of layers that can be offloaded from the server and deployed in-network is limited. By comparing front layers of different object detection models, such as YOLOv2, SSD, VGG, and Faster R-CNN, the common structures that all have in common are different combinations of convolutional layers followed by pooling layers, as in Table 19.1.

Choosing a proper split point of a model needs to take into consideration that i) the part before the split point should be capable of running on network devices and ii) split point should result in bandwidth savings to improve congestion. Consequently, the number of layers before the split point should not be too high, and to realize bandwidth savings, the output data of the front part should be smaller than the original input image size. In the example of this chapter, YOLOv2 is applied and analyzed for an explanation of model split strategies. As YOLOv2 has structural similarity with other commonly employed feature extractors, the model split approach can easily be transferred to those.

Table 19.1 Structure of the first ten layers in different object detection models.

Model	Structure of first ten layers
YOLOv2	Conv. + Pool. + Conv. + Pool. + 3 Conv. + Pool. + 2 Conv.
SSD	2 Conv. + Pool. + 2 Conv. + Pool. + 3 Conv. + Pool.
VGG16	2 Conv. + Pool. + 2 Conv. + Pool. + 3 Conv. + Pool.
Faster R-CNN	2 Conv. + Pool. + 2 Conv. + Pool. + 3 Conv. + Pool.

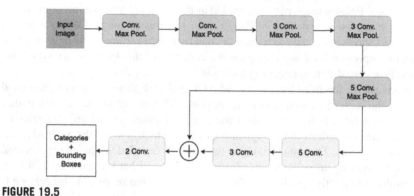

FIGURE 19.5

Model structure of YOLOv2.

19.2.3 Inside YOLO

As highlighted in Table 19.1, YOLOv2 is mainly constructed of convolutional layers and max-pooling layers. Convolutional layers are strong feature extractors in which the convolutional filters are capable of finding features of images. The function of max-pooling layers is reducing the size of feature maps and solve overfitting problems. The design of a model structure is performed through many experiments, which is a complex undertaking out of the scope of this chapter. The overall structure of YOLOv2 is illustrated in Fig. 19.5, noting that model weights and configuration files can be downloaded from the YOLOv2 website [307]. Considering that the computing capabilities of the server are assumed to be significantly increased when compared to those of the network nodes, the split point should be located in the front part of the model. The input image size of YOLOv2 is $1 \times 608 \times 608 \times 3$ when the batch size equals 1. Intuitively, the output shape of the chosen split point should be smaller than the input data shape. Furthermore, the front part cannot be too far into the layers to enable the timely execution as VNF. The output shapes of each layer are calculated as illustrated in Fig. 19.6 to determine a proper split point.

In Fig. 19.6 the x-axis represents the function of all layers in YOLOv2 with layer index, whereas the y-axis provides the output shape of each layer. The red line (dark gray in print version) corresponds to the shape of the input (i.e., $1 \times 608 \times 608 \times 3$). Furthermore, *conv* represents a convolutional layer, and *max* represents a max-pooling layer in Fig. 19.6. The outputs of the culminating layers are very small; for example, the output of *conv_27* is only 8% of the input size. In the front ten layers

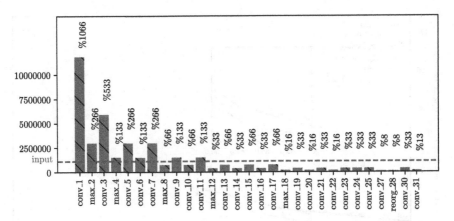

FIGURE 19.6

Output size of each layer in YOLOv2.

the output size of *max_8* and *conv_10* are both 66% of the input size. This can be considered as a split point, because they provide the potential possibility to compress the feature maps (which should result in smaller sizes than the input images). For this example, let the split point be at *max_8*: the layers before this point will be deployed as VNF, and the remaining layers will be deployed on the server.

19.2.4 Feature map compression

In object detection, feature maps refer to the output of convolutional or max-pooling layers; they are intermediate representations of the original image. The input image is processed by several layers, and those layers are constructed by convolutional filters. The attributes of filters are learned during training with CNN. The resulting weights or parameters of convolutional filters determine their attributes and also the feature maps (feature maps are extracted by different filters that represent different features of an image). Although the contents in feature maps depend on the input image, the feature extractors (filters) in all convolutional layers are not modified. Instead, the extracted feature of a filter only depends on its weights. For example, if the function of one filter is emphasizing the edge of objects, then the input content will not affect the ability of this filter to obtain the edge information.

The data types of feature maps are commonly of type float, which can be 16-bits, 32-bits, or 64-bits, as the initial weights of convolutional filters are randomly chosen float numbers defined by the model. During training, the weights will be updated by using optimization algorithms (one example algorithm is the gradient descent where the gradients are real numbers; see Chapter 8 for more details). After each convolution operation, activation functions such as ReLU [308] are applied on the output of the prior operation (which will not change the data type). The data type of an image will be initially preprocessed, for example, in YOLOv2 the image dimensions need to match the required input ranges of the model. This resizing should be noticed, be-

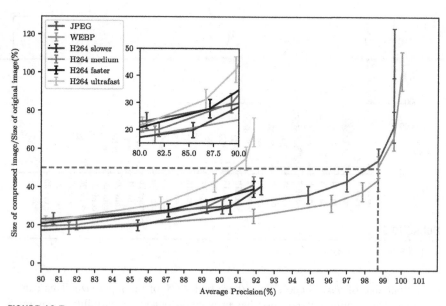

FIGURE 19.7

Image-based compression methods for JPEG input assumption.

cause the required input size of object detection models are not identical. In YOLOv2 the input shape is $1 \times 608 \times 608 \times 3$ when the batch size is equal to 1. In our example the input images are normalized to the range [0–1], that is, the data type of all feature maps will be float. The default setting for the weights data type is 32-bits float, namely FLOAT32.

Our preliminary discussion provides the main direction for the compression of feature maps, which is applying existing image compression and video compression methods to feature maps. Both approaches will be evaluated in the following, including compression ratio and the average precision on the COCO dataset [309].

For compression ratio evaluation, we assume that users send JPEG videos/images to the server over the network, as typically the data from cameras is encoded, for example, through hardware codecs in cameras. Based on this assumption, the compression ratio is defined as

$$R_{\text{jpg}} = \frac{L_{\text{jpeg}}}{l}, \tag{19.1}$$

where L_{jpeg} refers to the number of bytes for the Joint Photographic Experts Group (JPEG) image resized to shape $608 \times 608 \times 3$, and l represents the number of bytes for the compressed feature maps.

Fig. 19.7 illustrates all image-based feature maps compression methods including JPEG, WebP, and H.264 in different modes. JPEG and WebP have higher average precision, which are more than 92% of baseline average precision. H.264 cannot achieve such a high precision, because feature maps are not natural continuous video frames. The compression performance of WebP is better than JPEG and at AP range of 80% to 90% WebP also outperforms H.264 methods. In this range the compression performance of JPEG is a bit worse than H.264 in slower and medium modes and a bit better than H.264 in ultrafast mode. H.264 in ultrafast mode maintains low latency for encoding and causes lower compression ratios. The range under 80% of baseline precisions will not be considered, as it is very dangerous to sacrifice too much prediction quality in favor of higher compression ratios. Based on compression performance and computational complexity, JPEG-based compression is chosen and used in the application example.

19.3 Examples

Several `Python` programs are located in the dedicated example directory with *Com-NetsEmu* [310]:

- `Dockerfile.yolov2`: The Dockerfile to build the container with Distributed YOLOv2 framework installed.
- `topology.py`: The test topology is a simple chain with three nodes connected directly to a single switch: client–vnf–server.
- `preprocessor.py`: The program to send the test image to the remote server for object detection service. It has two running modes: 0: Send the raw image (in bytes) to the remote server without processing; 1: Preprocess the raw image and send the intermediate results to the remote server.
- `server.py`: The program to receive the image data from the client, determine detection results, and send them back to the client.
- `vnf.py`: Program to forward packets between client and server.
- `pedestrain.jpg`: A test image for detection from the pedestrian direction dataset [299].

This example setup can be used to perform a straight end-to-end detection latency comparison between store-and-forward and compute-and-forward when the second hop VNF (running `vnf.py`) is under congestion. Before running tests in each node terminal (Xterm by default), the interested reader is required to perform the initial setup steps by executing the commands in List 19.3. These commands build the YOLOv2 Docker image and start the network with CLI mode. We note that 10 GB disk space are required to build the YOLOv2 image and a minimum 4 GB RAM are required to run all tests smoothly. If the memory space is not sufficient, then the detection program (using Tensorflow) will automatically terminate (be killed by the Out Of Memory (OOM) killer).

```
$ sudo bash ./build_docker_images.sh
$ sudo python3 ./topology.py
```

Five terminals will be created when the network topology is initialized. The following two hands-on examples require executing commands interactively in these terminals. In the following, steps are marked with a *(node name 1, node name 2, ...) description* format. The commands of each step should be executed inside the corresponding terminals.

We note that the produced results are for the included example scenario. However, we encourage the interested reader to modify the different parameters, for example, the topology, service chain, or the detailed characteristics, as they are described in the related chapters of this book.

19.3.1 Infinite forwarding VNF

In this example, the VNF is able to forward an infinite number of packets, for example, as would be the case in unlimited or noncongested networks. The client runs the preprocessor.py initially in raw mode and then in processed mode. In the following example, we have combined the commands in the respective terminals and their produced output and interpretation of results.

```
# 1. (vnf) Run VNF program with default arguments.
$ python ./vnf.py


*** Maximal forwarding number: -1 (-1: infinite)
*** Packet socket is bind, enter forwarding loop

# 2. (server) Run server.py and wait for it to be ready.
$ python ./server.py


... Logs of tensorflow
*** Wait for data from client.

# 3. (client) Run preprocessor.py with raw image mode (mode 0).
$ python ./preprocessor.py 0


*** Processing delay: 0.52 s, receive timeout:14.48 s
*** Get response from server,
response: [{"object": "person", "score": 0.8786484003067017, "position": [164,
    121, 257,416]},
{"object": "person", "score": 0.803264856338501, "position": [145, 138, 185,
    345]}, {"object": "backpack", "score": 0.5143964886665344, "position":
    [223, 185, 246, 280]}]
*** Total time used: 11.43 s

# The client can get the detection result from the server and the total delay
    (including transmission and image processing for detection) is 11.43
    second.
```

```
# 4. (client) Run preprocessor.py with preprocessed mode (mode 1). The server.
     py should run on the server side. Restart it if the program crashes.
$ python ./preprocessor.py 1

*** Processing delay: 1.05 s, receive timeout:13.95 s
*** Get response from server,
response: [{"object": "person", "score": 0.9002497792243958, "position": [165,
     120, 256,416]},
{"object": "person", "score": 0.8104279637336731, "position": [145, 140, 185,
     343]}]
*** Total time used: 5.56 s

# The client can get the detection result from the server and the total delay
     (including transmission and image processing for detection) is 5.56
     second.
```

19.3.2 Limited forwarding VNF

In this example, the VNF can forward a maximum of 200 packets. This number is chosen based on the required number of data packets required to send the image to the server. Specifically, we consider the two modes of i) sending the raw image requiring 235 packets and ii) employing the preprocessing mode requiring only 137 packets to be sent. When the VNF is limited to forward maximal 200 packets, the last 35 packets in raw image mode will be lost, essentially showcasing the impacts of network congestion. The client runs the preprocessor.py initially in raw mode and then in processed mode.

```
# 1. (vnf) Run VNF program with maximal 200 packets.
$ python ./vnf.py --max 200

*** Maximal forwarding number: 200 (-1: infinite)
*** Packet socket is bind, enter forwarding loop

# 2. (server) Run server program on server node like step 2 in Test~\ref{sub:
     vnf_fwd_inf}.

# 3. (client) Run preprocessor program with raw mode like step 3 in Test~\ref{
     sub:vnf_fwd_inf}.
# The VNF program terminates
Reach maximal forwarding number, exits

# Both client and server will trigger timeout
# Output of client's terminal
*** Processing delay: 0.51 s, receive timeout:14.49 s
*** Failed to get response from server.
*** Total time used: 15.02 s
```

```
# Output of server's terminal
Server recv timeout! exist.

# 4. Restart VNF and server program with step 1 and 2.

# 5.(client) Run preprocessor program with preprocessed mode (mode 1).
$ python ./preprocessor.py 1

*** Processing delay: 1.00 s, receive timeout:14.00 s
*** Get response from server,
response: [{"object": "person", "score": 0.9002497792243958, "position": [165,
    120, 256,416]},
{"object": "person", "score": 0.8104279637336731, "position": [145, 140, 185,
    343]})]
*** Total time used: 5.41 s

# The pre-processing can reduce the required number of packets for
    transmission.
# This reduction can avoid potential buffer overflow of VNFs in the middle (
    emulated by maximal forwarding number).
```

Network coding for transport

20

Justus Rischke, Zuo Xiang

Technische Universität Dresden, Dresden, Germany

God does not play dice with the universe.
Albert Einstein

But we do.

20.1 Introduction

The current Internet is dominated by content delivery applications, such as video streaming, which are tolerant to packet loss. Future applications, such as the real-time steering and control of robots, will require a novel networking infrastructure that can reliably deliver data. For example, a robot would stop working if 3–5 contiguous packets were lost.

However, in the current Internet, there can be packet losses of up to 1%, even in the network core [311]. Multimedia applications can tolerate losses due to caching or the use of error concealment mechanisms. On the other hand, many other applications, such as data transfer, are critically affected and thus rely on reliable data transfer services from underlying layers, such as TCP. The trade-off is a higher latency and reduced throughput due to the flow and error control algorithms of TCP, and the TCP throughput already suffers with a loss ratio as low as 1%.

To reduce packet losses in the network, FEC has been proposed in the earlier days of the Internet [312]. Newer techniques have been proposed, for example, NC by Ahlswede et al. [170] and the extension RLNC [313], which tries to achieve more in terms of error correction and low latency. The idea of RLNC is creating new packets, including redundant packets, from linear combinations of the original ones using randomly generated coding coefficients. This makes RLNC applicable for a distributed networking environment, such as the Internet. Furthermore, RLNC allows us to recode, that is, to form newly encoded packets, at an intermediate network node to avoid retransmissions from the source. This reduces latency and losses, as demonstrated later by hands-on examples in Section 20.3. In addition, multiple coding schemes have been proposed, namely CRLNC [314], PACE [315], and systematic coding [186], to overcome the limitations of traditional block coding in terms of

Computing in Communication Networks. https://doi.org/10.1016/B978-0-12-820488-7.00035-9

339

latency and the ability to code on-the-fly. RLNC requires a paradigm shift from conventional *store-and-forward* to *compute-and-forward*; for that, network nodes need the computing capability to code packets. For a more in-depth discussion of network coding, we refer the interested reader to Chapter 9.

Several approaches using SDN have been proposed to bring RLNC into reality [316,317]. SDN-capable network nodes can flexibly redirect an incoming flow of packets to a compute node to perform encoding before forwarding the encoded packets to the destination.

The *NC-as-a-Service* approach as proposed by Szabo et al. [196] allows us to encode TCP and UDP communications without changing existing technologies and thus to improve throughput, reliability, and QoS.

On the other hand, modifications of existing protocols have been proposed to include RLNC, namely for TCP [194] and QUIC [318]. Fig. 20.1 shows the difference between traditional TCP and TCP with network coding (NC-TCP).

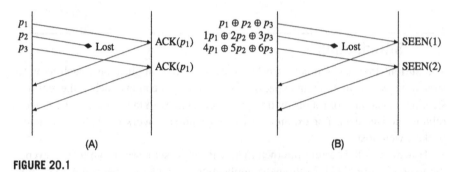

FIGURE 20.1

Difference between (A) traditional TCP and (B) NC-TCP.

Traditional TCP individually acknowledges each packet, that is, if a packet is lost, then the receiver retransmits the acknowledgment of the last packet until the sender repeats the lost packet. As a result, the lost packet ultimately stalls all consecutive packets, which reduces the throughput. NC-TCP, on the other hand, will acknowledge linear independent packets, which can be used to decode the original source packets. In this way, NC-TCP is less susceptible to losses when compared to traditional TCP.

Another application of network coding for transport is multicast. Multicast communication is essential for live video streaming or scalable file delivery.

The problem in multicast communication is the missing ARQ mechanism for repairing losses. FEC with a rateless code, such as RLNC, can improve the resilience without the need for costly feedbacks of the receiver. Fig. 20.2 illustrates the traditional multicast, which sends the individual packets p_1, p_2, in comparison with coded multicast, in which linear combinations $\alpha p_1 \oplus \beta p_2$ of the original source packets are sent instead.

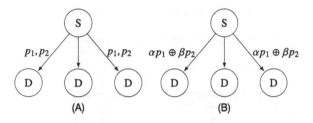

FIGURE 20.2

Difference between (A) traditional multicast and (B) coded multicast.

20.2 Network coding as virtualized network function

The traditional approach of implementing network coding is to directly include it into applications. The result is that NC inclusion has to be done for every application, such as a web browser or a video streaming software. Because the Open Systems Interconnection (OSI) model expects lossless communication at higher layers, the error recovery, which can be done with network coding, should be performed by the network, not the applications. Additionally, abstraction is improved because the coding does not have to be integrated into every application.

As already mentioned, NC can either be directly implemented into the application or, alternatively, as a VNF. The VNF approach allows us to code traffic without modification of existing protocols.

20.2.1 Virtualization approaches

The remaining challenge is the deployment of RLNC as Virtual Network Function (VNF) in the network. Several approaches exist, with varying levels of readiness for real world implementations:

Virtual Machines and Containers: The approach proposed in [317] uses a VM connected to an Open vSwitch. Fig. 20.3 illustrates this principle of general purpose compute nodes hosting the VNF.

The idea is that the VM includes the coding application or any other VNF. The VM is connected to the host machine via a virtual interface with virtual switch, which handles the routing. The benefit of this approach is that it is generic, because it can provide any VNF and can run on any system that can provide a virtual machine. As multiple machines can be spawned using deployment systems like *Vagrant* or *OpenStack*, this approach is also highly flexible. On the downside, the virtual machine and the guest operating system cause an overhead in terms of processing power and delay.

Data Plane Development Kit: As shown by Xiang et al. [26], modern packet processing libraries. such as Data Plane Development Kit (DPDK), are capable

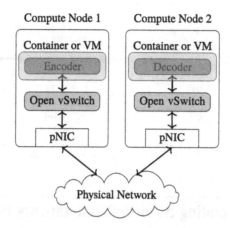

Compute Node 1   Compute Node 2

Container or VM   Container or VM

Encoder   Decoder

Open vSwitch   Open vSwitch

pNIC   pNIC

Physical Network

FIGURE 20.3

Virtualized encoder and decoder on general purpose compute nodes.

of implementing network coding as VNF. This approach helps in reducing the latency, because it avoids the Linux network stack.

Virtual Interface: The third approach is to use a virtual interface, which receives the packets, removes the header, and codes the payload. This method was proposed, for example, in [316] and in [319].

In the scope of this book, we use the first approach, whereby we use Docker containers provided by *ComNetsEmu* as hosts for the coding VNFs.

20.2.2 Coding the traffic

In the following hands-on examples, we consider only the encoding of UDP communications. Specifically, we only consider the payload of the UDP packet. The Ethernet header, the IP header, and the UDP header are stripped from the original packet and stored for later usage. After encoding the payload, coding coefficients and metadata (such as the generation number) are assembled with the original header as a new UDP packet with a coded payload. The structure of the newly generated UDP packet is illustrated in Fig. 20.4. After encoding and assembling, the packet is mirrored back from the coding instance. Recoding is implemented in a similar fashion, whereby the payload is recoded while the headers are stripped. For decoding, the payload of the coded UDP packet is decoded and reattached to the original UDP header.

20.3 Multihop recoding example

As explained in greater detail in Chapter 9, RLNC has the advantage of enabling recoding. However, to effectively recode, the recoder needs to be placed at the right

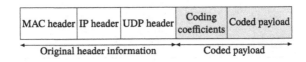

FIGURE 20.4

Coded UDP packet structure.

node. This example evaluates the placement of a recoder in a multihop topology with losses with the *ComNetsEmu* environment.

The example is located in the `comnetsemu/app/network_coding_transport` directory and requires one terminal. The example can be started with the command in Listing 20.1, which consists of a *Mininet* script and starting an additional *Iperf* UDP stream.

```
$ cd comnetsemu/app/network_coding_transport
$ sudo python3 ./multihop_topo.py
```

Listing 20.1: Running the multihop recoding example.

The script creates the topology depicted in Fig. 20.5, consisting of seven nodes. In this example, the first and last nodes are the client and server, respectively. The second and second last nodes are the encoder E and decoder D. The remaining three middle nodes consist of two forwarders F and one recoder R. All connections between the nodes are lossy.

FIGURE 20.5

RLNC recoder in a multihop network. Server and client have been omitted to improve readability.

The example performs an *Iperf* measurement three times, where the recoder is placed each time on a different node. An example log output of the *Iperf* measurement is shown in Listing 20.2.

```
...
------------------------------------------------------------
Server listening on UDP port 9999
Receiving 1470 byte datagrams
UDP buffer size:  208 KByte (default)
------------------------------------------------------------
...
[  3] 26.0-27.0 sec  5.12 Kbits/sec  175.485 ms    3/   19 (16%)
[  3] 27.0-28.0 sec  4.48 Kbits/sec  139.573 ms    3/   17 (18%)
```

```
[  3] 28.0-29.0 sec  3.84 Kbits/sec  195.148 ms    6/   18 (33%)
[  3]  0.0-29.9 sec  3.25 Kbits/sec  210.568 ms  178/  482 (37%)
...
```

Listing 20.2: Terminal 1: Example log output of *Iperf* server (formatted).

The losses measured by *Iperf* vary depending on the placement of the recoder. Theoretically, placing the recoder at the middle node, node 4, gives the best average results in terms of latency and losses. However, to show this, we multiple runs are required.

20.4 Adaptive redundancy example

An advantage of NC-TCP is that feedbacks or acknowledgments can be used to determine the amount of redundancy needed to repair losses. However, with SDN and network coding as VNF, the global knowledge of the SDN controller about the network can be exploited to adapt the redundancy created by the VNFs at runtime. This principle is illustrated in Fig. 20.6.

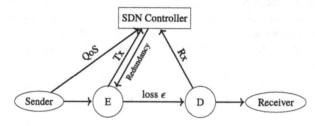

FIGURE 20.6

The SDN controller measures losses in the network and adapts redundancy according to desired QoS as proposed in [319].

The sender communicates its required QoS to the controller. The controller subsequently requests statistics from the compute nodes, which are hosting the coding instances (encoder E and decoder D). These statistics include the number of sent (Tx) and received (Rx) packets. This information is then used to estimate the losses ϵ between sender and receiver with $\epsilon = \frac{\text{Tx-Rx}}{\text{Tx}}$. The SDN controller then calculates the needed redundancy.

20.4.1 Delivery probability of packets

An important factor for adapting the degree of redundancy is the amount of coded packets needed to repair losses. For a block code, where all coding vectors are nonzeros, the decoding probability of a generation of size G can be calculated with the following probability mass function of the binomial distribution:

$$P_{G,\text{Block}}(n) = \sum_{G=0}^{n} \binom{n}{G} p^G (1-p)^{n-G} \qquad (20.1)$$

with $p = 1 - \epsilon$. Here ϵ represents the channel loss ratio, and n represents the number of packets sent per generation. The probability of decoding a single packet is the same as for decoding the whole generation. We need to consider the possibility to receive linearly dependent packets, which cannot be used for decoding. This probability can be neglected if a high field size (e.g. $q = 2^8$) is used.

The drawback of full vector coding is that at least G packets need to be received to decode all the packets. This head-of-line blocking can be avoided if the first original G packets are sent systematically, as proposed in [186]. In systematic full vector coding the generation decoding probability is the same as that of (unsystematic) full vector coding. However, the decoding probability of each packet is very different. This is because the packets received systematically may be delivered even if the entire generation cannot be decoded. The probability of decoding a single packet can be calculated as follows:

$$P_{p,\text{Systematic}}(n) = p + \big((1-p) \cdot P_{G,\text{Block}}(n-1)\big), \qquad (20.2)$$

that is, the probability that a packet is decoded is the sum of two probabilities: i) the probability of the packet being systematically delivered and ii) the probability that the packet is not delivered systematically, but the whole generation still is decoded. Fig. 20.7 depicts this coherence, where the green leaves (light gray in print version) represent the cases where the packets can be decoded, whereas the red leaf (dark gray in print version) depicts the case where the packets cannot be decoded.

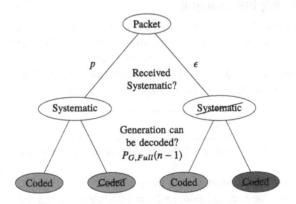

FIGURE 20.7

Tree diagram of decoding probability for packets with systematic coding.

The benefit that can be achieved with this approach is that a certain packet decoding probability can be achieved if enough packets n are sent. This can be used, for instance, if a service requires a certain transmission QoS. The service in this case can demand a high delivery probability. For example, with a delivery probability of 90%,

$$P_{p,\text{Systematic}}(n) \overset{!}{\geq} 0.9 \ . \tag{20.3}$$

This demand is sent to the SDN controller, which in turn calculates the required amount of redundancy with $r = n - G$. The challenge is to find the amount of packets n required to achieve the desired delivery probability. We calculate $P_{p,\text{Systematic}}(n)$ by incrementing n, starting from G, until satisfying condition (20.3). This can be done with the code snippet shown in Listing 20.3.

```
from scipy.stats import binom

def systematic_redundancy(k, p, qos=0.9):
assert type(k) == int
assert type(p) == float and 0 < p <= 1
n = k
while dec_prob_systematic_packet(k, n, p) <= qos:
n += 1
return n

def dec_prob_systematic_packet(k, n, p):
return p + ((1 - p) * binom.sf(k, n - 1, p, loc=1))
```

Listing 20.3: Redundancy calculation algorithm (Python).

20.4.2 Running the example

This example shows how the SDN controller collects statistics, calculates the losses, and adapts the redundancy with aforementioned approach. The example is located in comnetsemu/app/network_coding_transport and requires one terminal. In addition, the example needs additional libraries for statistical calculations. These can be installed with the command in Listing 20.4.

```
$ sudo ./install_dependencies.sh
```

Listing 20.4: Terminal 1: Installation of missing dependencies.

The example can then be started with the command in Listing 20.5, which consists of a *Mininet* script, and, in addition, an *Iperf* UDP stream is started.

```
$ sudo python3 ./adaptive_redundancy.py
```

Listing 20.5: Terminal 1: Starting the *Mininet* topology with *Iperf* UDP stream.

The script automatically starts the SDN controller in an xterm Terminal. An Example log output of the controller is shown in Listing 20.6.

```
...
Link  diff-tx-pkts diff-rx-pkts diff-lost-pkts loss_rate pred_loss
----- ------------ ------------ -------------- --------- ---------
s2-s3         580          411            169     0.291     0.291
pred_loss: 0.291 qos_level: 0.95
SYMBOLS: 10  REDUNDANCY:  7
Setting redundancy to: 7
Create OAM packet
Send OAM packet to encoder
...
```

Listing 20.6: xterm: Example log output of the SDN controller (formatted).

Note that the log outputs are formatted and truncated to improve readability. As shown in Listing 20.6, the controller measures the sent (`diff-tx-pkts`) and receives (`diff-rx-pkts`) packets each second. The controller then calculates the loss rate, and with the algorithm given in Listing 20.3, it calculates the redundancy for the desired delivery probability (`qos_level`). Listing 20.7 shows the log output of the *Iperf* server after finishing.

```
...
------------------------------------------------------------
Server listening on UDP port 9999
Receiving 1470 byte datagrams
UDP buffer size:  208 KByte (default)
------------------------------------------------------------
[  3] local 10.0.0.5 port 9999 connected with 10.0.0.1 port 59767
[ ID] Interval       Bandwidth       Jitter      Lost/Total Datagrams
[  3]  0.0- 1.0 sec   80.3 Kbits/sec  1.258 ms    72/  323 (22%)
[  3]  1.0- 2.0 sec   73.0 Kbits/sec  5.000 ms    92/  320 (29%)
[  3]  2.0- 3.0 sec   74.9 Kbits/sec  4.144 ms    85/  319 (27%)
[  3]  3.0- 4.0 sec   80.6 Kbits/sec  4.822 ms    60/  312 (19%)
[  3]  4.0- 5.0 sec   63.4 Kbits/sec 11.391 ms    18/  216 (8.3%)
[  3]  5.0- 6.0 sec  108 Kbits/sec   11.547 ms    48/  386 (12%)
[  3]  6.0- 7.0 sec   53.1 Kbits/sec 27.513 ms    18/  184 (9.8%)
[  3]  7.0- 8.0 sec  112 Kbits/sec    8.960 ms    18/  368 (4.9%)
...
```

Listing 20.7: Terminal 1: Example log output of *Iperf* server (formatted).

The last column contains the measured loss rate, which immediately stands out as lower than the default 30% losses set in the *Mininet* script. The measured losses decrease even further over time, because the controller averages multiple measurements. The effective losses seen by *Iperf* ϵ_{Iperf} then settles at the desired delivery probability or $\epsilon_{Iperf} \approx 1 - P_{p,\text{Systematic}}$, to be more specific.

The controller provides a REST API to change any of its variables. This can be used for further experimentation, for example, to change the delivery probability on the fly. Example commands are given in Listings 20.8 and 20.9 to get and set variables, respectively.

```
$ curl http://127.0.0.1:8080/simpleswitch/params/QOS_LEVEL
{"QOS_LEVEL": 0.95}
```

Listing 20.8: Usage of controller REST API: Get variables.

```
$ curl -X PUT http://127.0.0.1:8080/simpleswitch/params/QOS_LEVEL -d
'{"QOS_LEVEL": 0.99}'
{"QOS_LEVEL": 0.99}
```

Listing 20.9: Usage of controller REST API: Set variables.

In addition, the example also allows the user to change the loss rate to test different loss-delivery-probability combinations. To do this, the script `adaptive_redundancy.py` must be modified at the location shown in Listing 20.10.

```
$ cat adaptive_redundancy.py
...
# Connect switches
if switch.name == "s4" and last_sw.name == "s3":
        net.addLinkNamedIfce(switch, last_sw, use_htb=True, bw=10, delay="1ms"
            ,
        loss=30)
...
```

Listing 20.10: Usage of controller REST API: Set variables.

20.4.3 Example results

If different loss-delivery-probability combinations are tried, then the following plot shown by Fig. 20.8 can be produced, which shows that a certain delivery probability can be achieved independently of the channel losses. For example, if a delivery probability of 0.9 is desired, then independently of the channel loss ratio ϵ, the *Iperf* stream will see no more than 10% losses. However, a slight deviation is normal, since the number of packets is discrete, and therefore the redundancy can only be adapted with a certain precision.

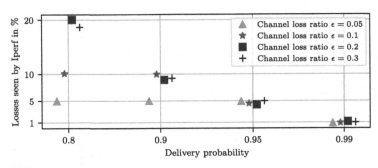

FIGURE 20.8

Losses seen by the application layer (e.g., *Iperf*) under different channel loss ratios and for different desired decoding probabilities.

Network coding for storage

21

Robert-Steve Schmoll
Technische Universität Dresden, Dresden, Germany

Everything not saved will be lost.
Nintendo Title Screen

21.1 Introduction

Ever since the creation of the Gutenberg printing press, wide-spread media creation experienced an explosion in numbers [320]. The advent of digital mass media allows for more users to consume content at an increased rate than with traditional, physically bound media. Its societal impact has been recognized by awarding the 2007 Nobel Prize in Physics for the discovery of the giant magnetoresistance effect, which laid the foundation for modern hard drive technology. Additionally, users of technology increasingly become content creators themselves. This creates completely new business models for storing, distributing, and finding data.

21.2 Distributed storage

The importance of data throughout society leads to several strategies concerning the handling, preservation, and integrity of digital data. Already on the block level of a Hard Disk Drive (HDD), simple measures such as Cyclic Redundancy Check (CRC) are employed to rule out errors while reading or writing bits from the physical medium. Unfortunately, every physical medium has an intrinsic failure probability, that – despite modern manufacturing methods – is still not negligible.

Fig. 21.1 illustrates the general failure probability of devices over their lifetime and the parameters influencing such probability. At the beginning of device usage, the failure rate is mostly determined by physical errors introduced beforehand, for example, through manufacturing faults, temperature fluctuation, or even transport damages. Additionally, the general probability of random failures leads to many devices failing unexpectedly even when they are brand new. After this initial period, considerably lower numbers of devices fail, statistically speaking. Nearing the end of the common lifetime of a physical device, the failure rate again increases due to the wear of the components, which in turn leads to the device becoming out-of-service.

Computing in Communication Networks. https://doi.org/10.1016/B978-0-12-820488-7.00036-0

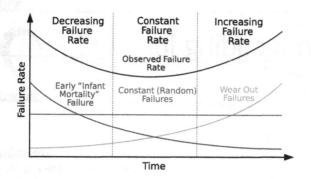

FIGURE 21.1

Bathtub curve of failure probability over life time.

With those processes in mind, the question is not if an HDD will fail, but the point in time of the failure and the strategy to either recover the data or keep the integrity of the data intact. This is where Distributed Storage Systems (DSSs) ensure that data are not lost – even in the event of a hardware failure.

One of such systems is the Redundant Array of Independent Disks (RAID) [321] (originally described with the attribute *inexpensive*). In its simplest form, it duplicates the data over two different storage disks (so-called level 1 RAID), thus allowing failure of either one without affecting the availability of the original files. More advanced schemes, such as RAID level 5, distribute the original data and its parity bits over numerous drives, leading to more effective storage, that is, a higher ratio of usable capacity to redundancy than merely duplicating information. Although this redundancy method is widely used in data centers, one intrinsic drawback is the process of repairing a failed node. In the case of level 1 a new hard drive replacing a failed one needs to be rewritten with the whole amount of the original data of the single remaining HDD, a process that is time-intensive with capacities reaching up to 20 TB, whereas interface speeds top out at approximately 500 MB/s. The introduction of a parity operation, for example, in RAID level 5, suffers similar limitations, but with the added benefit of arbitrary node replaceability. Additionally, a problem arises in the similarity of the remaining source HDDs and their potential failure while performing the recovery.

When DSSs go beyond the number of nodes common for RAID systems, the repair of failing nodes becomes a more complex task to the point where it turns into a research subject [322,323]. Here we have to differentiate between *exact* repair, where the data that were on the failed node are reconstructed, and the *functional* repair, where the new node should fulfill the same ability to preserve the integrity of the whole system as the previous failed node.

Replication exact repair requires finding the same data chunk that went missing when the storage node failed, leading to the overhead of searching the whole storage system. On the other hand, a functional repair can rely on simply copying a new data chunk from the remaining nodes in the system. This saving of overhead for the repair process is likely to cause issues after additional repairs, as the probability of losing unique but essential data chunks increases with each round of a failing node.

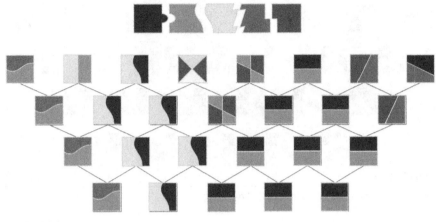

FIGURE 21.2

Distributed storage system with data coded in a simple fashion.

Fig. 21.2 shows an example of a simple coding scheme of five data packets, which results in a DSS of eight nodes. With randomly selecting a new node from the previous set of nodes, it only takes two rounds until the original data cannot be retrieved, since uniquely coded packets got lost in the process. This showcases that in terms of efficiency of a DSS, there is a trade-off between overhead and repair reliability for a replication scheme or even a simple coding scheme.

21.3 Network coding in distributed storage

As discussed in Chapter 9, RLNC has several desirable properties for transporting data through networks. Several of these properties are explained in Section 9.2 and show potential for the use in distributed storage, leading to a promising dual use of the same code for different applications. Being a rateless code, RLNC can create considerably more coded packets from the original data. This allows for more storage nodes without resorting to replicating packets. Moreover, each coded packet is not *unique* to the decoding process, that is, each coded packet can be used to restore the initial packets – in contrast to the example in Fig. 21.2. Additionally, employing coded packets on different storage nodes introduces an element of privacy, as access to a single storage node does not reveal parts of the original data, in contrast to the mere replication of data packets [324].

Fig. 21.3 illustrates a similar scenario to that described in Section 21.2, but with packets coded in an RLNC-like scheme. Here we can see that the random choice of nodes from one round to the next does not impair the ability of the system to

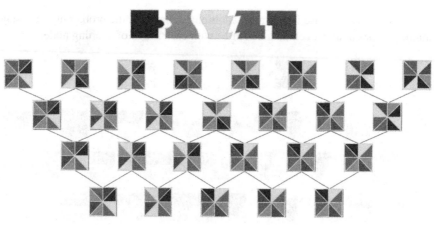

FIGURE 21.3

Distributed storage system with RLNC-like coding.

recover original data – even if only five coded packets are left. The ability of recoding with an insufficient rank grants RLNC an advantage over similar codes such as Reed-Solomon (RS), which display similar properties as previously mentioned [325]. However, where RS needs to retrieve a certain number of coded packets to create more coded packets, RLNC can do so with less transmitted coded packets, albeit with the possibility of not achieving full rank.

21.4 Running the example

The following example program is based on the research by Fitzek et al. [326]. It analyzes the efficiency of a DSS while employing different repair strategies with a focus on reliable repair. Three approaches are to be compared: i) uncoded packets, ii) coded packets without recoding, and iii) coded packets with recoding.

To provide a quick overview over the simulation scenario, we highlight the overall steps in the following. A simulator is started and uses a DSS with C nodes, each of which contains Q different packets of the encoded data. The encoding uses generation size G, resulting in G packets being required for successful decoding as a minimum. From this initial situation the scenario starts to iterate several rounds; at the beginning of each round, L storages get cleared of any data. Afterwards, the repair is started by randomly selecting $P \leq C - L$ (for *parent*) nodes and recoding over all $P \times Q$ packets. A decoder then creates $L \times Q$ new encodings for the previously erased nodes. At the end of each round the simulator confirms the system integrity through decoding over all C storage nodes. If decoding is successful, then the iteration continues with the next round. The number of consecutive rounds until decoding fails is an indicator for the resilience of the whole storage system.

The reference publication and this example use the values provided in Table 21.1 as configuration parameters.

Table 21.1 Values for the parameters in the distributed storage simulation.

Number of storage nodes C	Values
Generation size G	15
Number of lost nodes per round L	15
Number of packets Q per node	2 … 9
Number of parents P for repair	1 … 9

Listing 21.1 shows the help page output of the simulator located in the ComNet-sEmu example directory. For ease of use, parameter symbols were chosen similar to those displayed in the previous paragraph. The simulator allows the reader to run a reliability estimation for a specific repair strategy with a certain number of packets per storage and parents for repair. Here the number of parents are a measure for the cost of transfer, and the amount of packets per node defines the storage costs for a DSS.

```
usage: rlnc_storage_example.py [-h] [-p PARENTS] [-q PACKETS] [-r ROUNDS]
                               [-m {rlnc,uncoded,rs}] [-s]

Parameters for RLNC-DS-Simulation

optional arguments:
  -h, --help            show this help message and exit
  -p PARENTS, --parents PARENTS
                        Packets per storage node
  -q PACKETS, --packets PACKETS
                        Number of parents for reparation
  -r ROUNDS, --rounds ROUNDS
                        Number of deletion-and-repair rounds
  -m {rlnc,uncoded,rs}, --mode {rlnc,uncoded,rs}
                        Coding mode
  -s, --results         Show number of rounds for each iteration
```

Listing 21.1: Output of help function of example program.

21.4.1 Uncoded repair

In the case of uncoded repair, packets are generated with a systematic encoder, that is, symbols are created in clear text with only an overhead header telling the coders about systematic encoding. This leads to merely splitting the original data in $G = 15$ chunks, meaning that with only one packet per node, one failure is enough to break the system. Listing 21.2 shows this example.

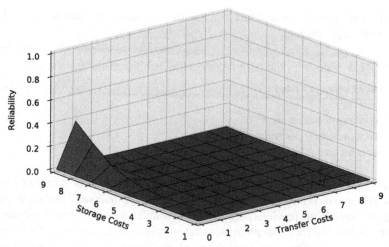

FIGURE 21.4

Simulation of 100 rounds with repair through replicating uncoded packets.

```
$ ./rlnc_storage_example.py -r 1 -p 3 -q 1
Success probability for reaching 1 rounds: 0 %
```

Listing 21.2: Output of simulator with one packet per node.

With each repair, the system chooses P parents, each of which randomly selects enough packets until Q packets are copied to the failed node. This repair scheme is not very robust, since unhelpful redundancy can easily occur. Fig. 21.4 illustrates that for 100 deletion rounds, no combination of P and Q can reliably keep the DSS integrity. This showcases that replication is not sufficient for many consecutive failed and repaired nodes.

Were a lower number of rounds considered, the simulator would show that the number of parents, that is, the cost of transport, is actually irrelevant for the reliability of the DSS. Since there is no advantage in having more packets available for the repair, due to the choice being random, the simulator can verify this behavior. Listing 21.3 shows example outputs.[1]

```
$ ./rlnc_storage_example.py -r 10 -q4 -m uncoded -p3
Success probability for reaching 10 rounds: 78 %
$ ./rlnc_storage_example.py -r 10 -q4 -m uncoded -p4
```

---

[1] Variations in values for the success probability stem from the lower number of iterations for the simulator to keep low computation times in virtual environments. The interested readers can dive into the source code to change values to attain statistical sound results.

```
Success probability for reaching 10 rounds: 79 %
$ ./rlnc_storage_example.py -r 10 -q4 -m uncoded -p5
Success probability for reaching 10 rounds: 80 %
$ ./rlnc_storage_example.py -r 10 -q4 -m uncoded -p6
Success probability for reaching 10 rounds: 81 %
```

Listing 21.3: Output of several simulator runs while increasing the number of parents for the repair process.

In the uncoded repair scenario the amount of packets per storage node is a far more important parameter, directly increasing the resilience of the system with respect to failures. A few more simulator runs, exemplary outputs shown in Listing 21.4,[1] allow the reader to test this process.

```
$ ./rlnc_storage_example.py -r 10 -p4 -m uncoded -q2
Success probability for reaching 10 rounds: 5 %
$ ./rlnc_storage_example.py -r 10 -p4 -m uncoded -q3
Success probability for reaching 10 rounds: 53 %
$ ./rlnc_storage_example.py -r 10 -p4 -m uncoded -q4
Success probability for reaching 10 rounds: 77 %
$ ./rlnc_storage_example.py -r 10 -p4 -m uncoded -q5
Success probability for reaching 10 rounds: 97 %
```

Listing 21.4: Output of several simulator runs while increasing the number of packets per storage node.

This scenario shows the inefficiency both in terms of reliability and in terms of repair of a distribution scheme that relies on replication. It can not only survive several deletion rounds with higher storage costs, but it also does not scale with increasing transfer costs, which could help mitigate limits in storage. As Fig. 21.4 shows, 100 deletions rounds are infeasible with replication of uncoded packets.

21.4.2 Simple network code with replication

This section evaluates how applying a rateless code to the original data can improve a DSS. Here the DSS encodes original data into packets with a rateless code, creating $C \times Q$ packets in total. In each deletion round, the system chooses randomly from $P \times Q$ packets and copies them to the previously failed node. With the added benefit of coded packets, any combination of them can recreate the original data. Fig. 21.5 shows the reliability of the system for different parameters and, when compared to the uncoded scenario, the advantage in reparability. The figure also indicates a difference in dependency, that is, the influence of the number of parents available for repair on the reliability.

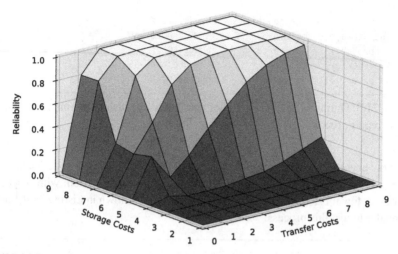

FIGURE 21.5

Simulation of 100 rounds with repair through copying coded packets.

21.4.3 Network coding with recoding

Here the ability of RLNC to recode is used as a mechanism to repair a previously deleted node. Due to the attempt at optimization of storage and transfer costs, two different strategies are possible [326]. The *prerecoding* approach would start the process at each parent node, recoding over the Q available packets and relaying the minimum amount of packets to the new node to be repaired. This approach leads to the minimum amount of packets transferred. For a more reliable repair, a *postrecoding* approach can be applied. Here all parents send all their packets to the new node, leading to a transfer of $T = P \times Q$ packets. Subsequently, the new node recodes over the available T packets and creates Q recoded packets. For simplicity, in this section, we only discuss the postrecoding approach. As Fig. 21.6 illustrates, this strategy is far more efficient in keeping the DSS reliable while minimizing the amount of traffic and storage.

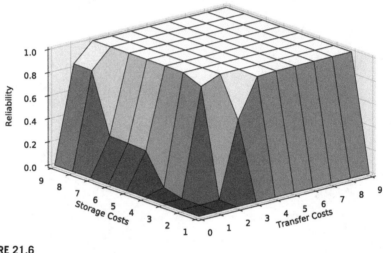

FIGURE 21.6

Simulation of 100 rounds with repair through recoding.

In-network compressed sensing

22

Maroua Taghouti, Malte Höweler
Technische Universität Dresden, Dresden, Germany

I like density, not volume.
Maureen Howard

22.1 Introduction

Compressed sensing has emerged as one of the key technologies providing a solution to handling massive amounts of data in the context of next generation networks. It is capable of exploiting the inherent sparsity in signals to achieve a lossy compression that reduces the required bandwidth usage of communication networks to a fraction of the original transmission requirements.

This sparsity, in practice, can be due to the deployment of wireless sensor network setups where the nodes are located close to each other, as in the case of industrial monitoring systems. Since these devices are in the same physical environment, the readout of the sensors at a given time produce similar values (e.g., room temperature varies only slightly in neighboring areas). Consequently, sampling of the generated data will show similar tendencies (i.e., only gradual temperature changes).

In statistical terms, this results in spacial and temporal correlation in the observed data, which can be exploited by compressed sensing methods to reduce the amount of data during transmission. The data, in turn, can be reconstructed accurately on the receiver side. For the theoretical details on compressed sensing, we refer to Chapter 10.

In this chapter, we introduce the reader to the compressed sensing toolset, which can be employed in the ComNetsEmu framework employed throughout this book. To access the compressed sensing features, the `python-numpy` and `python-sklearn` dependencies have to be available inside the emulator. In the following, we assume that the reader is already familiar with how ComNetsEmu is used and that the environment is already running. Inside the `compressed_sensing` directory of the practical examples, all the required resources to run compressed sensing on the dataset provided by Intel lab (see below) are available. First, we have to generate the compressed sensing input and build the `Docker` container as in Listing 22.1.

Computing in Communication Networks. https://doi.org/10.1016/B978-0-12-820488-7.00037-2

```
$ python3 generate-seed.py
$ sudo bash build_docker.sh -a
```

Listing 22.1: Setup of the compressed sensing scenario.

Note that the execution of these two commands have to be repeated each time the dataset is modified or in case we want to change the randomization of the dataset. In the following two sections, we show how to run compressed sensing in point-to-point and cluster-based scenarios.

As for the reconstruction algorithm, we opted for using one of the most complete and stable implementations of the OMP algorithms, which is intended for general machine learning in Python [327]. Other implementations of various compressed sensing algorithms can also be found, for example, in the KL1p library [328].

As in the other practical chapters of this book, the simulations are performed using Docker containers. In this particular exercise, we simply consider a sensor node as an independent Docker container, which has high computational capabilities to perform compression independently of the adopted topology. This is mainly for harmonization with the rest of the discussed topics found in this book, even though – as it will be seen later on – using specific networks or Docker containers is not necessary for the employment of compressed sensing.

We employ the entire data set of 54 sensors provided by Intel Lab [329] for the input used in this example. The sensor measurements took place in the Intel Berkeley Research Lab in 2004 and made publicly available with topology information together with humidity, light, voltage, and temperature values, which were sampled every 31 seconds (including timestamps). Without loss of generality, we tested our framework on the temperature readings. Under the `sensors` directory, we can find 54 CSV files comprised of one column of the temperature readings in Celsius. Note that to the best of our knowledge, this is a solid and reliable library, which can be employed for reproducibility, despite the fact that some of its data are missing or truncated (according to the Intel Lab Data webpage).

22.2 Point-to-point scenario

In this section, we show an elementary example of a point-to-point topology illustrated in Fig. 22.1. We assume that a perfect connection is established and no losses occur. Since the temperature readings are not sparse, a sparsification step should be inserted before the compression of the readings becomes possible. In Chapter 10, we thoroughly explained the possible mechanisms for sparsification: either i) by projecting the readings into a different potential basis or frame or ii) by training a dictionary using a large set of readings. As commonly found in the literature, sensor data have been mainly related to the DCT transform and trained dictionaries. We therefore chose these two methods as examples to sparsify the sensor readings.

FIGURE 22.1

Example of a point-to-point topology where only one sensor is transmitting its compressed readings.

22.2.1 Using DCT for data sparsification

For this example, initially, the reader has to execute the command shown in Listing 22.2.

```
$ sudo python3 topo.py 1 --dct 1
```

Listing 22.2: Setup of the compressed sensing scenario using the DCT transform.

The output of this command should be:

```
*** Adding controller
*** Adding switch
*** Adding dockerhost and links
head_host: kwargs {'ip': '10.0.0.21', 'mac': '00:00:00:00:00:01', 'cpu_quota':
    100000, 'cpuset_cpus': '0', 'volumes': ['/var/run/docker.sock:/var/run/
    docker.sock:rw']}
head_host: update resources {'cpu_quota': 100000, 'cpu_period': 100000, '
    cpuset_cpus': '0'}
(10.00Mbit 1ms delay 0.00000% loss) (10.00Mbit 1ms delay 0.00000% loss)
    node_host1: kwargs {'ip': '10.0.0.1', 'mac': '00:00:00:00:00:02', '
    volumes': ['/var/run/docker.sock:/var/run/docker.sock:rw']}
node_host1: update resources {'cpu_quota': -1, 'cpu_period': 100000}
(10.00Mbit 1ms delay 0.00000% loss) (10.00Mbit 1ms delay 0.00000% loss)
*** Starting network
*** Configuring hosts
head_host node_host1
*** Starting controller
c0
*** Starting 1 switches
s1 (10.00Mbit 1ms delay 0.00000% loss) (10.00Mbit 1ms delay 0.00000% loss)
    ...(10.00Mbit 1ms delay 0.00000% loss) (10.00Mbit 1ms delay 0.00000% loss
    )
*** Ping: testing ping reachability
head_host -> node_host1
```

```
node_host1 -> head_host
*** Results: 0% dropped (2/2 received)
*** Adding docker in docker
*** Wait for head to finish

Head: Setup socket on ('10.0.0.21', 8003)

Head: Now connected to:  ('10.0.0.1', 46886)

Head: Closing connection to:  ('10.0.0.1', 46886)

*** Head has finished
Head: Shape received data buffer:  (1, 20, 1)
Head: Decompression with DCT
Head: Shape reconstructed data buffer:  (1, 100, 1)
Head: MSE for Sensor1 : 2.4971498629114364
```

Listing 22.3: Details of the output from Listing 22.2.

Under multiple measurements vectors with compressed sensing, the sensor compresses a collection of $S = 100$ samples, each containing $n = 80$ temperature readings in $m = 40$ measurements – that is, the volume of the data is reduced to half – and the result is in turn sent to the sink. The sink is expected to be already aware of the type of sparsification the original data has undergone, which in this case is a DCT transformation. The sink in return applies the OMP algorithm to obtain an approximation of the original data. To evaluate the accuracy of the reconstructed data at the sink, we employ MSE scores, as they provide a good approximation of the reconstruction error. This scenario guarantees a reconstruction with MSE 2.497, which is not considered optimal for a compression ratio of 50%.

22.2.2 Using a trained dictionary for data sparsification

The difference between this example and the previous DCT-based one is the sparsification method, which relies on a trained overcomplete dictionary obtained using a large set of the sensor readings in advance. The framework provides a simple implementation of the K-SVD algorithm using the OMP method provided by the Scikit-learn API. For this second example, the reader needs to execute the command shown in Listing 22.4.

```
$ sudo python3 topo.py 1
```

Listing 22.4: Setup of the compressed sensing scenario using overcomplete dictionary learning.

The relevant part of the output from the aforementioned command is as follows.

```
*** Head has finished
*** Node1 logs
Node1: Connected to:  ('10.0.0.21', 8003)
Node1: Compression with dictionary
Node1: Shape original data:  (100, 80)
Node1: Shape sparse data:  (200, 100)
Node1: Shape compressed data:  (100, 40)
Node1: MSE:  0.1829645146171032
Node1: closed connection to:  ('10.0.0.21', 8003)
```

Listing 22.5: Details of the output from Listing 22.4.

The original data is reconstructed with an MSE of 0.182, which is considered minimal compared to the previous example using the DCT. This result is expected, because the *dictionaries are generated for specific data types* and are not as standardized as the DCT.

22.3 Single-cluster scenario

In this section, we use the cluster-based topology as it is a rather typical for wireless sensor network deployments. We repeat the same variation of scenarios performed in the previous section using six sensors in total for the single-cluster as illustrated in Fig. 22.2. Note that this topology can take up to 16 docker containers, that is, sensors.

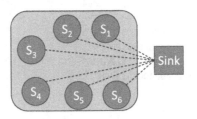

FIGURE 22.2

Example of a single-cluster topology consisting of six sensor nodes that transmit their compressed readings to a common sink.

22.3.1 Using DCT for data sparsification

This scenario can be generated using the command found in the following listing:

```
$ sudo python3 topo.py 6 --dct 1
```

Listing 22.6: Setup of the compressed sensing scenario using overcomplete dictionary learning.

In turn, the following listing shows the connections established between the containers and the sink:

```
Head: Setup socket on ('10.0.0.21', 8003)
Head: Now connected to:  ('10.0.0.1', 51770)
Head: Now connected to:  ('10.0.0.2', 38404)
Head: Now connected to:  ('10.0.0.3', 35576)
Head: Now connected to:  ('10.0.0.4', 33658)
Head: Now connected to:  ('10.0.0.5', 60990)
Head: Now connected to:  ('10.0.0.6', 40766)
```

Listing 22.7: Details of the output from Listing 22.6.

We further give the extension of the output from Listing 22.6. Similarly to the compressed sensing setup in the previous example, the size of the original data in its multivector shape is 100×80, and that of the compressed data or measurement is 100×40. The output shows the MSE of the reconstructed data, which are received from all six sensors.

```
Head: Shape received data buffer:  (6, 20, 1)
Head: Decompression with DCT

Head: Shape reconstructed data buffer:  (6, 100, 1)
Head: MSE for Sensor1 : 2.4971498629114364
Head: MSE for Sensor2 : 3.205683707754018
Head: MSE for Sensor3 : 1.4509731095228484
Head: MSE for Sensor4 : 2.5766685412602643
Head: MSE for Sensor5 : 3.3931169852261265
Head: MSE for Sensor6 : 2.497400591976876

*** Head has finished
*** Node1 logs
Node1: Connected to:  ('10.0.0.21', 8003)
Node1: Compression with dct
Node1: Shape original data:  (100,)
Node1: Shape compressed data:  (40,)
Node1: closed connection to:  ('10.0.0.21', 8003)

*** Node2 logs
Node2: Connected to:  ('10.0.0.21', 8003)
Node2: Compression with dct
```

```
Node2: Shape original data:  (100,)
Node2: Shape compressed data:  (40,)
Node2: closed connection to:  ('10.0.0.21', 8003)

*** Node3 logs
Node3: Connected to:  ('10.0.0.21', 8003)
Node3: Compression with dct
Node3: Shape original data:  (100,)
Node3: Shape compressed data:  (40,)
Node3: closed connection to:  ('10.0.0.21', 8003)

*** Node4 logs
Node4: Connected to:  ('10.0.0.21', 8003)
Node4: Compression with dct
Node4: Shape original data:  (100,)
Node4: Shape compressed data:  (40,)
Node4: closed connection to:  ('10.0.0.21', 8003)

*** Node5 logs
Node5: Connected to:  ('10.0.0.21', 8003)
Node5: Compression with dct
Node5: Shape original data:  (100,)
Node5: Shape compressed data:  (40,)
Node5: closed connection to:  ('10.0.0.21', 8003)

*** Node6 logs
Node6: Connected to:  ('10.0.0.21', 8003)
Node6: Compression with dct
Node6: Shape original data:  (100,)
Node6: Shape compressed data:  (40,)
Node6: closed connection to:  ('10.0.0.21', 8003)
```

Listing 22.8: Details of the output from Listing 22.6.

Note that despite all temperature readings being similar at the sensors – due to their geographical placement – and sharing sensing times, the reconstruction shows a different MSE for each sensor. This is a result of several potential errors within the data sets.

22.3.2 Using a trained dictionary for data sparsification

Now we turn to data sparsification with dictionaries. This scenario can be generated using the command

```
$ sudo python3 topo.py 6
```

Listing 22.9: Setup of the compressed sensing scenario using overcomplete dictionary learning.

We omit the presentation of the established connections between each container and the sink, as they are identical to that of the DCT case. The following listing shows the MSE of the reconstructed data for each of the nodes:

```
Head: Shape received data buffer:  (6, 100, 40)
Head: Decompression with dictionary

Head: Shape reconstructed data buffer:  (6, 100, 80)

Head: MSE for Sensor1 : 0.18296451461710317
Head: MSE for Sensor2 : 0.5187587073441643
Head: MSE for Sensor3 : 0.17176943538612072
Head: MSE for Sensor4 : 0.18735776249882208
Head: MSE for Sensor5 : 0.3512422779473302
Head: MSE for Sensor6 : 0.16979520531023373

*** Head has finished
*** Node1 logs
Node1: Connected to:  ('10.0.0.21', 8003)
Node1: Compression with dictionary
Node1: Shape original data:  (100, 80)
Node1: Shape sparse data:  (200, 100)
Node1: Shape compressed data:  (100, 40)
Node1: MSE:  0.1829645146171032
Node1: closed connection to:  ('10.0.0.21', 8003)

*** Node2 logs
Node2: Connected to:  ('10.0.0.21', 8003)
Node2: Compression with dictionary
Node2: Shape original data:  (100, 80)
Node2: Shape sparse data:  (200, 100)
Node2: Shape compressed data:  (100, 40)
Node2: MSE:  0.5187587073441643
Node2: closed connection to:  ('10.0.0.21', 8003)

*** Node3 logs
Node3: Connected to:  ('10.0.0.21', 8003)
Node3: Compression with dictionary
Node3: Shape original data:  (100, 80)
Node3: Shape sparse data:  (200, 100)
Node3: Shape compressed data:  (100, 40)
```

```
Node3: MSE:  0.17176943538612072
Node3: closed connection to:  ('10.0.0.21', 8003)

*** Node4 logs
Node4: Connected to:  ('10.0.0.21', 8003)
Node4: Compression with dictionary
Node4: Shape original data:  (100, 80)
Node4: Shape sparse data:  (200, 100)
Node4: Shape compressed data:  (100, 40)
Node4: MSE:  0.18735776249882208
Node4: closed connection to:  ('10.0.0.21', 8003)

*** Node5 logs
Node5: Connected to:  ('10.0.0.21', 8003)
Node5: Compression with dictionary
Node5: Shape original data:  (100, 80)
Node5: Shape sparse data:  (200, 100)
Node5: Shape compressed data:  (100, 40)
Node5: MSE:  0.3512422779473302
Node5: closed connection to:  ('10.0.0.21', 8003)

*** Node6 logs
Node6: Connected to:  ('10.0.0.21', 8003)
Node6: Compression with dictionary
Node6: Shape original data:  (100, 80)
Node6: Shape sparse data:  (200, 100)
Node6: Shape compressed data:  (100, 40)
Node6: MSE:  0.1697952053102337
Node6: closed connection to:  ('10.0.0.21', 8003)
```

Listing 22.10: Details of the output from Listing 22.9.

Listing 22.10 shows an average MSE varying from 0.169 for sensor 1 to 0.518 for sensor 2. Such a difference was also noticeable in the scenario where DCT was employed; see Section 22.3.1. However, this scenario shows remarkable improvements in the reconstruction of the original data compared with the previously mentioned one.

22.3.2.1 Overcomplete dictionary robustness

Previously, we applied the same overcomplete trained dictionary on all temperature readings from six different sensors. To test its robustness, we generated an overcomplete dictionary for each data set, applied compressed sensing under the same assumptions and conditions considered in this chapter, and obtained the following results:

```
Head: Shape received data buffer:  (6, 100, 40)
Head: Decompression with dictionary

Head: Shape reconstructed data buffer:  (6, 100, 80)

Head: MSE for Sensor1 : 0.18296451461710317
Head: MSE for Sensor2 : 0.3831813907447271
Head: MSE for Sensor3 : 0.17085131674783086
Head: MSE for Sensor4 : 0.18582421022630732
Head: MSE for Sensor5 : 0.2531560266775236
Head: MSE for Sensor6 : 0.16862887303364887
```

Listing 22.11: Details of the output from Listing 22.9 with a specific trained dictionary for each sensor.

Listing 22.11 shows the average MSE for each of the six sensors of the single-cluster topology. Note that the results are almost identical to those obtained in Listing 22.10 for sensors 1, 3, 4, and 6, whereas the MSE dropped by approximately 0.13 for sensor 2 and 0.1 for sensor 5. This could be due to the fact that some of the available sensor readings are not entirely exact. Nevertheless, one single trained dictionary remains sufficient and robust under data sets with questionable accuracy. Moreover, additional dictionary processing and storage can be omitted.

22.4 Next steps

As we have seen in this chapter, the compressed sensing framework for the ComNetsEmu provides some basic examples for getting hands-on experience with compressed sensing for temperature measurement. The exposed API itself provides the opportunity to fine-tune the general parameters involved in the compression process, as studied in Chapter 10 in detail, as well as those related to the proposed dictionary learning with the K-SVD algorithm.

In general, it is a rather challenging task to find an appropriate sparsification transform for a specific type of data, not to mention the selection of the most efficient one. That is one of the reasons why training overcomplete dictionaries can be an efficient and robust alternative, especially when large data sets can be provided in advance.

For motivated readers interested in evaluating the impact of DCS in single-cluster topologies, the JSM model can be potentially utilized to capture not only the intrasignal correlations but the intersignal correlations as well. Moreover, it guarantees more compression with fewer overall measurements.

Security for mobile edge cloud

23

Simon Hanisch, Amr Osman, Tao Li, Thorsten Strufe
Technische Universität Dresden, Dresden, Germany

The problem with quotes on the Internet is that it is hard to verify their authenticity.
Abraham Lincoln

23.1 Introduction

The first step in protecting the security of a system is to determine what to protect and from whom. Without knowing what possible threats could be encountered, one cannot build adequate defenses. For computer networks, the confidentiality, integrity, and availability of traffic that moves through the network are required. Adversaries in computer networks are diverse, because they can be located in various places in the network. This provides them with different levels of access and control over the traffic. One example is an attacker located at the edge of the network. This location enables the attacker to communicate with other clients on the network but does not provide control over traffic directed elsewhere. A more sophisticated attacker may have control over a router in the network, which would enable manipulation of the network traffic of others.

Because of this manifold of different attackers, it is necessary to classify them according to their defining aspects. Traditionally, these aspects are their intention, area of control, capabilities, and behavior. The intention describes what the attackers want to achieve, such as stealing specific information, whereas the behavior can be either active or passive. Capabilities refer to resources, including computation power, time, and money, that the attacker can utilize. The area of control lastly describes which parts of the system the attacker can access.

For the hands-on examples in this chapter, we work with the following two adversaries who have similar behaviors and intentions but differ with respect to their area of control in the network and their capabilities:

Malicious Endpoint: The Malicious Endpoint is an active adversary connected to the network and can communicate with other hosts. The intention is to steal information from services on the network. Due to the position in the network (as an endpoint), the area of control is restricted to network packets that are

Computing in Communication Networks. https://doi.org/10.1016/B978-0-12-820488-7.00038-4

directly sent to the adversary. This adversary can, however, use IP addresses that were not assigned to fool services (IP-Spoofing).

Man-in-the-Middle (MitM):   The Man-in-the-Middle (MitM) adversary is a classic Dolev-Yao adversary that has direct control over the network. This access allows us to view, drop, and alter traffic, making this a very strong adversary. This adversary is polynomially bound in time, that is, breaking cryptographic schemes that are currently deemed secure is considered infeasible.

In the remainder of this chapter, we first discuss network isolation to deny the Malicious Endpoint from accessing resources on the network. To implement network isolation, we utilize a packet filter. Secondly, we present how to protect network traffic from a Man-in-the-Middle attacker by using a secure network tunnel. Each section starts with an introduction of the core concepts, followed by some technical details, and conclude with practical exercises.

23.2 Network segmentation

In a plain network without any regulation, every endpoint can talk to every other endpoint. This is great for connectivity and exposes the network traffic to all tenants operating on the same edge cloud network. Even without malicious intent, mistakes, such as wrong configurations or broadcasts to all endpoints, can lead to an involuntary information disclosure. The threat is magnified when an attacker similar to the Malicious Endpoint is considered, who will try to access unauthorized information and services. Hence it is desirable to separate tenants and groups of endpoints from each other. This separation is called network segmentation.

The segmentation is achieved by placing a packet filter on a network device between the two groups. By filtering the network traffic we can regulate which endpoint of the first group can talk to which endpoint in the second group. Additionally, even the types of allowed and denied traffic can be configured. The segmentation of the network enables the establishment of different trust zones. In the following, we introduce the core concepts of packet filtering and how they can be applied to regulate network traffic.

23.2.1 Concepts

At its core, a packet filter uses a rule set to decide if network packets are allowed or denied (see Fig. 23.1). The employed rules consist of a condition that must be matched (e.g., the packet must have a specific source address) and an action (e.g., the packet gets dropped) that is invoked if the condition is met. To understand what can be filtered, we initially review the layers of the network protocol stack. Network packet filters are most frequently used to filter the network and transport layer protocols, providing the packet filter with access to IP addresses and port numbers. However, packet filters for higher-layer protocols such as HTTP exist as well. The

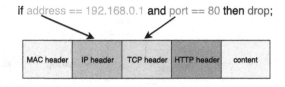

FIGURE 23.1

An example rule filtering an IP packet.

general rule of thumb is that filtering at a higher layer in the protocol stack allows for more fine-grained rules but comes with the additional overhead of parsing all the underlying protocol layers.

Matching only on the properties of a single packet is known as static filtering, as no knowledge of previous packets is required to decide the fate of the packet under consideration. Stateful filtering, on the other hand, additionally takes previous packets into account for filtering, for example, to filter all packets that belong to a specific connection. This stateful approach requires packet filters to maintain a record of the connection state between endpoints. In addition to their core functionality of filtering, packet filters can perform additional tasks, such as traffic shaping, network address translation, and the marking of packets.

When writing a rule set for a packet filter, there are two fundamental design approaches, *blacklisting* and *whitelisting*. Essentially, this refers to the difference of what to do in the default case: accept or deny. Backlisting accepts all packets by default and specifies rules for which packets to drop. This approach has the drawback that operators can miss malicious traffic that should be dropped, leaving holes in the firewall. Whitelisting denies all packets by default and only accepts those that match a rule. This way, operators can filter malicious traffic they did not expect but also have to add rules for every type of allowed connection. In general, whitelisting is considered more secure but less flexible than blacklisting.

There are three general data structures that could be used to implement a rule set. The naive approach is to simply maintain an ordered list of the rules and to evaluate them in sequential order. Though simple, this approach requires to check all rules in the list if none of them matches the traffic (i.e., this approach is $O(n)$). An alternative approach is to use a tree structure, whereby each node represents a condition. The benefit of this approach is that the worst case results in a rule checking complexity of $O(\log(n))$. An added benefit is provided through knowing the position in the tree, which implicitly results in knowledge of all previous conditions. The third option is representing the rule set as a hash table. A hash table offers a rule checking complexity of $O(1)$ but has the drawback that only one type of condition combination can be matched per table. Consider an example of rules matching the IP address and port as well as rules matching the IP address and transport protocol type. To implement the rule set, the packet filter must maintain two different hash tables. Modern packet filters utilize a combination of all three data structures giving the operator flexibility in structuring a rule set.

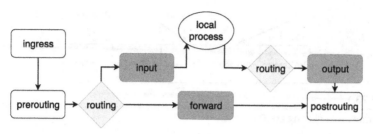

FIGURE 23.2

The Netfilter hooks of the Linux network stack.

23.2.2 Implementation

Packet filters are usually implemented as part of the kernel of the OS because of their coupling to the network stack. At different stages in the network stack, so-called *hooks* (see Fig. 23.2) allow the packet filter to intercept and filter traffic. Since the network stack requires time to process each packet, dropping packets as early as possible is desirable to save resources. For example, the Berkley Packet Filter (BPF) can filter traffic right before the kernel receives the packet from the network driver.

In the past the kernel modules of packet filters were mostly tailored to a fixed suite of protocols, making it necessary to have multiple tools and extensions to filter all protocols (e.g., iptables and ip6tables). The modern approach is using a virtual machine in the kernel that can be loaded with arbitrary bytecode. The rule set becomes a program having an end-state of either accept or deny. Rules are translated to comparison instructions with conditional jumps to the next rules. This enables writing filters for arbitrary protocols without the need to change the kernel module. Since this approach is so generic and efficient, it has already been adapted to filtering tasks outside the networking domain.

Whereas packet filters are implemented in the kernel, the configuration utility is placed in userspace. This utility takes the rule set specification, parses it, and programs the kernel module accordingly.

Since the connection tracking information is also used by other network operations, the connection tracking is not performed by the packet filter. Instead, it is most often implemented as its own kernel module. Due to a high number of connections in today's networks, the size and update speed of this table is often a bottleneck for the overall packet filter performance. This makes the maximum number of connections and the connection establishment rate two significant performance metrics for packet filters.

23.2.3 nftables

We will further use `nftables` [330] as an example to provide practical exercises on how to use a packet filter for network security. Main improvements over iptables are

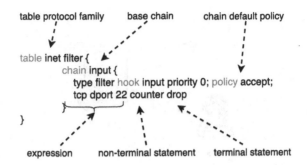

FIGURE 23.3

An example `nftables` table to filter all incoming SSH traffic.

an easier rule syntax, which simplifies reading and writing rules and native support for sets and traffic shaping. The `nftable` kernel module is a generic virtual machine and therefore can filter arbitrary protocols. However, the userspace utility only supports IPv4, IPv6, and ARP.

`nftables` organizes its rule set as *tables*, which are bound to a specific address family. The address family determines the type of packets processed in the table; the *inet* family is the most relevant here, as it processes both IPv4 and IPv6 packets.

Inside a table, the rules are kept in containers called *chains*. *Chains* are simply a list of rules that are sequentially evaluated (i.e., top to bottom). By binding a chain to a *Netfilter* hook the chain receives the packets from the hook making the chain a base chain. Base chains differ from normal chains, because they need to have a default policy (i.e., accept or deny), which is triggered if none of its rules match the packet under consideration. Packets can be transferred from one chain to another by so-called jumps.

`nftables` rules always consist of two things, expressions and statements. Expressions are conditions that have to be met, for example, that a packet contains the address 192.178.0.1. Multiple expressions can be combined by using logical operators. Statements describe actions that are triggered when the expressions evaluate to true. Statements can be either terminal or nonterminal. Terminal statements, such as jump, drop, or accept, end the rule evaluation, and hence each rule can only have one terminal statement. Nonterminal statements, on the other hand, do not end the rule evaluation, and each rule can have an arbitrary number of these statements. (See Fig. 23.3.)

23.3 Network isolation exercise

The following exercises should provide an introduction on how `nftables` can be used to implement network isolation. We start with a simple exercise, which demonstrates

client

server

attacker

FIGURE 23.4

Network scenario of the first nftables exercise.

the basic concepts of static blacklisting and whitelisting. We then move on to stateful filtering and advanced rule organization. Each exercise has a prepared network implementation in the network_security folder in *ComNetsEmu*.

23.3.1 Blacklisting and whitelisting

In the first exercise, we use nftables to create blacklist and whitelist filter rule sets to block the *attacker* from accessing the web *server* without blocking the legitimate *client* (see Fig. 23.4). All three actors are located in the same network, and therefore we employ nftables directly on the *server* to block incoming requests. The first step is creating a new table to filter IP traffic. nft add table inet filter creates a new table, called *filter*, that filters the protocol family *inet*, which is a combination of IPv4 and IPv6 traffics. Filtering both protocols with one table provides the advantage of less complex rule set management, as only one table has to be maintained. Generic rules are applied to both protocols (e.g., filtering a tcp port), and protocol-specific rules are only applied to the particular protocol under consideration.

The next step is creating a base chain connected to the *input* Netfilter hook. We choose the *input* hook because it is passed by all traffic directed toward the local processes on the host. A chain that is connected to a Netfilter hook becomes a base chain and therefore requires us to specify its type, priority, and default policy. The priority defines the order in which the base chains connected to the same hook are executed. Since lower priorities are executed first, we choose priority 0. The type of the base chain can be filter, nat, or route. Since we want to filter traffic, we select the *filter* chain type. Lastly, the base chain requires a default policy that describes what to do with the traffic if none of the rules in the chain matches. For this example, we choose *accept* by default.

```
$ nft add chain inet filter input \{ type filter hook input priority 0 \;
    policy accept \; \}
```

Our current setup already receives packets, but since we accept all by default, we see no different behavior. A rule that actually blocks traffic from *attacker* is still missing. The following rule drops all traffics that come from the *attacker*'s IP address.

```
$ nft add rule inet filter input ip saddr 10.0.0.3 drop
```

Now the traffic from *attacker* to *server* is dropped, and *attacker* can no longer access the Web server. To verify that the packets are being dropped, we can use ping from *attacker* to *server* or the other way round. Since we block all traffic from *attacker*, an ICMP echo request from *server* to *attacker* will also fail because the response of *attacker* is filtered.

In the next part of the exercise, the *attacker* changes its IP address to a random one, and our approach of blacklisting the *attacker* does not work anymore. Instead of blacklisting, every time the *attacker*'s IP changes, we use whitelisting to only allow *client* to connect to *server*. We start by removing the blacklist from our filter table and then change the default policy for the base chain from accept to drop. We subsequently add a rule that allows incoming traffic from *client*.

```
$ nft flush rule inet filter input
$ nft add chain inet filter input \{ type filter hook input priority 0 \;
    policy drop \; \}
$ nft add rule inet filter input ip saddr 10.0.0.1 accept
```

The reader can check if *client* can reach *server* and if *attacker* is blocked. This concludes the first exercise with nftables, and the reader should know the difference between a whitelist and a blacklist, as well as some basic nftables operations.

23.3.2 Stateful filtering

In our second exercise, we introduce more advanced filtering techniques that do more than simply blocking traffic. First, we allow our *server* to initialize a connection to the Internet and receive the responses, even though the Internet is blocked from accessing *server*.

The scenario already has the whitelist rule set from the first exercise deployed on *server*. In turn, the *server* can send requests to an endpoint on the Internet but does not receive the response because only the traffic of *client* is whitelisted. Since whitelisting every endpoint on the Internet is not a good idea, we use a stateful rule that allows endpoints to respond if they have been contacted by *server* first. This is accomplished by matching on the connection tracking state of a packet instead of matching on its IP address. We want only to pass connections that are *established* or *related*. Related are all connections that relate to an established connection, for example, ICMP traffic that belongs to a TCP connection.

```
$ nft add rule inet filter input ct state established,related accept
```

With *server* being able to talk to the Internet, we now want to focus on the *client* accessing *server*. The *client* is a legitimate user that is using too much of the bandwidth; we want to restrict him to no more than 1 MB per second.

We again want to insert the rule into our existing rule set, but for the rate limit to work, it has to be evaluated before the traffic is accepted by one of the other rules, and therefore it has to be the first rule in the filter chain. So far we added rules to chains using `nft add rule`, which appends rules at the end of the list of rules by using `nft insert rule` we insert at the top of the list.

```
$ nft insert rule inet filter input limit rate over 1 mbytes/second drop
```

Additionally, rules can be inserted in the middle of a chain by using the handle of the rule you want to insert above or below as the position argument when creating the rule. The rule handles can be printed by using `nft list table inet filter -a`.

Both examples should have provided an introduction to how stateful packet filter rules can be used for dynamic filtering.

23.3.3 Chains and jumps

In our last exercise, we restructure a rule set with additional chains to keep the rules for different networks separate from each other. The packet filtering are performed on the *router* connecting the three networks illustrated in Fig. 23.5 with each other.

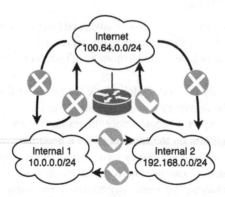

FIGURE 23.5

Three networks that are isolated from each other with a central firewall

The first step is creating a table and a base chain. The base chain is connected to the forward hook because we want to filter the network traffic that is passing through the router and not the one that is directed to it. Additionally, we create a chain for each of the network interfaces of the router.

```
$ nft add table inet filter
$ nft add chain inet filter forward \{ type filter hook forward priority 0 \;
    policy drop \; \}
$ nft add chain inet filter forward-s1
$ nft add chain inet filter forward-s2
$ nft add chain inet filter forward-s3
```

The second step is redirecting the traffic from the network interfaces to their respective chains. Instead of accepting or dropping packets, we use the jump statement to switch from our base chain to the interface-specific chain. At the end of each `nftables` rule, there is a statement that is executed if the rule matches. Statements equivalent to accept, drop, and jump are terminal and end the rule evaluation, and therefore there can only be one terminal statement in a rule. Nonterminal statements do not end the rule evaluation and allow actions, such as logging or tagging traffic. Different from terminal statements, there can be arbitrary many nonterminal statements in a rule, allowing rules where traffic is first logged and then accepted. In the rules below, we integrated the *counter* statement to count the number of bytes that pass through each network interface before we jump into the next chain.

```
$ nft add rule inet filter forward iif router-s1 counter jump forward-s1
$ nft add rule inet filter forward iif router-s2 counter jump forward-s2
$ nft add rule inet filter forward iif router-s3 counter jump forward-s3
```

The counters can be evaluated by listing the table with `nft list table inet filter`. To have the same rule set as before, we now need to add the individual rules for the networks to their chains.

```
$ nft add rule inet filter forward-s1 ip saddr 10.0.0.0/24 accept
$ nft add rule inet filter forward-s2 ip saddr 192.168.0.0/24 accept
$ nft add rule inet filter forward-s3 ct state established,related ip daddr
    10.0.0.0/24 accept
```

Whereas the number of rules in this example is still manageable without structuring them into chains, this approach is invaluable when dealing with large rule sets. Besides the maintainability, it also improves the evaluation speed of the rule set –

when a packet is received via interface *s1*, only the rules in the forward-s2 chain have to be evaluated.

With the forwarding neatly organized, we have to address the open ports on the router and on a host in network *s1*. For the host in *s1*, we can simply add a rule to the forward-s1 chain dropping all traffic that is directed to port 22 and 1337. We add a base chain that denies all incoming traffic to the router itself to protect the *router*.

```
$ nft add rule inet filter forward-s1 tcp dport {ssh, 1337} drop
$ nft add chain inet filter input \{ type filter hook input priority 0 \;
    policy drop \; \}
```

23.4 Secure network tunnels

The goal of secure network tunnels is to protect the confidentiality and integrity of a network traffic that is in transit. Network tunnels achieve this goal by encrypting network packets and encapsulating them in a tunnel protocol. Additionally, a cryptographic checksum, a message authentication code, is added to detect manipulation of the transported packets. A well-known representative of a secure network tunnel is a VPN.

23.4.1 Concepts

Our initial scenario is that of two edge cloud services, labeled *service A* and *service B*, that wish to communicate with each other via a network. The adversary is a Man-in-the-Middle attacker located in between *A* and *B* and has control of the network traffic. This position provides the attacker with the ability to read, alter, or drop every message that is exchanged between the two services.

To keep the attacker from reading, the messages are encrypted with a cipher to conceal their content. Ciphers can be symmetric or asymmetric. Symmetric ciphers use the same key for encryption and decryption, whereas asymmetric ciphers employ two separate ones. The benefit of asymmetric ciphers is that the encryption key cannot be used to decrypt the traffic, and therefore it can be announced to the world as a public key. The drawback of asymmetric ciphers is that they are based on expensive mathematical operations, which require significant computational power. This results in asymmetric encryption being slower than symmetric encryption. For network traffic, symmetric ciphers are more beneficial, because they induce less overhead on the traffic.

In order for *service A* and *service B* to use a symmetric cipher for their secure network tunnel, they first have to agree on a secret key for the cipher. If they exchange the key via the network, then the attacker learns the key, and the encryption becomes useless. Therefore they need a secure channel to exchange keys. Since asymmetric

FIGURE 23.6

Operations of a cipher.

ciphers do not require their encryption key to remain secret, they can be used to build a secure tunnel to exchange the key for the symmetric cipher. With the keys exchanged and the encryption in place, A and B can conceal their traffic from the attacker. (See Fig. 23.6.)

Encryption, however, does not keep the adversary from modifying the content of the message. Even though the attacker might not know what is changed (because the attacker cannot read the message). Since the attacker is assumed to be able to alter the messages (the attacker has full control of the network traffic), at least any modification should be detected.

Message Authentication Codes (MACs) are similar to network protocol checksums used to detect errors during the transmission of packets. However, normal checksums are ineffective against deliberate attacks, because the attacker can simply recalculate the checksum after altering the message. To keep the attacker from performing the recalculation, the MAC uses an additional secret key as input to calculate the checksum. As long as the attacker does not possess the key, the checksum cannot be recalculated. Hence the receiver would notice that the message was altered.

If A and B have an established secure network tunnel between them, then the attacker now can neither read nor undetected manipulate the content of the traffic. However, there remains one attack vector to render the tunnel useless. During the setup of the tunnel, the attacker can intercept the traffic and claim to A that he is B and vice versa. The outcome would be two tunnels with the attacker in the middle. To prevent this attack, A and B need to authenticate each other. Remote authentication can be achieved by using a signature algorithm. Signature algorithms are asymmetric ciphers that use the encryption key to sign a message and the decryption key to verify a message. If the verification is successful, then it proves that the signer of the message is in possession of the secret key. For A to authenticate B, A needs to know the public key of B. This can be achieved by exchanging public keys beforehand or by using a trusted third party.

Secure network tunnels operate by encapsulating the entire packet inside a tunnel protocol. This is different from just encapsulating the payload, since the metainformation of the packet is also protected.

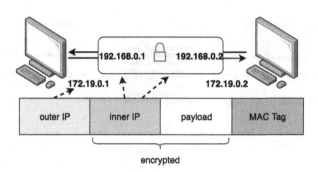

FIGURE 23.7

The structure of a network tunnel and the corresponding network packet.

23.4.2 Implementation

Secure network tunnels typically provide their own network protocol for encapsulating packets. These tunnel protocols can operate on every layer of the protocol stack. However, because most of the time the intended communication should be secured end-to-end between two parties, they are often implemented on the Internet or transport layer. (See Fig. 23.7.)

As described before, the initial setup of a tunnel can be challenging due to the required key exchanges. Additionally, the parties often have to agree on a mode of operation and a cipher suite (a set of cipher algorithms) to use. This is the reason why most secure network tunnels have an additional protocol or phase that realizes these parameter negotiations between the endpoints.

23.4.3 Wireguard

Traditional secure network tunnels, like IPsec or OpenVPN, offer multiple modes of operations and parameter negotiations, which make them complex. Hence they are difficult to implement and use without making mistakes that lead to security problems. Wireguard [331], on the other hand, is designed to be simple to implement and use by only offering a minimal feature set. Instead of allowing for flexibility, the designers chose a fixed stream cipher for encryption (ChaCha20 [332]) and a specific MAC (Poly1305 [333]), which have been designed for high throughput at low computational cost. The security of the Wireguard protocol was formally verified in [334].

23.5 Secure network tunnel exercise

In the following two exercises, we demonstrate how to set up a Wireguard tunnel between two or more endpoints and provide an example of how to automate the process.

23.5.1 **Man-in-the-middle**

In our first exercise, we show the benefit of having a secure network tunnel at the example of a simple Man-in-the-Middle attacker. The scenario is that the *client* wants to access a file on the *server* via the File Transfer Protocol (FTP). Since security was not considered during the design of FTP, it does not offer any form of protection against our Man-in-the-Middle *attacker*. All messages are exchanged without encryption, including passwords and usernames.

Because we consider a MitM *attacker*, we have to assume that it can read and manipulate every message that is exchanged between *client* and *server*. One example way of performing such an attack is to compromise a networking device on the path between *client* and *server*. However, in this exercise the attacker achieves its position by manipulating (spoofing) the Address Resolution Protocol (ARP), which is used to find the correct MAC addresses for IP addresses. The idea of ARP spoofing is that when some host on the local network asks *Who has IP address 192.168.0.5?*, the *attacker* simply responds with *I have!* (even though it has not). This way, all traffic destined for 192.168.0.5 gets send to the *attacker* and not the actual recipient. By forwarding the packets (via the recipient MAC address) to the real recipient the attacker now becomes a Man-in-the-Middle. This way the attacker becomes an additional hop between *client* and *server* without them knowing.

Although we cannot stop the *attacker* from spoofing ARP (again security was not considered during protocol design), we can protect the confidentiality and integrity of our network traffic from the *attacker*.

The first step is generating key pairs for the *client* and *server* by using Wireguard command-line tool `wg`. A good rule of thumb is that keys should always be generated where they are being used, and this way keys cannot be forgotten or end up where they are not supposed to be. Therefore we generate the keys on the hosts. With `umask 077; wg genkey > privatekey`, the private key is generated and stored in the file `privatekey`, which can be read only by the owner of the file. To derive the public key from the private one, we use `wg pubkey < privatekey > publickey`. The public key needs no additional file access restrictions because we will anyway announce this key to the world. We could even send this key to our *attacker*.

After the key generation phase, we now set up the Wireguard interfaces on our hosts. This is done by using the iproute2 tool since Wireguard has its one network interface type. After the interface is created, we add an IP address to it as we would with a normal interface, this address will be used inside the tunnel.

```
$ ip link add dev wg0 type wireguard
$ ip address add dev wg0 192.168.0.1/24
```

Now all that is left is configuring the interface with the generated private key and a configuration for the peer (endpoint) we want to connect to. We require three pieces of information from our peer: i) The public key of our peer, ii) the IP address our peer uses inside the tunnel, and iii) the external address of the tunnel (IP address plus port). We give an example how to configure the interface for a peer with the public key PUBLIC_KEY, the internal tunnel address 192.168.0.2/32, and the external tunnel address 10.0.0.2:1337. Our tunnel is configured to use the private key in the file *privatekey* and to listen on port 1337 using interface wg0. After wg0 is configured, we use iproute2 to set the interface up.

```
$ wg set wg0 listen-port 1337 private-key ./privatekey peer PUBLIC_KEY allowed
    -ips 192.168.0.2/32 endpoint 10.0.0.2:1337
$ ip link set up dev wg0
```

After the Wireguard tunnel is configured on *client* and *server* (with their respective keys and addresses), the tunnel is operational and can be used. The command wg show can be used for troubleshooting and to display information about peers and interfaces. When the tunnel is up and working, we can try to connect to the FTP server through it and see if the attacker can still snoop on the passwords!

23.5.2 Tunnel network

In the second exercise, we showcase how to set up Wireguard with multiple peers and a configuration file using wg-quick to automate the process. Inside the secure-tunnel-2.py file, there are four hosts: *center*, *client1*, *client2*, and *client3*. We want *client1*, *client2*, and *client3* to establish a Wireguard tunnel to *center*.

The generation of the key pairs is already done, and the respective keys are stored in files on the hosts. We now can write a configuration file for each of the hosts that sets up the Wireguard interface. The configuration files are stored in /etc/wireguard and follow the naming scheme INTERFACE_NAME.conf. For example, for the wg0 interface, we get /etc/wireguard/wg0.conf.

The first part of the configuration describes the interface and can only appear once in the configuration. To replicate our setup from exercise 1, we need to specify the internal tunnel address, the listening port, and the private key of the interface. Additional options allow specifying DNS server and network configurations. Wireguard will also create its own routing table for the interface, which can be disabled via the options. All options can be found in the manpage of wg-quick.

```
[Interface]
Address = 192.168.0.1/24
ListenPort = 1337
PrivateKey = oK56DE9Ue9zK7fgfggfgdopphsdfsd1sdsdfsdcXXsQKrQM=
```

The second part of the configuration specifies the peers of the interface. There can be multiple peers per interface, and all that is required to specify is the public key and the AllowedIPs. The endpoint is optional, but at least one side of the tunnel must specify one. The AllowedIPs specify which IPs are allowed to traverse the tunnel and are used to determine which peer is the recipient of a packet. Just like regular routing Wireguard looks up the shortest prefix match of its peers to decide which peer should receive the packet. This requires the AllowedIPs prefixes to be unique per peer, and hence you cannot have two peers with the same AllowedIPs.

```
[Peer]
PublicKey = GtL7fZc/bLnqZldpVofMCD6hDjrK28SsdLxevJ+qtKU=
AllowedIPs = 192.168.0.2/32
Endpoint = 10.0.0.2:1337
```

With a complete configuration in hand, the creation of the interface boils down to calling `wg-quick up wg0`. The interested reader can now write a single configuration file for each host and check if all they can reach *center* via their tunnels.

An additional benefit of interface configurations is that the creation of a Wireguard interface can be triggered on system boot by using the *systemd* service as `systemctl enable wg-quick@wg0` (note that this does not work in the *ComNetsEmu* containers).

Extensions

7

Outline

In this part, we highlight via examples how to extend the ComNets Emulator with new features and how to connect it to the external world or to SDR technologies.

Connecting to the outer world

24

Fabrizio Granelli
University of Trento, Trento, Italy

*There was a time when people felt the Internet was another world, but now
people realise it's a tool that we use in this world....*
Tim Berners-Lee

24.1 Introduction

The purpose of the Virtual Machine (VM) proposed in this textbook is enabling anyone to apply the concepts described in the different chapters through the usage of an
emulation environment on a generic PC platform. Indeed, the proposed scenarios and
scripts can be run within a single VM.

Nevertheless, we might want to extend the size of the emulated test bed and to
run it on different machines, even across the Internet. This chapter describes how
to enable *ComNetsEmu* to connect to the Internet and how to interconnect different
ComNetsEmu instances across the Internet. This allows us to study systems that go
beyond the limitations of a single VM in terms of processing and storage power and
to incorporate and send traffic to the Internet.

The possibilities offered by this paradigm are virtually endless, as endless are the
potential services supported by the Internet.

In the next sections, we propose a step-by-step procedure to set up a virtual network interface on Open vSwitch inside the Mininet component of *ComNetsEmu* to
connect to the Internet. By exploiting Network Address Translation (NAT) we also
present a simpler approach to finally interconnect two (or more) *ComNetsEmu* instances across the Internet.

24.2 Connecting ComNetsEmu to the Internet

In this chapter, we describe the steps required to connect the network emulated within
ComNetsEmu using Mininet to the global Internet. There are two basic ways to inter-
connect the emulated network to the Internet: i) by manually setting the NICs of the
Mininet hosts and ii) by exploiting the NAT service. In the next sections, we step-by-
step describe how to perform such configurations and finally explain how to use the
Domain Name System (DNS) address resolution.

Computing in Communication Networks. https://doi.org/10.1016/B978-0-12-820488-7.00040-2
Copyright © 2020 Elsevier Inc. All rights reserved.

24.2.1 Manual host configuration

This section provides the step-by-step instructions for manual configuring Mininet to connect to the Internet. We assume that the *ComNetsEmu* virtual machine is hosted by a PC running the VirtualBox virtualization software and that the hosting OS has already been configured to provide Internet connectivity, that is, the hosting OS is presumed to have Internet connectivity as a baseline.

24.2.1.1 Checking connectivity and NIC of the host

In the VirtualBox network settings for the VM, a NAT interface needs to be enabled that allows connection to the Internet. Typically, an IP address is similar to 10.0.2.15, that is, an IPv4 class A address.

The proper connectivity can be evaluated by pinging www.google.com within the VM (or any other Internet address) to make sure that the guest OS, the one provided by *ComNetsEmu*, is connected to the Internet.

We can check the actual Network Interface Control (NIC) used by the VM to connect to the Internet by issuing the following command and identifying the NIC with the correct IP:

```
$ ifconfig
```

We should take note of the NIC name, for example, eth0 or other. To fix ideas in the following sections of this chapter, we assume that the example NIC name is eth0.

24.2.1.2 Running an example network

With the general network setup preliminaries, we can now start a Mininet network with a switch and a host or any other preferred topology. To continue our example, we issue

```
$ sudo mn  --switch ovsk --mac --topo single,2
```

inside the terminal. This command creates a network with single switch and two hosts, that is, a switch *s1* and two hosts *h1* and *h2*.

24.2.1.3 Connecting the guest interface to the OVS bridge

The command used to enable a guest interface on Open vSwitch is the ovs-vsctl command, which is used for querying and configuring openvswitchd (the daemon of openvswitch). An xterm window is required for programming *s1*, as this command does not run directly on Mininet. One example is opening a new secure shell connection from the host OS into the guest OS by using the -X or -Y command switches of ssh. The Open vSwitch configuration can now be evaluated using the command

```
$ sudo ovs-vsctl show
```

The resulting output should look similar to:

```
$ sudo ovs-vsctl show
d27a9060-3edf-4ee7-a4cf-09e705c93f56
    Bridge "s1"
        Controller "ptcp:6634"
        Controller "tcp:127.0.0.1:6633"
            is_connected: true
        fail_mode: secure
        Port "s1-eth1"
            Interface "s1-eth1"
        Port "s1-eth2"
            Interface "s1-eth2"
        Port "s1"
            Interface "s1"
                type: internal
    ovs_version: "2.2.9"
```

Next, the interface `eth0` needs to be connected to switch *s1* by running the following command:

```
$ sudo ovs-vsctl add-port s1 eth0
```

The command `ovs-vsctl show` can now be employed to verify the configuration again:

```
$ sudo ovs-vsctl show
```

The new interface should show up as in the example output:

```
$ sudo ovs-vsctl show
d27a9060-3edf-4ee7-a4cf-09e705c93f56
    Bridge "s1"
        Controller "ptcp:6634"
        Controller "tcp:127.0.0.1:6633"
            is_connected: true
        fail_mode: secure
        Port "eth0"
            Interface "eth0"
        Port "s1-eth1"
            Interface "s1-eth1"
        Port "s1-eth2"
            Interface "s1-eth2"
        Port "s1"
            Interface "s1"
                type: internal
    ovs_version: "2.2.9"
```

24.2.1.4 Update IP addresses on the hosts

The next configuration step requires two terminal windows for *h1* and *h2*. These can be created with the command

```
mininet> xterm h1 h2
```

The following commands are now issued on the first host *h1*:

```
h1> ifconfig h1-eth0 0
h1> dhclient h1-eth0
h1> ifconfig
```

The first command removes the IP address from h1-eth0, and the second command reassigns the IP address for h1-eth0 by querying the built-in Dynamic Host Configuration Protocol (DHCP) server. The third command shows the renewed network interface description, which now should be similar to the following example output:

```
h1-eth0   Link encap:Ethernet  HWaddr 00:00:00:00:00:01
          inet addr:10.0.2.16  Bcast:10.0.2.255  Mask:255.255.255.0
          inet6 addr: fe80::200:ff:fe00:1/64 Scope:Link
          UP BROADCAST RUNNING MULTICAST  MTU:1500  Metric:1
          RX packets:24 errors:0 dropped:0 overruns:0 frame:0
          TX packets:12 errors:0 dropped:0 overruns:0 carrier:0
          collisions:0 txqueuelen:1000
          RX bytes:3304 (3.3 KB)  TX bytes:1764 (1.7 KB)

lo        Link encap:Local Loopback
          inet addr:127.0.0.1  Mask:255.0.0.0
          inet6 addr: ::1/128 Scope:Host
          UP LOOPBACK RUNNING  MTU:65536  Metric:1
          RX packets:1252 errors:0 dropped:0 overruns:0 frame:0
          TX packets:1252 errors:0 dropped:0 overruns:0 carrier:0
          collisions:0 txqueuelen:0
          RX bytes:151432 (151.4 KB)  TX bytes:151432 (151.4 KB)
```

The Internet connectivity of the setup can again be verified using ping:

```
h1> ping 8.8.8.8
```

Similarly, for host *h2*:

```
h2> ifconfig h2-eth0 0
h2> dhclient h2-eth0
h2> ifconfig
```

(We omit the output as it is similar to that for *h1*.) With these configuration steps completed, the Mininet emulated network is now capable of connecting to the Internet.

24.2.2 **Using NAT service**

In several intranets today, Network Address Translation technology is used for both enabling the connection of multiple devices sharing the same public IP address and improved security. Typically, this is also a service offered by virtualization environments such as VirtualBox or VMWare, which provides a local NAT service to the VMs. Therefore it would be useful to connect the emulation Virtual Machine with the Internet exploiting the already available NAT service.

Indeed, it is possible to enable NAT connectivity by using a Mininet primitive addNAT(), as is demonstrated in the following example. In this case, a tree topology is implemented with one switch and provided Internet connectivity:

```python
!/usr/bin/python
from mininet.cli import CLI
from mininet.log import lg, info
from mininet.topolib import TreeNet

if __name__ == '__main__':
    lg.setLogLevel( 'info')
    net = TreeNet( depth=1, fanout=4 )
    # Add NAT connectivity
    net.addNAT().configDefault()
    net.start()

    info( "*** Hosts are running and should have internet connectivity\n" )
    CLI( net )
    # Shut down NAT
    net.stop()
```

24.2.3 **Using DNS resolution**

To use Internet host *names*, a properly configured DNS server is required. To perform this task, we can open a terminal on host *h1* and then use the following command to edit the corresponding configuration file (note that it is a good practice to remember to always make a backup copy, as in the first line of the following example):

```
h1> sudo cp /etc/resolv.conf /etc/resolv.conf.backup
h1> sudo nano /etc/resolv.conf
```

Then we can enter the desired DNS information in the configuration file of the local DNS resolver in the following format:

```
nameserver 8.8.8.8
```

This example uses the public Google DNS server, which is located at the known IP address 8.8.8.8. This configuration can be adapted to use any reachable DNS server that might be most suitable for the reader, for example, the address of a company DNS server or of another public DNS server.

Note that the configuration of the hosts is shared with the hosting machine, and therefore the changes to the DNS server(s) also affect the host. For this reason, if something goes wrong or if one needs to revert to the original configuration, then the command

```
$ sudo cp /etc/resolv.conf.backup /etc/resolv.conf
```

can be issued to revert back to the original configuration.

24.3 Connecting different test bed VMs

Another important way to build larger test beds is interconnecting different Virtual Machines running SDN/NFV emulation across the Internet. This enables an increase of the size of the test bed by using several VM hosting services while managing them as a self-contained environment.

One way to interconnect hosting VM instances that contain Mininet software is through a tunnel across the Internet. In this case, we propose the usage of Generic Routing Encapsulation (GRE) to build such a tunnel.

This section provides an overview of the methodology to directly interconnect two or more VMs. The example is inspired by http://csie.nqu.edu.tw/smallko/sdn/vm2vm_gre.htm.

A GRE tunnel is used when packets need to be sent from one network to another over the Internet or, more generally, an insecure network. With GRE, a virtual tunnel is created between the two endpoints, and packets are sent through the GRE tunnel. Fig. 24.1 presents a conceptual diagram describing how a GRE Tunnel works.

When the sending router decides to send a packet into the GRE Tunnel, it *wraps* the whole packet into another IP packet with two headers: i) the GRE header used to

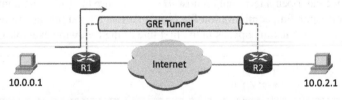

FIGURE 24.1

Overview of the concept of the GRE Tunnel.

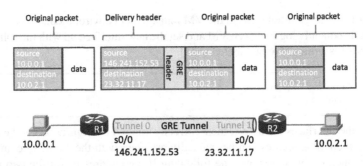

FIGURE 24.2

Detailed function of the GRE Tunnel.

manage the tunnel (4 bytes) and ii) the *Delivery header* (20 bytes), which includes the new source and destination IP addresses of two virtual interfaces of the tunnel (called tunnel interfaces). This process is called *encapsulation*. An example of detailed setup and usage of the GRE Tunnel in provided in Fig. 24.2.

In the example of Fig. 24.2, when R1 receives an IP packet, it wraps the whole packet with a GRE header and a delivery header. The delivery header includes new source IP address of 146.241.152.53 (the IP address of the physical interface of R1 that is used to create the tunnel) and the new destination IP address of 23.32.11.17 (the IP address of the physical interface of R2 that is used to create the tunnel). These two IP addresses are purely random and do not present actual IP addresses of routers.

It is important to note that the GRE tunnel does not encrypt the packet, but it only encapsulates it. In case encryption is required, IPSec must be used. Moreover, since GRE is an encapsulating protocol, we might need to adjust the Maximum Transmission Unit (MTU) to 1400 bytes and Maximum Segment Size (MSS) to 1360 bytes. This is possible through the following commands in a common Linux environment:

```
$ ip mtu 1400
$ ip tcp adjust-mss 1360
```

For our example, we consider a simple scenario where we want to interconnect a Virtual Machine running a default Mininet distribution (which we will call VM#1) to a Virtual Machine running the *ComNetsEmu* provided with this book (which we will VM#2). The VMs can run on the same PC or on different PCs. However, for simplicity and to enable easy deployment of this example, we assume that both VMs are running on the same host. Moreover, we deploy a single controller to administrate the merged topology. The controller will run on VM#1.

The first step of the procedure is enabling the two VMs to offer a proper IP network interface. To achieve this goal, if we use VirtualBox environment, then we can define a *host only* interface on each VM and activate it through the configuration interface. Now that the interfaces are set up, we need to evaluate their status.

After the configuration steps in the VM management environment are performed, we can subsequently log into each VM and set the new interface up with the following commands (assuming it to be labeled `eth1`):

```
$ sudo ifconfig eth1 up
$ sudo dhclient eth1
```

Once the interfaces are active, their correctly applied configuration can be evaluated, and their IP addresses can be manually assigned with the `ifconfig` command. The continued example assumes the following IP two addresses for the two virtual machines:

1. VM#1 IP address: 192.168.56.101
2. VM#2 IP address: 192.168.56.104

Now we should build the scripts for activating two Mininet instances and interconnect them. On VM#1 (i.e., the default Mininet VM), we create the following Python file and save it as `vm1_script.py`. Note that the file is in Python v.2, as per system specifications:

```python
#!/usr/bin/python
from mininet.net import Mininet
from mininet.node import Controller, RemoteController, OVSKernelSwitch
from mininet.cli import CLI
from mininet.log import setLogLevel
from mininet.link import TCLink, Intf

def topology():
    print "Create a network."
    net = Mininet( controller=Controller, link=TCLink, switch=OVSKernelSwitch
        )
    print "*** Creating nodes"
    s1 = net.addSwitch( 's1')
    h1 = net.addHost( 'h1', ip="10.0.0.1" )
    # controller will run on VM#1 at IP 192.168.56.101
    c0 = net.addController('c0', controller=RemoteController, ip='
        192.168.56.101', port=6633 )

    print "*** Adding Link"
    net.addLink(h1,s1)

    print "*** Starting network"
    c0.start()
    s1.start( [c0] )
    # set up GRE tunnel between 192.158.56.101 and 192.168.56.104
    s1.cmd("ip link add s1-gre1 type gretap local 192.168.56.101 remote
        192.168.56.104 ttl 64")
```

```
s1.cmd("ip link set s1-gre1 up")
Intf("s1-gre1", node=s1)

print "*** Running CLI"
net.start()
CLI( net )
print "*** Stopping network"
s1.cmd("ip link del dev s1-gre1")
net.stop()

if __name__ == '__main__':
    setLogLevel( 'info' )
    topology()
```

On VM#2 (i.e., the *ComNetsEmu* VM), we create the following file and save it as
vm2_script.py. In this case, we use Python v3:

```
#!/usr/bin/python
from mininet.net import Mininet
from mininet.node import Controller, RemoteController, OVSKernelSwitch
from mininet.cli import CLI
from mininet.log import setLogLevel
from mininet.link import TCLink, Intf

def topology():
    print("Create a network.")
    net = Mininet( controller=Controller, link=TCLink, switch=OVSKernelSwitch
        )
    print("*** Creating nodes")
    s2 = net.addSwitch( 's2')
    h2 = net.addHost( 'h2', ip="10.0.0.2" )
    # controller will run on VM#1 at IP 192.168.56.101
    c0 = net.addController('c0', controller=RemoteController, ip='
        192.168.56.101', port=6633 )

    print("*** Adding Link")
    net.addLink(h2,s2)

    print("*** Starting network")
    c0.start()
    s2.start( [c0] )
    # set up GRE tunnel between 192.158.56.101 and 192.168.56.104
    s2.cmd("ip link add s2-gre1 type gretap local 192.168.56.104 remote
        192.168.56.101 ttl 64")
    s2.cmd("ip link set s2-gre1 up")
    Intf("s2-gre1", node=s2)
```

```
    print("*** Running CLI")
    net.start()
    CLI( net )

    print("*** Stopping network")
    s2.cmd("ip link del dev s2-gre1")
    net.stop()

if __name__ == '__main__':
    setLogLevel( 'info' )
    topology()
```

The system is now configured and ready for running the experiment. At this point, we should log on VM#1 and run the POX SDN controller on a separate terminal with the following command:

```
VM#1$ ./pox/pox.py pox.forwarding.l2_learning
```

This will instruct switches to operated as Layer 2 learning switches. To generate the network infrastructure, we should run the scripts we created before on each machine:

```
VM#1$ sudo python vm1_script.py
VM#2$ sudo python3 vm2_script.py
```

At this point, the merged Mininet environment is up and running. In particular, the terminal running the POX controller should report that the two switches on both VMs are connected.

We can now check the correct configuration by pinging from Host 1 (on VM#1) to Host 2 (on VM#2):

```
mininet> h1 ping 10.0.0.2 -c 3
```

We can check the switch configuration and identify the GRE tunnel by logging, for example, on switch *s2* on VM#2:

```
mininet> s2 ovs-vsctl show
1627d3fe-f88f-44aa-b651-9f3a2916f9af
    Bridge "s2"
        Controller "tcp:192.168.56.101:6633"
            is_connected: true
        fail_mode: secure
        Port "s2"
            Interface "s2"
                type: internal
        Port "s2-eth1"
```

```
        Interface "s2-eth1"
    Port "s2-gre1"
        Interface "s2-gre1"
ovs_version: "2.9.2"
```

Similar configurations can be realized across the world. This enables the incorporation of actual software (e.g., external controllers) and devices (e.g., actual SDN switches) into *ComNetsEmu* environments, ranging from teaching over designing and prototyping to implementations of solutions for current and future computing in communication networks – the possibilities are endless!

24.4 Exercises

This section provides some exercises aimed at developing interconnected test beds.

24.4.1 Exercise 1

Following the example in Section 24.3, build a distributed test bed consisting of three or more Virtual Machines running the emulator of the book or Mininet, managed by a single controller.

24.4.2 Exercise 2

Build a Mininet topology in the provided Virtual Machine and enable an external SDN controller to connect to the SDN switches.

Integrating time-sensitive networking

25

Marian Ulbricht, Javier Acevedo
Technische Universität Dresden, Dresden, Germany

The only reason for time is so that everything doesn't happen at once.
Albert Einstein

25.1 Introduction

Ethernet is the most important technology for wired and wireless networking. Consequently, it is the first choice when data need to be transferred between computer systems. Adapting heterogeneous ecosystems to a common information technology saves resources and reduces costs. Due to its robust design, Ethernet allows hot *plug and play* of devices, regardless of dropped packets and transmission delays. Nevertheless, for some applications, this undeterministic behavior is not acceptable. In industrial implementation scenarios, for example, applications based on bus systems, for example, CAN, Ethercat, or Profinet, must fulfill time-critical constraints to provide real-time communication between devices. This chapter describes the TSN technology, an extension of the Ethernet standard, which enables the deployment of time-critical applications through real-time data transmissions. The main motivation for the development of TSN is adapting Ethernet technology to fulfilling latency and redundancy requirements in industrial applications. TSN is a generic term used for many of the IEEE standards to describe time-sensitive extensions of the Ethernet technology. Many device vendors employ the term TSN-ready to promote their products, but they support only a subset of the standards described within this chapter. A mandatory feature for managing time-aware network devices is a common time base. The methods described in the IEEE802.1AS standard provide the time synchronization requirements to select and distribute the best clock reference through a time-sensitive network. The functionality of TSN is based on the Time-Aware Shaper (TAS), which is fully described in the IEEE802.1Qbv standard. The TAS supports a time-controlled and cyclic opening and closing device ports. This makes a network deterministic as it can be programmed when a device opens or closes the gate of a transmit data queue. Then, for a known network routing implementation, the maximum latency of each packet is defined if the TAS configuration of each device is known. The TAS handles only forwarding of outgoing packets. Consequently, in theory, TSN employing only this technology can be flooded by any device connected to the input ports of any TSN switch. The flooding packets simply fill the transmit queues, whereas the deterministic behavior of the network is not guaranteed anymore. To tackle this issue, the Per-Stream Filtering and Policing (PSFP) is employed, which is defined in the IEEE802.1Qci standard. The PSFP implements a gatekeeper mechanism that protects the TSN network nodes against packets that arrive at a time out of their assigned time slot. Time slot reservation generates gaps during the transmission of packets, especially if full-sized Ethernet frames are considered. An additional time slot, named guard-band, is needed to separate them from packets belonging to other queues. The Frame preemption, defined in the IEEE802.1Qbu standard, introduces the possibility to stop the transmission of a large Ethernet frame until the time slot of the frame is opened again in the next cycle. This chapter provides an overview of the main TSN standards and their relationships. At the end of the chapter, a hands-on experiment will show the time shaping protocols in action within the *ComNetsEmu*.

25.2 IEEE802.1AS – if timing matters

A key feature of TSN is its highly accurate clock distribution in the whole network. The Precision Time Protocol (PTP), which is defined in IEEE1588 [335], and its evolution in IEEE802.1AS [336], defines a mechanism to synchronize the clocks of several network devices at the scale of microseconds. In the IEEE802.1AS [336] standard the concept of clock distribution was simplified to a network structure containing time-aware endpoints and time-aware bridges. Fig. 25.1 illustrates an example network comprised of time-aware network devices. As depicted, the clock distribution is defined for several MAC architectures, for example, Ethernet and IEEE802.11 (WLAN). Every time-aware node has a hardware or software clock, which is synchronized using L2 or L3 PTP frames. In theory, each time-aware device is able to distribute its clock over the network. Practically, one single reference clock is selected in the network, which is named *clock-grandmaster*. For each bidirectional clock relationship, master and slave clocks are selected. For clock selection, the clock quality information is included in every PTP message. This clock description, consisting of clock priority and class information, encodes the type of the clock, for example, crystal oscillator, Global Positioning System (GPS)-based clock, or atomic clock. Each device selects the best priority and accuracy values. In case of a time-aware bridging device, this clock is distributed continuously over the network. To transfer the current clock value from master to slave peer through a PTP Peer to Peer (P2P) connection, a measurement about the path delay of the used link is required. Figs. 25.2 and 25.3 illustrate the path delay measurement procedure for Ethernet and WLAN technologies, respectively. Both figures show a two-step delay measurement procedure. The network stack of both timing-aware devices is able to report the current transmission point in time for each packet back to the transmitting application. The receiving time is reported from the hardware or software stack to that application as well. Hardware-based timestamping is much more accurate than software-based timestamping, because the timestamp is reported directly by the network hardware

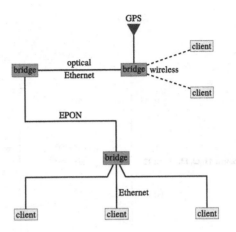

FIGURE 25.1

IEEE802.1AS clock distribution network structure [336].

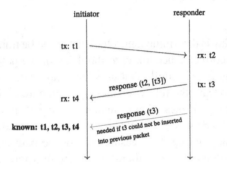

FIGURE 25.2

Path delay measurement for Ethernet MAC.

and represents the point in time where the packets are delivered from the Physical Layer Chip (PHYchip) and transmitted over the line. With this feature, the effective transmission time can be reported in a follow-up message. As shown in Fig. 25.2, the procedure contains three messages. The responder submits the receiving time of the first message back to the initiator using the data space of the second message. Finally, the transmission time of the second message is transmitted in a third message from the responder to initiator. At this point the initiator has all the information about the timestamps t1, t2, t3, and t4, which represent the sending and receiving times of the first two messages.

The path delay can now be calculated using the formula

$$d = \frac{(t_2 - t_1) + (t_4 - t_3)}{2}.$$

(25.1)

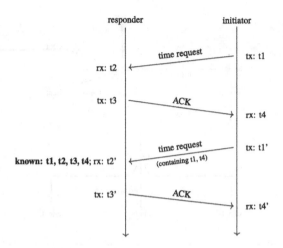

FIGURE 25.3

Path delay measurement for WLAN MAC.

Once the path delay is determined, the clock value can be transmitted from master to slave and will be corrected according to the delay of the path. The procedure on WLAN shown in Fig. 25.3 is similar but uses the delay request message of the next measurement cycle to leave the follow-up message. Time-aware bridges use a special correction field in the PTP messages to announce their own message propagation delay. Theoretically, PTP messages could be forwarded via nontime-aware hops. Due to the lack of accuracy during the measurement of the path delay, these paths are omitted for the selection of the grandmaster and are downgraded in comparison to native time-aware links.

25.3 Different shapes of packets – IEEE802.1Qav and IEEE802.1Qbv

For time-sensitive transmission between a TSN-talker and a TSN-listener, the communication is classified into streams and traffic classes. Depending on the QoS requirements, different shaping techniques are applied to the packets. Sections 25.3.1 and 25.3.2 introduce the Credit-based Shaper (CBS) and the TAS. The Stream Reservation Protocol (SRP) can be used to automatically announce the required QoS during a time-aware transmission between the talker and listener.

25.3.1 Credit-based shaper

The CBS selects the frames to be transmitted based on credits. Credits are accumulated during waiting times and are decreased when a frame of the corresponding

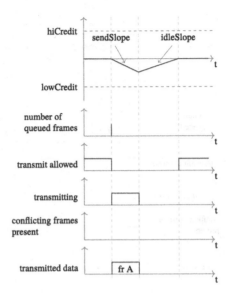

FIGURE 25.4

CBS scenario 1 [337].

traffic class is transmitted. Figs. 25.4, 25.5, and 25.6 illustrate the behavior defined in the IEEE802.1Qav [337] standard for the frames to be transmitted (fr A..C). Each traffic class owns an amount of credits. The rules to handle these credits are the following:

1. If the amount is not negative, then a frame that is queued for transmission will be transmitted. This decision is erroneous when the output resource is empty. In turn, a running transmission will not be interrupted or aborted.
2. If a frame is waiting for transmission, then the amount of credits is increased with a rate determined by the parameter *idleSlope*.
3. If a frame is transmitted, then the amount of credits is decreased with a rate of *sendSlope*.
4. If there are no further frames to transmit from the handled traffic class, then the amount of credits will be set to zero. The amount of credits is limited by upper and lower bounds called *highCredit* and *lowCredit*, respectively.

Fig. 25.4 shows the behavior of the CBS if the system is in idle state and a single frame enters into the transmitting queue. If a frame is blocked by higher prior traffic, then the credit stock is increased by the *idleSlope* until the queue could be opened again as in Fig. 25.5. Fig. 25.6 illustrates a scenario with several queued frames. The IEEE802.1Qav standard recommends the CBS for scheduling video and audio traffic. By tuning the *Credit and *Slope parameters accordingly to the available channel bandwidth and allowed frame sizes, the CBS can be configured to guarantee a particular data rate and maximum hop delay for each traffic class. (We refer the interested

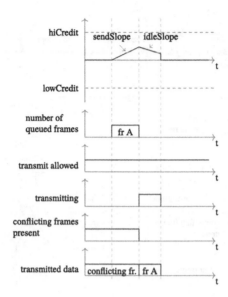

FIGURE 25.5

CBS scenario 2 [337].

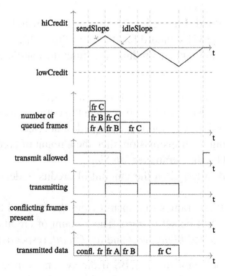

FIGURE 25.6

CBS scenario 3 [337].

reader to [337] for further details.) As shown in Fig. 25.6, the QoS parameters cannot be guaranteed if several streams of the same traffic class are concentrated inside a network node.

25.3.2 Time-aware shaper

Whereas other shapers handle packets and process queues as soon as new packets arrive, the TAS, in contrast, is able to control the queue processing with high accuracy and ensures a deterministic behavior. The TAS can be used to organize the packet forwarding in an Ethernet network through TDMA. Similarly to the CBS, the TAS handles one queue per traffic class. As illustrated in Fig. 25.7, a gate controls the

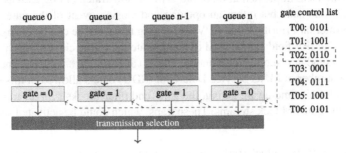

FIGURE 25.7

TAS block view [338].

frame transmission at the end of each queue. The Gate Control List (GCL) rules the states of all gates inside the shaper. Each GCL entry consists of a gate state vector that is a binary mask encoding the gate states and an activation time, which determines the length of the validity of each specific entry. If a GCL entry is activated, then the gate state vector is applied to the gates until the activation time period expires. Subsequently, the next GCL entry is selected. The IEEE802.1Qbv [338] standard additionally allows the reordering of the frames inside each queue based on other shaping algorithms, including the CBS. To archive a deterministic behavior in the whole time-sensitive network, the configuration process of the TAS must be time-sensitive. The parameters *Basetime* and *Cycletime* control the handling of the GCL. The *Cycletime* is defined as the sum of all GCL time slots. The base time defines the point in time where a new GCL handling starts. During the configuration process, a distinction between `oper` times and `admin` times is made. `oper` times contain the actual running configuration, whereas the `admin` time values contain new configuration information that will be activated during the configuration procedure. Once the configuration is activated, the GCL is handled within a loop, which is restarted automatically if the Oper-Cycle-Time expires. The combination of PTP and TAS allows the network administrator to plan a deterministic network that can guarantee a maximum handle time per traffic class.

25.4 IEEE802.1Qci – you shall not pass!

The TAS manages the scheduling of the already queued frames. In contrast, the IEEE802.1Qci [339] standard defines a gatekeeper mechanism preventing frames to be queued if they arrive at a wrong time slot or trigger other filtering thresholds. This instance is designed to organize packets inside streams. Differently put, a flow of packets is matched into a stream, and a stream filter is applied as a first step of the PSFP-block. As shown in Fig. 25.8, each stream is bound to several parameters specifying the priority, gate ID, and meter ID. Several streams can be routed to pass through a specific gate, defined in the next stage of the PSFP block. The gates operate similar to the GCL implementation of the TAS. As illustrated in Fig. 25.8 as well, the GCL defines no state but rather an opening and closing event of the gate. Furthermore, the Internal Priority Value (IPV) belonging to each gate can be changed using this GCL. The IPV field can be used to manipulate the traffic class of the stream, for example, to send it to another TAS queue. The `IntervalOctetMax` field represents the maximum number of MAC Service Data Unit (SDU) octets allowed in this timeslot. In addition to this time-aware gating, the gate- and filter-blocks support traps, which can close the gate or block streams. These traps, which could be activated separately, are:

SDU oversize trap: block traffic due to recognizing an oversized frame.
Invalid RX trap: block traffic routed through a gate due to frames arriving during while the gate was closed.
Interval octet trap: block traffic due to an exceeding number of octets per timeslot.

The traps, which can be enabled one by one, offer a very strict way to block network traffic that is not aligned to the network time slicing scheme or to prevent flooding the network with oversized frames. On the other hand, a skilled network administration is needed to not overblock important communication. The last PSFP stage, illustrated in Fig. 25.8, is used for collecting flow information. The measured pa-

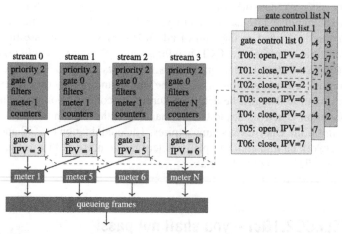

FIGURE 25.8

PSFP block view [339].

rameters, for example, CBS, Commited Information Rate (CIR), Excess Burst Size (EBS), and Excess Information Rate (EIR), can be used to calculate and control the PSFP settings.

25.5 IEEE802.1Qbu, IEEE802.3br – filling the gaps

Using the TAS described in Section 25.3.2, a queued frame can be transmitted if the gate is open at the end of the corresponding queue. Once a transmission of a frame is started, it will be not interrupted, even if the gate closes during the transmission of the frame. Due to possible ongoing transmissions of low prior traffic, a guard band has to be introduced before a phase with high-priority traffic begins. This guarantees that the bandwidth, which is planned for high-priority frames, is not blocked by ongoing frames of the last time slot. The length of the guard band depends on the link speed and maximum size of the low-priority frames. With the frame preemption, the guard band can be minimized or removed. This fact enlarges the available link capacity. As shown in Fig. 25.9, the frame preemption allows the split of Ethernet frames into sev-

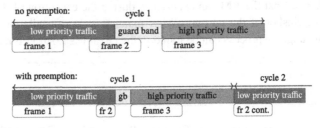

FIGURE 25.9

Frame preemption comparison.

eral pieces. In the current example the second frames are transmitted at the end of the low-priority time slot. Without frame preemption, a guard band is needed to finish the frame before the high-priority traffic is scheduled. With the frame preemption, the transmission of the second frame ends with the low-priority period and is continued at the beginning of the next low-priority time slot. A preempted frame is labeled by a special preamble and CRC sum. The IEEE802.3br [340] standard defines four different preamble types. A preempted frame can be split into four pieces. For the preemptive frames, the CRC is XORed with $0x0000FFFF$, making an inversion of the last part of the frame. Preempted frames are not transmitted via several hops; instead, each device with preemption capabilities reassembles the frames. If a frame is transmitted to a device that does not support preemption, then this frame is discarded because the CRC is changed.

25.6 Hands-on: time-sensitive queueing in the new Linux kernel 5.2

Versions of the Linux Kernel above release 5.0 introduce a new feature of time-sensitive packet handling, taprio and net/sched [341]. This feature includes time-scheduled queue selection, similar to the TAS described in Section 25.3.2, and the specification of a time-point transmission inside the packet socketbuffer structure. This section shows how to configure a simple time-gated frame transmission using the *ComNetsEmu*. We note that the TSN standard provides an accuracy in the range of nanoseconds. Nevertheless, software-based packet processing cannot match up to that level of accuracy. Consequently, to use TSN in real-world applications, a hardware-accelerated TSN switch should be used. One example can be found in [342].

25.6.1 ComNetsEmu setup

The deployment of this testbed setup uses the *ComNetsEmu* VM, with further details described in Chapter 13. To use the new features of the Linux kernel, some system upgrades are necessary. For this implementation, the system requires the installation of the Linux kernel version 5.2.15 and some additional tools, such as iproute2. To tackle changes that the VM can experience during the execution of different experiments, a snapshot of the current VM is mandatory. An installation script that automates the upgrade process can be found in the comnetsemu/util directory.

```
$ cd comnetsemu/util
$ ./install_TSN_testbed.sh
```

This script launches the first stage of the installation process. The VM is restarted at the end of the installation process. To initiate the second stage of the installation procedure, the execution of the following commands is necessary:

```
$ cd ~/TASim
$ ./setup.sh 2
```

After a successful installation, there should be four additional network interfaces, TN0, TN1, TN2, and TN3.

25.6.2 Using the TAS simulator

To use the *taprio* features, a network interface with several TX queues is required. Actually, only some Intel NIC hardware supports this property. The installed *TASim* provides four multiqueue interfaces, where a pair of two is connected to a bidirectional tunnel.

25.6.3 Preparing the TAS

The user can use the command in Listing 25.1 to configure the taprio queues on the virtual interface TN2. This example is based on the tutorial of [343]. A detailed description of the used tc command can be found in Section 27.5.

```
$ sudo tc qdisc add dev TN2 parent root handle 100 taprio \
num_tc 2 \
map 1 0 1 1 1 1 1 1 1 1 1 1 1 1 1 1 \
queues 1@0 1@1 \
base-time 0 \
sched-entry S 01 500000000 sched-entry S 02 500000000 sched-entry S 00
    500000000 \
clockid CLOCK_TAI
```

Listing 25.1: Taprio configuration.

In this example, two traffic classes are in use (*num_tc 2*). The Linux kernel can handle up to 16 priorities, which are mapped to two traffic classes, similarly to the TAS queues described in Section 25.3.2. The priority 0 is mapped to the traffic class 1, and priority 1 is mapped to the traffic class 0. The priorities from 2 to 15 are assigned to traffic class 1 as well. The next line maps two traffic classes to the TX queues of the networking device. Overall, traffic with priority 0 is in queue 0 and traffic with priority 1 is in queue 1. The next lines follow the definitions of the IEEE802.1Qbv standard. A basetime of 0 indicates that the new configuration should be applied immediately. The GCL contains three entries, which open queue 0 and queue 1 for 0.5 seconds, followed by 0.5 seconds of guard band. The assignment of a traffic class to the test packets is based on iptables:

```
$ sudo iptables -t mangle -A POSTROUTING -d 1.1.1.1 -j CLASSIFY --set-class
    0:1
$ sudo iptables -t mangle -A POSTROUTING -d 2.2.2.2 -j CLASSIFY --set-class
    0:0
```

This rule labels every traffic with the ip destination 1.1.1.1 to the traffic class 1.

25.6.4 Measurement and results

The TAS on port TN2 is now configured to schedule the outgoing packets. We will use tcpreplay to replay traffic of a prerecorded file. Since the tcpreplay traffic is not affected by iptable rules, a small trick is required by employing two network tunnels connected with a bridge. The following command configures a bridge to connect TN1 and TN2:

```
$ ./install_bridge.sh
```

The packet transmission is started with the following command:

```
$ sudo tcpreplay -i TN0 --loop=0 -p10 testpackets.pcap
```

With the command `tcpdump` described in Chapter 27, the user can inspect the traffic on the interfaces TN1 and TN3, which are the outgoing points of the virtual tunnels.

```
$ sudo tcpdump -i TNx -n
```

Fig. 25.10 depicts the expected output of the interfaces TN1 and TN2. On port TN1, tcpdump shows some IP addresses alternating their destinations. Fig. 25.10A shows that on port TN3 the packets are sorted by their destination addresses.

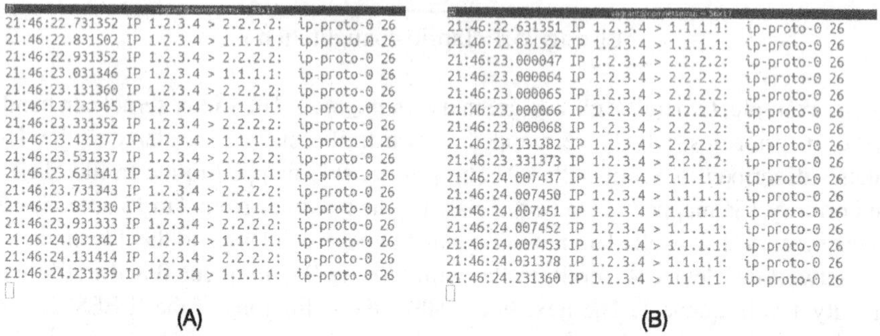

```
21:46:22.731352 IP 1.2.3.4 > 2.2.2.2:  ip-proto-0 26
21:46:22.831502 IP 1.2.3.4 > 1.1.1.1:  ip-proto-0 26
21:46:22.931352 IP 1.2.3.4 > 2.2.2.2:  ip-proto-0 26
21:46:23.031346 IP 1.2.3.4 > 1.1.1.1:  ip-proto-0 26
21:46:23.131360 IP 1.2.3.4 > 2.2.2.2:  ip-proto-0 26
21:46:23.231365 IP 1.2.3.4 > 1.1.1.1:  ip-proto-0 26
21:46:23.331352 IP 1.2.3.4 > 2.2.2.2:  ip-proto-0 26
21:46:23.431377 IP 1.2.3.4 > 1.1.1.1:  ip-proto-0 26
21:46:23.531337 IP 1.2.3.4 > 2.2.2.2:  ip-proto-0 26
21:46:23.631341 IP 1.2.3.4 > 1.1.1.1:  ip-proto-0 26
21:46:23.731343 IP 1.2.3.4 > 2.2.2.2:  ip-proto-0 26
21:46:23.831330 IP 1.2.3.4 > 1.1.1.1:  ip-proto-0 26
21:46:23.931333 IP 1.2.3.4 > 2.2.2.2:  ip-proto-0 26
21:46:24.031342 IP 1.2.3.4 > 1.1.1.1:  ip-proto-0 26
21:46:24.131414 IP 1.2.3.4 > 2.2.2.2:  ip-proto-0 26
21:46:24.231339 IP 1.2.3.4 > 1.1.1.1:  ip-proto-0 26
```
(A)

```
21:46:22.631351 IP 1.2.3.4 > 1.1.1.1:  ip-proto-0 26
21:46:22.831522 IP 1.2.3.4 > 1.1.1.1:  ip-proto-0 26
21:46:23.000047 IP 1.2.3.4 > 2.2.2.2:  ip-proto-0 26
21:46:23.000064 IP 1.2.3.4 > 2.2.2.2:  ip-proto-0 26
21:46:23.000065 IP 1.2.3.4 > 2.2.2.2:  ip-proto-0 26
21:46:23.000066 IP 1.2.3.4 > 2.2.2.2:  ip-proto-0 26
21:46:23.000068 IP 1.2.3.4 > 2.2.2.2:  ip-proto-0 26
21:46:23.131382 IP 1.2.3.4 > 2.2.2.2:  ip-proto-0 26
21:46:23.331373 IP 1.2.3.4 > 2.2.2.2:  ip-proto-0 26
21:46:24.007437 IP 1.2.3.4 > 1.1.1.1:  ip-proto-0 26
21:46:24.007450 IP 1.2.3.4 > 1.1.1.1:  ip-proto-0 26
21:46:24.007451 IP 1.2.3.4 > 1.1.1.1:  ip-proto-0 26
21:46:24.007452 IP 1.2.3.4 > 1.1.1.1:  ip-proto-0 26
21:46:24.007453 IP 1.2.3.4 > 1.1.1.1:  ip-proto-0 26
21:46:24.031378 IP 1.2.3.4 > 1.1.1.1:  ip-proto-0 26
21:46:24.231360 IP 1.2.3.4 > 1.1.1.1:  ip-proto-0 26
```
(B)

FIGURE 25.10

Terminal showing the packet shaping inside the ComNetsEmu. (A) Test traffic before TAS scheduling; (B) Test traffic after TAS scheduling.

In Fig. 25.10B a group of packets with the IP address 1.1.1.1 are followed by a group of packets with the IP address 2.2.2.2. Then a 0.5 seconds gap follows until the GCL repeats its pattern. The user can employ the -w argument to write the captured traffic of the TN3 interface into a file and inspect it outside the VM using the *IO graph* feature of Wireshark. A detailed description of Wireshark can be found in Chapter 27.

Integrating software-defined radios

26

Javier Acevedo, Marian Ulbricht, Dongho You
Technische Universität Dresden, Dresden, Germany

When wireless is fully applied the earth will be converted into a huge brain,
capable of response in every one of its parts.
Nikola Tesla

26.1 Introduction

With the rapid growth of the wireless/mobile communication markets and continuous developments of new communication technologies, users are demanding small-sized, lightweight, and low-cost terminals that can receive high-quality services. In turn, service providers require multistandard communication protocols capable of providing flexibility for implementations of various mobile communication and multimedia services. A realization of these services is practically impossible with traditional fixed hardware modes. Subsequently, a great need exists for technologies that are more flexible and more economical while supporting multimode, multiband, and multifunctional.

Software-Defined Radios (SDRs) are considered to be part of the technologies required to satisfy these demands. An SDR system is a device with physical components comprised of an antenna, an Analog-to-Digital Converter (ADC) and/or a Digital-to-Analog Converter (DAC), and a reprogrammable processor to execute different signal processing applications. SDR can support multiple wireless standards or service functions by a software module implemented in a high-speed processing element capable of programming most functional blocks, excluding the radio frequency domains. Therefore SDR enables service providers to add and remove various wireless standards and functionalities with only one device. As several universal SDR devices with different capabilities have emerged, users including students, engineers, and researchers can employ them for their respective endeavors.

At their core, wireless communication systems have been designed to transmit data by using a specific radio waveform. The specific radio waveform to be utilized depends on the wireless standard to be implemented, for example, WCDMA and LTE. In the past, generating and employing a prescribed waveform for one standard required matching dedicated hardware for that particular standard. Using SDR overcomes this drawback and provides the flexibility to dynamically select a radio waveform, as we will discuss in this chapter.

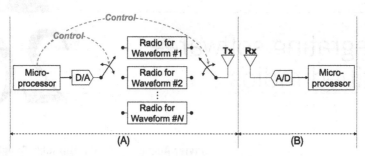

FIGURE 26.1

Basic concept of (A) SCR and (B) SDR. Whereas processing SCR enables switching between a limited number of predetermined discrete radio characteristics, the SDR approach results in full flexibility of the radio, where many radio components can be configured by software.

Fig. 26.1 illustrates the basic concept utilizing an SDR receiver and a traditional Software-Controlled Radio (SCR) sender. A microprocessor of an SCR device can dynamically control a radio waveform according to system requirements. The waveform that can be selected varies according to the configuration of the SCR hardware. Although intuitively, the sender radio overall seems to be software-controlled (and thus to be SDR), it is strictly an SCR. SDR *defines* the radio waveform to be used in the microprocessor rather than controlling something to select the waveform. In addition, one or more communication blocks, such as FEC or modulation, can be defined by software (we discuss further details in Section 26.2.1). Therefore, the digital data from the microprocessor are converted into the desired waveform and then sent to the antenna. The receiving process is the opposite of this, as illustrated in Fig. 26.1B for an SDR receiver.

Generally, significant time is required to adapt new technologies proposed by researchers and engineers for real-world implementations. One of the reasons is the abundance of unexpected issues arising in practical wireless environments, which cannot be determined ex ante. Employing SDR has the potential of unearthing and solving some of these issues before broader implementations, in addition to providing other advantages, such as:

Interoperability: SDR can be used not only for communication with multiple incompatible radios, but also for relaying between them. Therefore it is very suitable for various use cases, ranging from personal to military [344].

Compatibility: Wireless standards, such as WiFi and LTE, can be imported to an SDR device by only updating software. Furthermore, various necessary system functions can be inserted and utilized.

Lower Cost: Service providers can add many additional standards and functions to an SDR device through software modifications. This is economical, as even after deployment, service providers can add many standards and functions to an SDR device. Subsequently, the cost of maintenance and training is also reduced. As there are many universal SDR devices, students, engineers, and researchers can employ them for general purpose activities.

Spectrum Reuse: SDRs enable selecting the center frequency dynamically, and in turn underutilized spectrum can be exploited efficiently. This fact enables SDR users to significantly increase the overall spectrum access.

Energy Efficiency: SDR has the ability to generate the waveform in accordance to system requirements. For instance, a low-power waveform can be used for the IoT communication standards demanding low battery consumption.

These advantages for SDR have fueled its broad adoption. In the following, we describe the overall principles in greater detail before providing examples for how to employ SDR in the context of this book as an extension to the *ComNetsEmu* experiments introduced in earlier chapters.

26.2 Basic principles

26.2.1 What is programmable in SDR?

Fig. 26.2 illustrates a generic wireless communication system through typically employed building blocks for common functionalities. Most of these communication system blocks can be programmed in an SDR environment, as highlighted in Fig. 26.2. The data source is normally transmitted to the data sink. However, before transmitting and after receiving, the data source must be processed to make transmission and reception reliable and efficient.

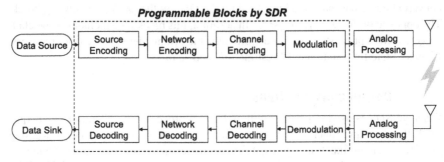

FIGURE 26.2

Modern wireless communication system blocks. By using SDR the baseband functions of each communication building block are implemented in software.

When a binary data source, such as image/video and audio, is introduced on the transmitter side, the source should be compressed to increase the transmission efficiency. This is generally performed in the *source encoder*. In the case of video data, for example, there are many redundancies between successive video frames, and the number of these can be reduced by the video encoder, such as with High Efficiency Video Coding (HEVC) [345]. Note that the inverse process is handled on the receiver side in the *source decoder*. Once the source data has been compressed, a *network encoder* can be employed to improve network throughput and performance. This can be considered an optional technique in traditional communication systems. Since its performance is widely proven in multihop wireless networks, it is considered a key enabling technology in 5G networks [346]. In general, network coding does not increase the data rate, and RLNC [173] and Fulcrum Network Coding (FNC) [347] are notable examples. A *network decoder* is normally used to recover the network-coded packets by Gaussian elimination. We describe network coding in greater details in Chapter 9. The next step, performed through the *channel encoder*, is adding special redundancy bits (i.e., parity bits) to increase the protection capability from the potential errors occurring in the transmission over wireless channels. This is especially important for wireless broadcast channels that do not consider retransmissions, and hence it is also referred to as the FEC. Turbo codes [348], Low-Density Parity-Check (LDPC) codes [349], and polar codes [350] are notable examples used in modern wireless communication systems. A channel decoder is used on the receiver side to return the binary information back to its original form by removing the parity bits. The channel-coded bits should be mapped into a certain electromagnetic waveform employing amplitude, frequency, and phase by a *modulator*. On the opposite side, a *demodulator* converts the electromagnetic waveform back into binary bits. However, sometimes the demodulator outputs soft-bits (not binary bits) such as the Log-Likelihood Ratio (LLR) values to use for iterative decoding process in the channel decoder. Since this case shows high bit error performances in the receiver, but the decoding complexity is high, the number of iterations should be controlled in accordance with terminal capabilities. Finally, the modulated waveform is sampled and converted into a baseband analog signal by the DAC in the *analog processing* block. The opposite of the DAC is the ADC. In addition, this block also includes some challenges, such as pulse shaping, bandwidth, data rate, frame detection, carrier recovery, Channel State Information (CSI) estimation, and so on.

26.2.2 Design considerations

When an SDR is used to perform a digital signal processing application, the majority of the baseband functionality is implemented in software throughout processing blocks. To achieve the design goals, the programmer has to consider some parameters that can constrain the functionality of a wireless application. We list some of those challenges:

Pulse Shaping: In real wireless communication systems, pulse shaping is an important factor to consider in the design of transmit and receive filters, because it defines the conditions to maximize spectrum utilization while minimizing the Inter Symbol Interference (ISI) to shrink bit error rates during the transmission of digital pulse bit streams. To fulfill those two properties, the pulse shape must be wide in time and satisfy the Nyquist condition to make the ISI zero [351,352].

Bandwidth and Data Rate: In wireless digital communications, it is important to establish the maximum bandwidth and data rate that a transceiver can achieve to prevent high bit error rates. The Nyquist formula calculates the upper bound in data rate at which data can be transmitted in noise-free channels without bit errors.

Frame Detection: Framing is essential for a receiver (or decoder) to determine the time of data detection and when data was delivered from the transmitter to initiate the decoding process. In practical wireless channels, it is a challenge to achieve synchronization between the receiver and transmitter due to propagation and computation delays. Therefore, in wireless applications, SDR designers should consider synchronization schemes, such as framing bit or syncwork framing.

Several additional design considerations exist for SDR systems, such as shadowing, multipath, multiantennas, multiple access, or indoor/outdoor environments. However, these considerations are out of the scope of this chapter. We refer the interested reader to [353] for more details.

26.2.3 Design constraints

SDR systems stand out for providing a high level of reconfigurability and trivial programmability: Users can implement custom functions using a graphical interface to create signal flow graphs. Although these features ensure flexibility in the design of hardware radios, they simultaneously constrain the throughput and provoke higher latency when processing a signal. SDR is based on a combination of a General Purpose Processors (GPP) and a Digital Signal Processor (DSP). Therefore a bus system is employed to transfer samples from the RF front-end to the GPP of the host computer, introducing latency that is not negligible. Furthermore, the computation of signal processing on the GPP results in low throughput due to the scheduling of buffers inside the operating system to compute incoming data. For instance, the Ettus Research Universal Software Radio Peripheral (USRP) N210 is connected to the host computer via Gigabit Ethernet, which has a bandwidth of 25 MHz. Gigabit Ethernet has a maximum data rate of 125 MBps, but the USRP cannot handle more than half of the Ethernet sample rate, 62.5 MBps. For 16-bit complex samples of 4 bytes each, the maximum sample rate is approximately 15 MS/s, which is significantly lower than the maximum sample rates of 50 MS/s or 25 MS/s for 8-bit or 16-bit samples, respectively (details as provided by the manufacturer). Subsequently, the connection interface appears as an additional bottleneck. To overcome these limitations, it is necessary to employ new parallel pipeline techniques in the FPGA for the physical layer. Furthermore, the separation of the data and control flow is implemented in hardware in the MAC layer. However, the development of hardware on the internal FPGA requires the design and verification of each hardware block using Hardware Description Languages (HDL).

26.3 Software stacks

To implement any signal processing application on an SDR system, it is necessary to consider the entire architecture of the device to understand its programming environment and hardware limitations. This section describes the Ettus Research SDR hardware and one of its programming frameworks, which will be later employed on the design of an OFDM transceiver in Section 26.4.

26.3.1 Universal Software Radio Peripheral (USRP)

Ettus Research has developed its own family of SDR systems and denominated them USRP. The USRPs are configured via software to be tunable transceivers for designing, prototyping, and deploying radio communication systems in several frequency bands. Although they can be paired with National Instruments' LabView as development framework, for the purpose of this chapter, they will be programmed through GNU Radio (which will be described in detail in Section 26.3.2). Several interfaces, such as Ethernet, USB, or even Thunderbolt, can be employed to connect a host computer to a USRP. Ettus Research has developed a free and open-source driver, called USRP hardware driver (UHD), which provides portability across the USRPs devices.

The design of the transceiver presented in Section 26.4 was based on the USRP N210 series. This USRP provides 50 MS/s bandwidth of complex samples to receive and transmit directions using a Gigabit Internet interface, which is ideal for physical layer prototyping and dynamic spectrum access applications. It has a built-in Spartan 3A-DSP 3400 FPGA, and a MIMO expansion port used for synchronization when connecting two devices of the same series.

26.3.2 GNU radio

GNU Radio is a free and open-source development framework, which provides the processing blocks and tools to design and implement software radios or signal processing applications either on physical RF hardware or simulation-based environments. The GNU Radio applications are based on flow graphs, in which an extensive library of processing blocks are interconnected to deploy signal processing functionalities. Each processing block contains parameters that can be set depending on the requirements of the application. At its beginning, GNU Radio was conceived for radio amateurs and enthusiasts, but it gathered interest from industry and academia as an alternative to proprietary frameworks and drivers that inhibited the comparisons and dissections of multiple solutions.

The GNU Radio application consists of a flow graph, where the vertices of each graph represent the processing blocks, such as signal sources and sinks, whereas the edges constitute the data flows between them. Each block contains attributes, such as the type of data that the block can handle and configuration parameters about the input and output ports. Then the signal sources are characterized by outgoing ports, whereas the sinks feature incoming ports. The signal processing blocks have a variable number of ports, depending on the purpose of the block. In addition to ports, blocks operate on different data formats, which can be adjusted by the user depending on the design requirements.

The GNU Radio framework employs C++ to ameliorate the performance of the DSP and Python to provide good programmability. Therefore the processing blocks are written in C++, whereas the signal flow graphs are designed in Python. Apart from the processing and signal blocks, GNU Radio is composed by a scheduler, which uses Python's built-in threading module to control the start, stop, and wait operations during the execution of the signal flow graph. GNU Radio utilizes C++ wrappers for Python to extract the signal processing functionalities. The link between C++ and Python is an interface called SWIG2. The GNU Radio programming stack is completed by the communication interface between the host computer and the USRP through the UHD driver, which was primarily developed on GNU/Linux and embedded Linux, but it has been extended in support to proprietary operating systems, such as macOS and Microsoft Windows, as illustrated in Fig. 26.3.

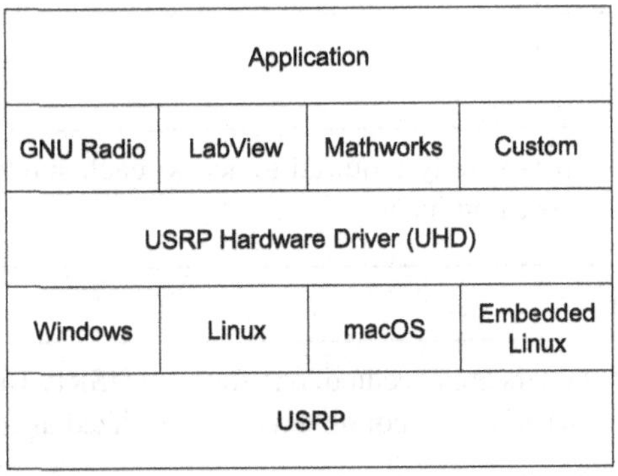

Application			
GNU Radio	LabView	Mathworks	Custom
USRP Hardware Driver (UHD)			
Windows	Linux	macOS	Embedded Linux
USRP			

FIGURE 26.3

Software stack for SDR. To transmit or receive data packets using an SDR-based application, the packets transverse various hardware and software layers initiating from high-level GNU Radio application to low-level physical mapping.

26.4 Examples

We present two basic examples to illustrate how to program real SDR devices. The overall implementation approach is based on the *ComNetsEmu*, which contains the Python execution code within the SDR's application example directory. For these specific examples, GNU Radio is the employed framework to program the Ettus Research's N210 USRPs through executable Python files. However, due to the UHD driver, the following programs can be ported to any USRP N-series device. The first implementation refers to the design of a practical OFDM transceiver using the

GNU Radio Companion, whereas the second implementation is based on the latency measurements realized during the transmission of ping packets between two USRPs through a virtual Ethernet interface.

26.4.1 Setup

The *ComNetsEmu* VM described in Chapter 13 can be used to execute the following examples. To prepare the VM to have access to the USRP hardware, a script containing all the setup parameters must be executed once using the following commands:

```
$ cd comnetsemu/app/integrating_software_defined_ratios/
$ ./setup.sh
```

As described before, this chapter presents two implementations. Each exercise is located in each of the subdirectories of examples:

```
$ ls examples
MODULATION TUNNEL
```

To run each example, it is simply required to access each subdirectory and run the respective docker-compose command

```
$ docker-compose up
```

At this point, two containers are executed, one for each USRP. To log into each of the running containers, Docker's exec command can be utilized as follows:

```
# MODULATION example
$ docker exec -it modulation_sdr1_1 /bin/bash # for the first USRP
$ docker exec -it modulation_sdr2_1 /bin/bash # for the second USRP
or
# TUNNEL example
$ docker exec -it tunnel_sdr1_1 /bin/bash
$ docker exec -it tunnel_sdr2_1 /bin/bash
```

26.4.2 OFDM transceiver exercise

In this example, we evaluate the implementation of an OFDM transceiver. The interested reader can refer, for example, to [354–358] for a broader introduction to the parameter space under consideration for the transmission and reception of data using this modulation scheme.

The implementation starts with the placement of signal processing blocks inside the GNU Radio Companion workspace. The GNU Radio Companion is a graphical interface employed for creating signal flow graphs and generating flow graph source code. To interact with the physical hardware radios, a USRP sink and a USRP source are inserted into the design of the transceiver using GNU Radio's UHD USRP sink and source blocks, respectively. Inside each block, it is necessary to assign some configuration parameters, such as IP address used to identify each USRP, radio frequency options (e.g., channel gains), and the antenna configuration for receiver and transmitter.

A simple OFDM example, provided by the GNU Radio project [359], is illustrated in Figs. 26.4 and 26.5. The figures depict the flow graphs in the GNU Radio

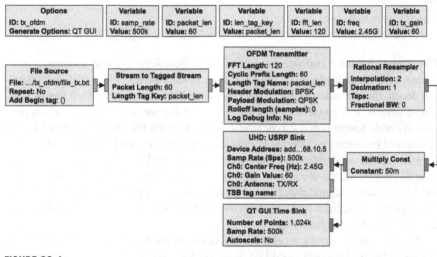

FIGURE 26.4

Transmitter flow graph in GNU Radio. The ODFM transmitter building block is a C++ program in charge of transforming bit streams into baseband modulated signals.

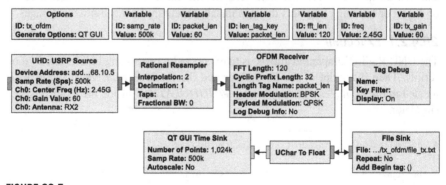

FIGURE 26.5

Receiver flow graph in GNU Radio. The OFDM receiver building block is a C++ program; its functionality is the conversion of a complex modulated signal into a bitstream.

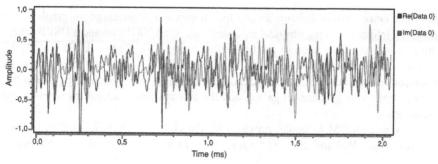

FIGURE 26.6

Transmitted output signal.

Companion for the transmitter and receiver, respectively. For the transmitter side, a file source block is placed on the left corner to load the text file, which is transmitted during the execution of the graph. As the OFDM transmitter block has as input a byte stream, it is required to add the *Stream to Tagged Stream* block, which periodically adds length tags for synchronization. Then the OFDM transmitter block is configured by setting its internal parameters regarding the FFT length, cyclic prefix length, and the modulation for the header and the payload. It provides a complex modulated signal at baseband as output. At the end of the flow graph, a USRP sink is used to transmit the modulated signal by adjusting its center frequency to 2.4 GHz. The signal is visualized by using the *Time Sink* QT GUI block. Fig. 26.6 illustrates the corresponding output signal.

On the receiver side the received signal is a complex baseband signal, which is received and passed through an OFDM receiver block, where the detected data packets are demodulated as a stream of packed bits and stored into a file sink. Depending on the distance at which the USRPs are placed and the configuration parameters, losses can arise during transmissions. In the received file, those losses are presented in terms of wrong or missing characters through misalignment in the transmitted messages.

26.4.2.1 Execution

The GNU Radio Companion allows us to export radio processing blocks into Python scripts. Then the script of the transmitter and the receiver, *transceiver_ofdm.py*, is located in the examples directory *example/MODULATION* inside the ComNetsEmu simulator.

```
$ cd comnetsemu/app/integrating_software_defined_ratios/examples/MODULATION
$ ls
docker-compose.yml file_tx.txt transceiver_ofdm.py
```

The attentive reader can have a look into the script files. To start the example, we execute the *docker-compose* command:

```
$ docker-compose up
```

In this example, two GNU Radio Docker containers are executed to accomplish the transmission of a text file called `file_tx.txt` from one container to the other using two SDRs connected through an external network to the *ComNetsEmu* environment.

```
$ ls
docker-compose.yml file_tx.txt  file_rx.txt transceiver_ofdm.py
```

26.4.2.2 Results and analysis

Once the *docker-compose.yml* file is executed to configure both SDRs, the user can manually gain access to each container using the following commands:

```
$ docker exec -it modulation_sdr1_1 /bin/bash
$ docker exec -it modulation_sdr2_1 /bin/bash
```

With the following command, we run the GNU Radio's Python script:

```
$ ./transceiver_ofdm.py
```

If the data transmission is overall successful, then a file called `file_rx.txt` appears in the working directory, containing the data stream of the `file_tx.txt` file. Thereby note the aforementioned potentials for differences between source and received files. After the Docker containers are executed, the terminal has to exhibit the following outcome, showing some parameters such as the input stream, the packet number, and the offset values from packet to packet:

```
Tag Debug:
Input Stream: 00
Offset: 50820  Source: n/a     Key: packet_num    Value: 1363
Offset: 50820  Source: n/a     Key: ofdm_sync_carr_offset Value: 0
Offset: 50820  Source: n/a     Key: packet_len    Value: 60
```

26.4.3 Latency measurement exercise

The second example demonstrates how two USRPs can communicate with each other via TCP/IP networking using GNU Radio. This implementation illustrates how GNU Radio tunnel code creates an Ethernet virtual interface, typically gr0, between two USRPs through the TUN/TAP Linux kernel modules to tunnel any kind of IP traffic. The main purpose of this implementation is providing the reader with an estimate of the latency during the transmission of OFDM symbols between two USRPs by measuring the RTT.

TUN/TAP enables virtual network devices (i.e., those that are not supported by hardware network adapters, but rather by software) to send and receive data packets to/from a user-space program, which is linked through an operating system to a hardware device. GNU Radio builds a TCP/IP tunnel among two USRPs to transport data packets between both devices. In this *ping* exercise, ICMP Echo packets are selected to be transmitted between the USRPs, as they can provide latency information about the RTT for messages from a source device to a destination device. The RTT is composed of four latency sources [360]: i) latency due to GNU Radio and OS kernel, ii) latency due to communication bus between the host computer and the USRP, iii) latency inside the USRP hardware, and iv) latency in the air interface.

26.4.3.1 Execution

To run the latency example, access to the directory *examples/TUNNEL*:

```
$ cd comnetsemu/app/integrating_software_defined_ratios/examples/TUNNEL
$ ls
docker-compose.yml
```

The attentive reader will have a look into the `docker-compose.yml` file. By launching docker-compose two containers will be created running the OFDM-tunnel example:

```
$ docker-compose up
```

The following Listing 26.1 shows the output of the startup procedure:

```
tunnel_sdr1_1 is up-to-date
tunnel_sdr2_1 is up-to-date
Attaching to tunnel_sdr1_1, tunnel_sdr2_1
sdr2_1  | [INFO] [UHD] linux; GNU C++ version 7.4.0; Boost_106501; UHD_3
    .14.1.0-release
sdr2_1  | [INFO] [USRP2] Opening a USRP2/N-Series device...
sdr2_1  | [INFO] [USRP2] Current recv frame size: 1472 bytes
sdr2_1  | [INFO] [USRP2] Current send frame size: 1472 bytes
sdr2_1  | [WARNING] [UDP] The send buffer could not be resized sufficiently.
sdr2_1  | Target sock buff size: 2500000 bytes.
sdr2_1  | Actual sock buff size: 1048576 bytes.
sdr2_1  | See the transport application notes on buffer resizing.
sdr2_1  | Please run: sudo sysctl -w net.core.wmem_max=2500000
sdr2_1  | [WARNING] [UDP] The send buffer could not be resized sufficiently.
sdr2_1  | Target sock buff size: 2500000 bytes.
sdr2_1  | Actual sock buff size: 1048576 bytes.
sdr2_1  | See the transport application notes on buffer resizing.
sdr2_1  | Please run: sudo sysctl -w net.core.wmem_max=2500000
sdr2_1  | [WARNING] [UDP] The send buffer could not be resized sufficiently.
sdr1_1  | [INFO] [UHD] linux; GNU C++ version 7.4.0; Boost_106501; UHD_3
    .14.1.0-release
```

```
sdr2_1 | Target sock buff size: 2500000 bytes.
sdr1_1 | [INFO] [USRP2] Opening a USRP2/N-Series device...
sdr2_1 | Actual sock buff size: 1048576 bytes.
sdr1_1 | [INFO] [USRP2] Current recv frame size: 1472 bytes
sdr2_1 | See the transport application notes on buffer resizing.
sdr1_1 | [INFO] [USRP2] Current send frame size: 1472 bytes
sdr2_1 | Please run: sudo sysctl -w net.core.wmem_max=2500000
sdr1_1 | [WARNING] [UDP] The send buffer could not be resized sufficiently.
sdr2_1 | [WARNING] [UHD] Unable to set the thread priority. Performance may
    be negatively affected.
sdr1_1 | Target sock buff size: 2500000 bytes.
sdr2_1 | Please see the general application notes in the manual for
    instructions.
sdr1_1 | Actual sock buff size: 1048576 bytes.
sdr2_1 | EnvironmentError: OSError: error in pthread_setschedparam
sdr1_1 | See the transport application notes on buffer resizing.
sdr2_1 | WARN: The gr::digital::ofdm_mapper_bcv block has been deprecated.
sdr1_1 | Please run: sudo sysctl -w net.core.wmem_max=2500000
sdr2_1 | Note: failed to enable realtime scheduling
sdr1_1 | No gain specified.
sdr1_1 | Setting gain to 19.000000 (from [0.000000, 38.000000])
sdr1_1 |
sdr1_1 | No gain specified.
sdr1_1 | Setting gain to 15.750000 (from [0.000000, 31.500000])
sdr1_1 | WARN: The gr::digital::ofdm_insert_preamble block has been
    deprecated.
sdr1_1 | WARN: The gr::digital::ofdm_sampler block has been deprecated.
sdr1_1 | WARN: The gr::digital::ofdm_frame_acquisition block has been
    deprecated.
sdr1_1 | WARN: The gr::digital::ofdm_frame_sync block has been deprecated.
sdr1_1 | /root/.gnuradio/prefs/vmcircbuf_default_factory: No such file or
    directory
sdr1_1 | vmcircbuf_createfilemapping: createfilemapping is not available
sdr1_1 | modulation:    bpsk
sdr1_1 | freq:          2.4
sdr1_1 | Carrier sense threshold: 30 dB
sdr1_1 |
sdr1_1 | Allocated virtual ethernet interface: gr0
sdr1_1 | You must now use ifconfig to set its IP address. E.g.,
sdr1_1 |
sdr1_1 |    $ sudo ifconfig gr0 192.168.200.1
sdr1_1 |
sdr1_1 | Be sure to use a different address in the same subnet for each
    machine.
```

Listing 26.1: Example output of the OFDM tunnel setup

Once the tunnel is created between the containerized applications running on each USRP, the user can gain access to each container through the following commands:

```
$ docker exec -it tunnel_sdr1_1 /bin/bash
$ docker exec -it tunnel_sdr2_1 /bin/bash
```

In each container, there is a virtual interface called *gr0*, which represents the endpoint of the OFDM tunnel. The virtual Ethernet IP address of each device can be set with the following command:

```
$ ifconfig gr0 192.168.200.1 # For the tunnel_sdr1_1
$ ifconfig gr0 192.168.200.2 # For the tunnel_sdr2_1
```

The information about each virtual Ethernet interface can be acquired with the `ifconfig` command:

```
$ ifconfig gr0
gr0: flags=4099<UP,BROADCAST,MULTICAST>  mtu 1500
        inet 192.168.200.1  netmask 255.255.255.0  broadcast 192.168.200.255
        ether 02:42:76:52:9a:cd  txqueuelen 0  (Ethernet)
        RX packets 0  bytes 0 (0.0 B)
        RX errors 0  dropped 0  overruns 0  frame 0
        TX packets 0  bytes 0 (0.0 B)
        TX errors 0  dropped 0 overruns 0  carrier 0  collisions 0
```

26.4.3.2 Results and analysis

The *ping* tool, described in Chapter 27, provides an easy solution to determine the delay between both docker containers over the OFDM link. The variation of packet size and data rate parameters of the ping packets allows us to visualize the available channel bandwidth during the communication.

To specify the packet size of a ping packet, the parameter s is used. The default packet size of a ping packet is 56 bytes. On the other hand, to modify the time interval between sending each packet, the parameter i is used. By default, Linux operating systems employ a time interval between sending each packet of one second. Therefore, to transmit ten packets per second, with each packet having a size of 2000 bytes, we need to employ the following command:

```
$ ping -i 0.1 -s 2000 192.168.200.1/2
```

By using different packet sizes we derive that the bigger the ping packet size, the larger the RTT. Nevertheless, the delay can increase exponentially when the data rate and the packet size increase at the same time. Fig. 26.7 depicts this relationship for small data rates. The RTT depends only on the size of the packet size, that is, bigger

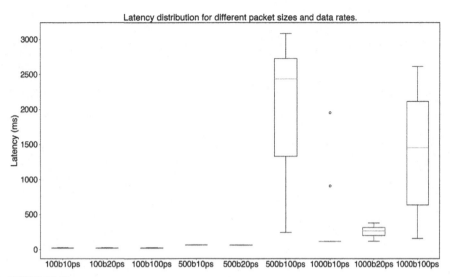

FIGURE 26.7

Latency distribution for different packet sizes and data rates. For small data rates, the RTT depends only on the size of the packet, i.e., bigger packet sizes increase latencies. For high data rates, the RTT increases with the number of packets and the packet size as the USRP cannot handle the amount of data transmitted at those rates.

packet sizes result in larger latencies. For high data rates, the RTT increases with the number of packets and the packet size as the data rate exceeds the processing capacity of the USRP.

PART

Tools

8

Outline

In this last part, we list several tools that are handy for the reader throughout the different chapters of the book. The main idea is to provide a readily available guide describing how to use common networking tools in the presence of the ComNets Emulator.

Networking tools

27

Sreekrishna Pandi, Alexander Kropp, Roland Schingnitz, Sebastian A.W. Itting

Technische Universität Dresden, Dresden, Germany

You cannot mandate productivity, you must provide the tools to let people become their best.
Steve Jobs

The previous parts of this book focused on providing a high-level overview, theoretical backgrounds, and several applications of the theory in a hands-on approach via examples. Whereas the provided examples provide different approaches to the novel networking techniques described throughout this book, several new tools were introduced, both in theory and through practice. These tools are oftentimes employed at the bleeding edge of network practicing and research and applied in the *ComNetsEmu* environment. A significant amount of background familiarity with Linux and the Linux networking tools commonly employed has been assumed. This chapter provides a review and basic knowledge for developing skills in the fundamental Linux networking stack and related tools. These will enable the reader to manage and troubleshoot a network and can be used as a reference as the interested reader works through the examples in this book.

The tools discussed in this chapter are the minimum set of tools that one must be familiar with to design, manage, and troubleshoot any generic network within the Linux environment. (Note that similar or even identical tools exist for other operating systems, but covering each one of them is out of the scope of this chapter.) All the tools explained in this chapter are also included in the *ComNetsEmu* Vagrant image, and hence all examples mentioned in this chapter may be tried on the *ComNetsEmu*. Each described tool contains a brief overview to provide a basic understanding of the purpose and design of the tool, together with a set of minimal usage manual.[1] In this chapter, we discuss the following tools:

ping: Connectivity testing and latency measurement [361,362]
iproute2: Basic network administration [361,363]
iperf: Traffic generation [364]
htop: Process monitoring [365]
tc: Traffic manipulation [251]

---

[1] One may even consider this chapter a poor man's man-page of the essential tools.

Computing in Communication Networks. https://doi.org/10.1016/B978-0-12-820488-7.00044-X
Copyright © 2020 Elsevier Inc. All rights reserved.

tcpdump: Traffic monitoring (terminal) [366]
Wireshark: Traffic monitoring (GUI) [367]
jupyter: Fast Python prototyping [368]

27.1 Connectivity testing – ping

Ping is presumably the most commonly used network administration tool. It is typically utilized to test the connectivity between two nodes in a network. It operates by sending an ICMP *echo-request* message to the specified destination host. Upon receiving the *echo-request*, the host sends an ICMP *echo-response* back to the original sender. A successful ping (echo request and response) signifies that the host is in fact reachable over the IP network. In addition, the time difference between the echo-request and the corresponding response denotes the RTT. The ping tool in the common Linux distributions is the implementation of RFC792 [369] and its respective updates.

The ping command follows the syntax: `ping [OPTIONS] <destination>`. The following is an example command to ping the localhost three times:

```
$ ping -c 3 localhost
PING localhost(localhost (::1)) 56 data bytes
64 bytes from localhost (::1): icmp_seq=1 ttl=64 time=0.027 ms
64 bytes from localhost (::1): icmp_seq=2 ttl=64 time=0.059 ms
64 bytes from localhost (::1): icmp_seq=3 ttl=64 time=0.088 ms

--- localhost ping statistics ---
3 packets transmitted, 3 received, 0% packet loss, time 2044ms
rtt min/avg/max/mdev = 0.027/0.058/0.088/0.024 ms
```

The RTT of each ping may be seen at the end of the line. The statistics about RTT and packet losses are accumulated at the end of the output to get a coarse understanding of the link characteristics. Some of the most useful options to the ping command are as follows [361]:

-c count   The number of pings to send to the destination and after which the statistic is printed. By default this is set to 0, implying that infinitely many pings will be sent until the user interrupts it (using `Ctrl+C`).

-i interval   The time interval between each successive ping in seconds. Default value: 1. This interval is usually reduced to get more precise RTT measurements in a short time. Note that reducing this value to less than 0.2 (200 ms) requires superuser permissions.

-s size   The size of each ping payload in bytes. Default value: 56. The ICMP header is eight bytes long, making the default ping size 64 (56 + 8). This can be increased to troubleshoot MTU issues in the network.

27.2 Basic network administration – iproute2

iproute2 is a suite of tools used to display and manipulate network devices, interfaces, and routing information in a Linux system. It may be considered the *Swiss Army Knife* of Linux networking. One of the most versatile command line tools packaged in *iproute2* is *ip*. The usage of the *ip* tool is as follows:

```
ip [OPTIONS] OBJECT [COMMAND]
```

where, OBJECT := { link |address |addrlabel |route |rule |neigh |ntable |tunnel |tuntap |maddress |mroute |mrule |monitor |xfrm |netns |l2tp |tcp_metrics |token |macsec }

OBJECT is the type of network element intended to be displayed or manipulated. COMMAND denotes the action that should be performed on the specified object. To limit the scope of this book to the most basic and fundamental tools, in this chapter, we introduce only three objects *addr*, *link*, and *route*.

27.2.1 ip addr

`ip addr`, shorthand for `ip address`, is a command commonly used to operate on the IP address of a network interface. It is frequently used to display the current IP address(es) of a specific interface (e.g., to check if the DHCP server on the network has published an IP address for the specific client). It can also be used to manually add/delete an IP address to/from the interface. Note that manipulating addresses require superuser privileges. Example use cases are given further.

```
$ ip addr
1: lo: <LOOPBACK,UP,LOWER_UP> mtu 65536 qdisc noqueue state UNKNOWN group
       default qlen 1000
link/loopback 00:00:00:00:00:00 brd 00:00:00:00:00:00
inet 127.0.0.1/8 scope host lo
valid_lft forever preferred_lft forever
inet6 ::1/128 scope host
valid_lft forever preferred_lft forever
2: eth0: <BROADCAST,MULTICAST,UP,LOWER_UP> mtu 1500 qdisc fq_codel state UP
       group default qlen 1000
link/ether 08:00:27:c2:be:11 brd ff:ff:ff:ff:ff:ff
inet 10.0.2.15/24 brd 10.0.2.255 scope global dynamic eth0
valid_lft 64793sec preferred_lft 64793sec
inet6 fe80::a00:27ff:fec2:be11/64 scope link
valid_lft forever preferred_lft forever
```

Listing 27.1: Display the address information of all interfaces.

As seen before, the `ip addr` command also displays other important metainformation about the network interface in addition to the ip address, such as state of the interface (UP/DOWN), queue length, and MTU. These settings can be manipulated using the `ip link` command (discussed later in this chapter).

```
$ ip addr show dev eth0
2: eth0: <BROADCAST,MULTICAST,UP,LOWER_UP> mtu 1500 qdisc fq_codel state UP
    group default qlen 1000
link/ether 08:00:27:c2:be:11 brd ff:ff:ff:ff:ff:ff
inet 10.0.2.15/24 brd 10.0.2.255 scope global dynamic eth0
valid_lft 64427sec preferred_lft 64427sec
inet6 fe80::a00:27ff:fec2:be11/64 scope link
valid_lft forever preferred_lft forever
```

Listing 27.2: Display the address of only one specific interface.

```
$ sudo ip addr add dev eth0 10.0.8.1/24
$ ip addr show dev eth0
2: eth0: <BROADCAST,MULTICAST,UP,LOWER_UP> mtu 1500 qdisc fq_codel state UP
    group default qlen 1000
link/ether 08:00:27:c2:be:11 brd ff:ff:ff:ff:ff:ff
inet 10.0.2.15/24 brd 10.0.2.255 scope global dynamic eth0
valid_lft 64072sec preferred_lft 64072sec
inet 10.0.8.1/24 scope global eth0
valid_lft forever preferred_lft forever
inet6 fe80::a00:27ff:fec2:be11/64 scope link
valid_lft forever preferred_lft forever
```

Listing 27.3: Add an IP address to an interface.

```
$ sudo ip addr del dev eth0 10.0.8.1/24
$ ip addr show dev eth0
2: eth0: <BROADCAST,MULTICAST,UP,LOWER_UP> mtu 1500 qdisc fq_codel state UP
    group default qlen 1000
link/ether 08:00:27:c2:be:11 brd ff:ff:ff:ff:ff:ff
inet 10.0.2.15/24 brd 10.0.2.255 scope global dynamic eth0
valid_lft 63953sec preferred_lft 63953sec
inet6 fe80::a00:27ff:fec2:be11/64 scope link
valid_lft forever preferred_lft forever
```

Listing 27.4: Delete an IP address from an interface.

27.2.2 **ip link**

The `ip link` command is used to operate on the properties of physical and virtual links created on the interface. These properties include MTU, link type (VLAN, message authentication code, bridge, etc.), ip address, and Transmission (TX) queue length. `ip link` is often used to create and maintain virtual links over physical interfaces. Virtual links are commonly used to support different higher-layer protocols as a virtual interface on top of the physical interface (e.g., IEEE802.1q tagged vlan interface). It is also used to create and maintain different network namespaces and assign the interfaces to them. We discuss this in greater detail in Section 27.8. All interfaces present in the host can be listed with the `ip link` command.

```
$ ip link
1: lo: <LOOPBACK,UP,LOWER_UP> mtu 65536 qdisc noqueue state UNKNOWN mode
    DEFAULT group default qlen 1000
link/loopback 00:00:00:00:00:00 brd 00:00:00:00:00:00
2: eth0: <BROADCAST,MULTICAST,UP,LOWER_UP> mtu 1500 qdisc fq_codel state UP
    mode DEFAULT group default qlen 1000
link/ether 08:00:27:c2:be:11 brd ff:ff:ff:ff:ff:ff
```

The general usage of ip link is `ip link [COMMAND]`. The commonly used commands are `add, delete, set, show`. The `add` and `delete` commands, as the names suggest, are used to create and remove virtual interfaces (links) to the device. The `set` command is used to modify the properties of the given link. Some of the commonly modified properties include *name*, *address*, *mtu*, *state* (up/down), *type* (vlan, macsec, bridge, etc.). For instance, to create an IEEE802.1q vlan interface with the vlan tag 100 on top of the `eth0` interface and set its MTU to 1450, we would do the following:

```
$ sudo ip link add link eth0 name eth0.100 type vlan id 100
$ sudo ip link set dev eth0.100 mtu 1450 up
$ sudo ip link show dev eth0.100
3: eth0.100@eth0: <BROADCAST,MULTICAST,UP,LOWER_UP> mtu 1450 qdisc noqueue
    state UP mode DEFAULT group default qlen 1000
link/ether 08:00:27:c2:be:11 brd ff:ff:ff:ff:ff:ff
```

All the packets going through the `eth0.100` interface will be appended with an IEEE802.1q header with a vlan tag of 100. Only vlan-aware network devices will be able to process these packets and will be dropped by other devices.

27.2.3 **ip route**

A typical Linux computer has multiple network interfaces installed. For instance, a standard laptop has an Ethernet port and a WiFi card on it. It is possible for the laptop to simultaneously be connected to both Ethernet and WiFi networks. Each interface typically has an IP address assigned from the DHCP server of the corresponding network. It is the function of the operating system to determine which packets should go out through which interface. Linux does so by using the routing table, which is essentially a database comprised of a series of rules, called routes, governing whether and how to process the different network packets. `ip route` is a tool used to view, create, and delete rules (routes) in the routing table:

```
$ ip r
default via 10.0.2.2 dev eth0 proto dhcp src 10.0.2.15 metric 100
10.0.2.0/24 dev eth0 proto kernel scope link src 10.0.2.15
10.0.2.2 dev eth0 proto dhcp scope link src 10.0.2.15 metric 100
```

27.3 Traffic generation – iPerf

iPerf is one of the most important networking tools. It can generate traffic between two hosts and works with both TCP and UDP. The use cases of *iPerf* are very diverse. It can be used, for example, to measure the end-to-end bandwidth between two hosts and the required retransmissions (TCP) and losses (UDP) over time. Another use case is the generation of load in the network to simulate certain network conditions for another application. The traffic generation itself can be adjusted via numerous parameters, such as window size, buffer size, or target bandwidth. The current version of *iPerf* is *iPerf3* and can be installed from the default repositories of *Debian/Ubuntu*. *iPerf* uses the client–server model, which implies that it must be first started on the server, and then the client can connect to the server. The *iPerf* command follows the syntax: `iperf3 [-s|-c host] [OPTIONS]` [364]. The following example commands make a performance test for both server(s) and client on localhost with TCP for three seconds:

```
$ iperf3 -s
- - - - - - - - - - - - - - - - - - - - - - - - - - - - - - - - - - - - - - - -
Server listening on 5201
- - - - - - - - - - - - - - - - - - - - - - - - - - - - - - - - - - - - - - - -
Accepted connection from 127.0.0.1, port 50816
[  5] local 127.0.0.1 port 5201 connected to 127.0.0.1 port 50818
[ ID] Interval           Transfer     Bandwidth
[  5]   0.00-1.00   sec  2.87 GBytes  24.6 Gbits/sec
[  5]   1.00-2.00   sec  2.80 GBytes  24.0 Gbits/sec
[  5]   2.00-3.00   sec  3.08 GBytes  26.5 Gbits/sec
[  5]   3.00-3.04   sec  89.8 MBytes  18.7 Gbits/sec
- - - - - - - - - - - - - - - - - - - - - - - - - - - - - -
[ ID] Interval           Transfer     Bandwidth
[  5]   0.00-3.04   sec  0.00 Bytes   0.00 bits/sec        sender
[  5]   0.00-3.04   sec  8.83 GBytes  25.0 Gbits/sec       receiver
```

Listing 27.5: Test the performance of the server for three seconds.

```
$ iperf3 -c 127.0.0.1 -t 3
Connecting to host 127.0.0.1, port 5201
[  4] local 127.0.0.1 port 50818 connected to 127.0.0.1 port 5201
[ ID] Interval          Transfer     Bandwidth       Retr  Cwnd
[  4]   0.00-1.00   sec  2.95 GBytes  25.3 Gbits/sec    0   3.18 MBytes
[  4]   1.00-2.00   sec  2.79 GBytes  24.0 Gbits/sec    0   3.18 MBytes
[  4]   2.00-3.00   sec  3.09 GBytes  26.5 Gbits/sec    0   3.18 MBytes
- - - - - - - - - - - - - - - - - - - - - - - - -
[ ID] Interval          Transfer     Bandwidth       Retr
[  4]   0.00-3.00   sec  8.83 GBytes  25.3 Gbits/sec    0   sender
[  4]   0.00-3.00   sec  8.83 GBytes  25.3 Gbits/sec        receiver

iperf Done.
```

Listing 27.6: Test the performance of a specific client for three seconds.

As seen before, the server begins to listen and continues by accepting the connection request from the client, and, subsequently, the measurements are starting. The current bandwidth is shown after each interval and at the end; statistics are accumulated on both the server and client sides. Additionally, the client process shows information about retransmissions and contention window size when using TCP for transmissions, whereas in UDP mode, jitter and packet loss are displayed instead. Some of the most useful options to the *iPerf* command are as follows:

-s –server   Runs *iPerf* in server mode.

-c –client <host>   Runs *iPerf* in client mode and connects to the server <host>.

-p –port #   Sets the port. Defaults to 5201.

-i –interval #   Sets the output interval in seconds.

-B –bind <host>   Binds to a specific interface.

–logfile <file>   Writes the output to the log file <file>.

-u –udp   Uses UDP instead of TCP.

-t –time #   Sets the transmit time in seconds. Defaults to 10 s.

-b –bandwidth #[KMG]   Sets the target bandwidth in bits/s (0 for unlimited). Defaults to unlimited for TCP and 1 Mbit/s for UDP.

-w –window #[KMG]   Sets the window/buffer size.

-l –len #[KMG]   Sets the read/write buffer length. Defaults to 128 KB for TCP and 8 KB for UDP. The maximum buffer length is 1 MB for TCP and 65507 B for UDP.

-P –parallel #   Sets the number of clients to be connected in parallel.

-R –reverse   Changes the direction of the traffic from client–server to server–client.

27.4 Process monitoring – htop

htop is an interactive tool to monitor processes. It provides many methods not only to filter processes efficiently, but also to actively control processes. Fig. 27.1 illustrates the *htop* interface. Information about the total CPU and memory utilization is presented on the top left, whereas the number of running processes or threads is presented on the top right of the screen. Additionally, the average CPU load of the last 1, 5, or 15 minutes is shown. The table below this overview information is a detailed view of the individual processes and threads. It contains general information, such as process ID and user, as well as CPU and memory usage, runtime, state, and the process tree. The two most common states are **s** (sleeping) and **r** (running). The process tree represents two aspects: on the one hand, the command that is actually executed and, on the other hand, the relationship between parent/child processes. Additional features of the interface enable sorting and filtering the table and navigating it with the arrow keys or the mouse. htop allows us to adjust the priority of processes and also to terminate processes in various ways.

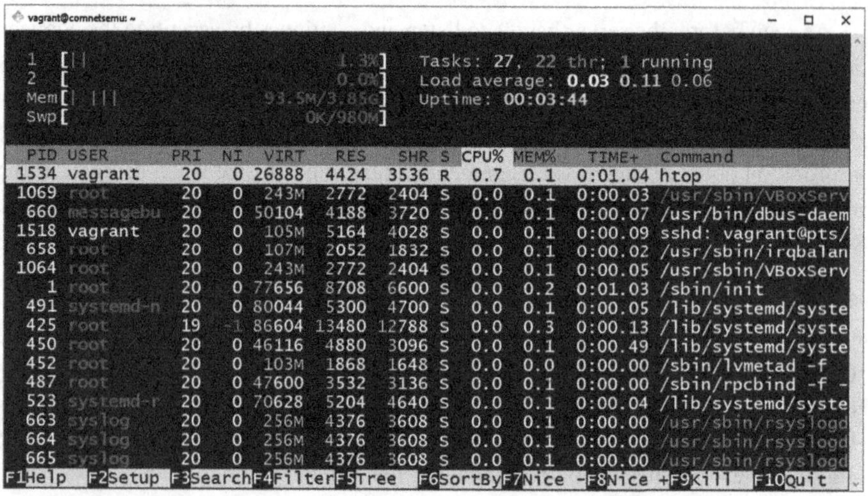

FIGURE 27.1

The interactive *htop* terminal.

Some of the most useful commands and interactive commands for htop are as follows [365]:

-p –pid=PID   Shows only the processes with the following PIDs.
-u –user=USERNAME   Shows only the processes of a specific user.
u   Filters processes by a specified user.
M   Sorts processes by memory usage.
P   Sorts processes by CPU usage.
T   Sorts processes by time

Selects the process with the corresponding PID.

Key[Space]   Tags the selected process. This is important to operate commands on multiple processes.

U   Untags all processes.

Key[F5]   Switches between normal view and tree view, which shows relation between parent and child processes.

Key[F7/F8]   Increases/Decreases the priority of the selected process (requires sudo).

Key[F9]   Terminates the selected process (requires further selection of the specific signal to terminate the process).

27.5 Network traffic manipulation – TC

Traffic Control (tc) is also one of important networking tools and can be used to shape or schedule network traffic. The applications of *tc* are very versatile and range from the prioritization of certain data streams to the emulation of large networks with corresponding link properties. It consists of three main components [251]:

Queueing Disciplines (qdiscs)   Before the kernel sends packets out to a specific interface, they are enqueued to the *qdisc* for that interface. Then the kernel requests packets from *qdisc* of this interface to hand them to the network adapter driver. The default *qdisc* is an adapted version of the common FIFO buffer. A *qdisc* itself can contain several classes and filters or even more *qdiscs*.

Classes   A *qdisc* can contain classes, which then contain further *qdiscs*.

Filters   A filter is used to determine in which class a packet will be enqueued.

The overall structure of the traffic control is a hierarchical tree, where the nodes are *qdiscs*, *classes*, and *filters*. *Classes* and *qdiscs* have IDs that consist of major and minor numbers, denoted as `major:minor`. The major of a *qdisc* is also called *handle*, and the minor in this case is always zero. Therefore it is usually only specified as `major`: *Classes* share the major of their parent *qdisc*. A special value is the *root*, which refers to the root *qdisc* of an interface. Fig. 27.2 represents an example schematic of such a tree. In this example the traffic is initially filtered based on the transport layer protocol (TCP, UDP). The UDP data rate is then capped at 1 Mbit/s, and the TCP traffic is further classified. Web traffic (ports 80 and 443) is restricted to 100 Mbit/s, and the rest to 10 Mbit/s.

The basic structure of the *tc qdisc* command is as follows [251]:

```
tc qdisc [add|change|replace|link|delete] dev <DEV> [parent <qdisc-id>|root] [
    handle <qdisc-id>] <qdisc> [qdisc options]
```

A *qdisc* can be either added, changed, replaced, linked, or deleted. The commands *change* and *replace* work similarly: Whereas *change* only executes on existing *qdiscs* and cannot modify the handle or the parent, *replace* performs *remove* and *add*, which

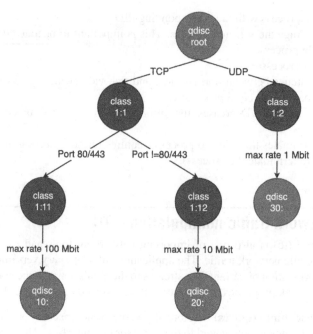

FIGURE 27.2

An example hierarchical tree structure of tc.

allows it to modify everything. *Link* is similar to *replace* but only works on existing nodes.

One of the most used *qdiscs* is *netem*, which is short for network emulator, for example, to add delays and packet losses or to create duplicates. We provide some examples on how to add delay with *netem* [251]. The tool *ping* from Section 27.1 is used to verify the delay.

```
$ ping 8.8.8.8 -c 2
PING 8.8.8.8 (8.8.8.8) 56(84) bytes of data.
64 bytes from 8.8.8.8: icmp_seq=1 ttl=50 time=15.0 ms
64 bytes from 8.8.8.8: icmp_seq=2 ttl=50 time=15.5 ms

$ sudo tc qdisc add dev eth0 root netem delay 200ms

$ ping 8.8.8.8 -c 2
PING 8.8.8.8 (8.8.8.8) 56(84) bytes of data.
64 bytes from 8.8.8.8: icmp_seq=1 ttl=50 time=217 ms
64 bytes from 8.8.8.8: icmp_seq=2 ttl=50 time=216 ms
```

Listing 27.7: RTT before and after adding delay.

It is also possible to add some variation to the delay or change the distribution function of the delay. The following example adds an average delay of 2000 ms and a jitter of 1500 ms. In the example the jitter even causes packet 5 to arrive after packet 4.

```
$ sudo tc qdisc change dev eth0 parent root netem delay 2000ms 1500ms

$ ping 8.8.8.8 -c 5
PING 8.8.8.8 (8.8.8.8) 56(84) bytes of data.
64 bytes from 8.8.8.8: icmp_seq=1 ttl=50 time=1730 ms
64 bytes from 8.8.8.8: icmp_seq=2 ttl=50 time=765 ms
64 bytes from 8.8.8.8: icmp_seq=3 ttl=50 time=1649 ms
64 bytes from 8.8.8.8: icmp_seq=5 ttl=50 time=1962 ms
64 bytes from 8.8.8.8: icmp_seq=4 ttl=50 time=3051 ms
```

Listing 27.8: RTT with delay and jitter.

netem also allows us to directly reorder packets with a certain percentage. In the following example, a delay of 1100 ms is added for around 50% of the packets, causing only packets 2 and 4 to arrive with delay:

```
$ sudo tc qdisc change dev eth0 parent root netem delay 1100ms reorder 50%

$ ping 8.8.8.8 -c 5
PING 8.8.8.8 (8.8.8.8) 56(84) bytes of data.
64 bytes from 8.8.8.8: icmp_seq=1 ttl=50 time=15.4 ms
64 bytes from 8.8.8.8: icmp_seq=3 ttl=50 time=15.6 ms
64 bytes from 8.8.8.8: icmp_seq=2 ttl=50 time=1115 ms
64 bytes from 8.8.8.8: icmp_seq=5 ttl=50 time=16.1 ms
64 bytes from 8.8.8.8: icmp_seq=4 ttl=50 time=1116 ms
```

Listing 27.9: RTT with reordering of packets.

netem furthermore supports the possibility to add packet losses or create duplicates with certain probabilities:

```
$ sudo tc qdisc change dev eth0 parent root netem loss 50%

$ ping 8.8.8.8 -c 5
PING 8.8.8.8 (8.8.8.8) 56(84) bytes of data.
64 bytes from 8.8.8.8: icmp_seq=1 ttl=50 time=15.5 ms
64 bytes from 8.8.8.8: icmp_seq=4 ttl=50 time=15.8 ms
64 bytes from 8.8.8.8: icmp_seq=5 ttl=50 time=16.9 ms
```

Listing 27.10: Packet loss with netem.

```
$ sudo tc qdisc change dev eth0 parent root netem duplicates 100%

$ ping 8.8.8.8 -c 3
PING 8.8.8.8 (8.8.8.8) 56(84) bytes of data.
64 bytes from 8.8.8.8: icmp_seq=1 ttl=50 time=14.9 ms
64 bytes from 8.8.8.8: icmp_seq=1 ttl=50 time=14.9 ms (DUP!)
64 bytes from 8.8.8.8: icmp_seq=2 ttl=50 time=15.5 ms
64 bytes from 8.8.8.8: icmp_seq=2 ttl=50 time=15.5 ms (DUP!)
64 bytes from 8.8.8.8: icmp_seq=3 ttl=50 time=15.8 ms
```

Listing 27.11: Duplicates with netem.

Advanced options for *netem* allow us to define a correlation, which can be employed for emulation of bursts for delay, losses, or duplicates. The following example demonstrates how an existing *qdiscs* can be deleted. Deleting a parent *qdisc* always results in the deletion of all child *qdiscs*, *classes*, and *filters*.

```
$ tc qdisc del dev lo root
```

Listing 27.12: Deleting qdiscs.

With *tc* also the bandwidth of a network can be limited. A suitable *qdisc* for this purpose is *tbf*, which is short for token bucket filter. The arguments for the *qdisc tbf* are `rate`, `buffer`, and `limit`. A buffer defines the maximum burst that can be sent out and limits the amount of queued bytes. Whereas a buffer that is too large can result in the data rate not being throttled, values for buffer and limit that are too small can have a negative effect on the data rate. The tool *iPerf* from Section 27.3 can be used to verify the rate limit as in the following example:

```
$ iperf3 -c 127.0.0.1 -t 3
Connecting to host 127.0.0.1, port 5201
[  4] local 127.0.0.1 port 50818 connected to 127.0.0.1 port 5201
[ID] Interval       Transfer     Bandwidth       Retr  Cwnd
[  4] 0.00-1.00 sec  2.95 GBytes  25.3 Gbits/sec    0   3.18 MBytes
[  4] 1.00-2.00 sec  2.79 GBytes  24.0 Gbits/sec    0   3.18 MBytes
[  4] 2.00-3.00 sec  3.09 GBytes  26.5 Gbits/sec    0   3.18 MBytes
- - - - - - - - - - - - - - - - - - - - - - - - -
[ID] Interval       Transfer     Bandwidth       Retr
[  4] 0.00-3.00 sec  8.83 GBytes  25.3 Gbits/sec    0   sender

$ sudo tc qdisc add dev lo root tbf rate 1Gbit buffer 10M limit 1M

$ iperf3 -c 127.0.0.1 -t 3
Connecting to host 127.0.0.1, port 5201
[  4] local 127.0.0.1 port 43398 connected to 127.0.0.1 port 5201
[ID] Interval       Transfer     Bandwidth       Retr  Cwnd
```

```
[ 4]  0.00-1.00 sec   124 MBytes  1.04 Gbits/sec   0    1.37 MBytes
[ 4]  1.00-2.00 sec   119 MBytes  999 Mbits/sec    0    1.37 MBytes
[ 4]  2.00-3.00 sec   119 MBytes  998 Mbits/sec    0    1.37 MBytes
- - - - - - - - - - - - - - - - - - - - - - - - - -
[ID]  Interval        Transfer    Bandwidth        Retr
[ 4]  0.00-3.00 sec   362 MBytes  1.01 Gbits/sec   0    sender
```

Listing 27.13: Bandwidth before and after rate control.

Several *qdiscs* can also be combined. For example, a bandwidth control *qdisc*, such as *tbf*, can be combined with *netem* to add delay. For this purpose, the first *qdisc* needs to obtain a referable handle. The second *qdisc* is declared to utilize the first *qdisc* as a parent. In the following example the additional delay results in less than the maximum specified rate of 1 Gbit/s, here due to TCP:

```
$ sudo tc qdisc add dev lo root handle 1:0 tbf rate 1Gbit buffer 10M limit 1M
$ sudo tc qdisc add dev lo parent 1:1 netem delay 30ms

$ iperf3 -c 127.0.0.1 -t 3
Connecting to host 127.0.0.1, port 5201
[ 4]  local 127.0.0.1 port 43450 connected to 127.0.0.1 port 5201
[ID]  Interval        Transfer    Bandwidth        Retr Cwnd
[ 4]  0.00-1.00 sec   36.3 MBytes 305 Mbits/sec    0    9.24 MBytes
[ 4]  1.00-2.00 sec   45.0 MBytes 377 Mbits/sec    0    9.24 MBytes
[ 4]  2.00-3.00 sec   48.1 MBytes 403 Mbits/sec    0    9.24 MBytes
- - - - - - - - - - - - - - - - - - - - - - - - - -
[ID]  Interval        Transfer    Bandwidth        Retr
[ 4]  0.00-3.00 sec   129 MBytes  362 Mbits/sec    0    sender
```

Listing 27.14: Rate control with additional delay.

The target bandwidth of 1 Gbit/s is reached when the *iPerf* test is performed with 5 parallel TCP connections.

```
$ iperf3 -c 127.0.0.1 -P 5
Connecting to host 127.0.0.1, port 5201
[ 4]  local 127.0.0.1 port 55534 connected to 127.0.0.1 port 5201
[ 6]  local 127.0.0.1 port 55536 connected to 127.0.0.1 port 5201
[ 8]  local 127.0.0.1 port 55538 connected to 127.0.0.1 port 5201
[10]  local 127.0.0.1 port 55540 connected to 127.0.0.1 port 5201
[12]  local 127.0.0.1 port 55542 connected to 127.0.0.1 port 5201

- - - - - - - - - - - - - - - - - - - - - - - - - -
[ID]  Interval        Transfer    Bandwidth        Retr
[ 4]  0.00-10.00 sec  236 MBytes  198 Mbits/sec    0    sender
[ 6]  0.00-10.00 sec  236 MBytes  198 Mbits/sec    0    sender
[ 8]  0.00-10.00 sec  236 MBytes  198 Mbits/sec    0    sender
```

```
[10]   0.00-10.00 sec   237 MBytes   199 Mbits/sec   0   sender
[12]   0.00-10.00 sec   235 MBytes   197 Mbits/sec   0   sender
[SUM]  0.00-10.00 sec   1.15 GBytes  989 Mbits/sec   0   sender
```

Listing 27.15: Rate control with additional delay and 5 TCP connections.

Qdiscs alone would not be very useful in practice, since the *aggregated* traffic of an interface would always be manipulated as a whole, that is, irrespectively of individual streams and their content. *Filters* allow us to influence only specific packets [251]. In the following example the *ping* command is used to measure the RTT to both DNS servers of Google. The filter is attached to the parent 1:0, which is a simple priority *qdisc*. If the destination IP address matches 8.8.8.8/32, then the packet is sent to the flow with Identifier (ID) 1:1 (which is the *netem qdisc*), adding 30 ms of delay. All other packets are unaffected.

```
$ tc qdisc add dev eth0 root handle 1: prio
$ tc qdisc add dev eth0 parent 1:1 netem delay 30ms
$ tc filter add dev eth0 protocol ip parent 1:0 prio 1 u32 match ip dst
    8.8.8.8/32 flowid 1:1

$ ping 8.8.8.8 -c 1
PING 8.8.8.8 (8.8.8.8) 56(84) bytes of data.
64 bytes from 8.8.8.8: icmp_seq=1 ttl=50 time=45.3 ms

$ ping 8.8.4.4 -c 1
PING 8.8.4.4 (8.8.4.4) 56(84) bytes of data.
64 bytes from 8.8.4.4: icmp_seq=1 ttl=50 time=15.5 ms
```

Listing 27.16: TC filter.

27.6 Traffic monitoring – tcpdump/Wireshark

Network traffic monitoring or analyzing is a method for deeply inspecting what is going on in a network. Two main parts are necessary to i) obtain the information and ii) present it and aid in its analysis. A close to hardware component (typically, a low-level driver) will actually listen for what is passing by at interfaces, for example, a network card, a Bluetooth device, or Universal Serial Bus (USB) hardware. It can monitor what happens in OSI layers 1 and 2 and either record packets destined for the interface itself or (in the so-called promiscuous mode) all packets passing by the interface under consideration. Afterwards, an additional software part is used to interpret the recorded frames/packets by reassembling and analyzing their content and presenting it to the user in an appropriate way.

There are several traffic monitors and analyzers, commonly differentiated according to their level of detail and the user interface they provide. In this chapter, we describe two prominent traffic monitoring tools, *tcpdump* and *Wireshark*. Whereas *Wireshark* is a powerful software with a huge amount of possible protocols to be analyzed and a sophisticated graphical user interface, *tcpdump* is a more lightweight command-line capturing and analyzing tool. *Wireshark* does not capture frames or packets directly, but provides the user-accessible interface to ongoing live captures or previously captured packet capture traces.

27.6.1 **tcpdump**

tcpdump offers interesting insights into network behavior. To have a quick look into the network traffic passing by the network interface of a computer, *tcpdump* is a good choice. It *dumps* packets directly from the network interface and displays it human-friendly in the terminal. In the *ComNetsEmu*, there are several examples and situations where GUI-programs cannot be used to display the traffic between the emulated hosts, so *tcpdump* can be useful for those scenarios. To be able to capture all traffic and not only traffic destined for the current computer network interfaces, the listening interface has to support the so-called monitoring (promiscuous) mode, which has to be established beforehand. For a first quick look, we can start `tcpdump` without any parameters. It will monitor the default interface for connecting to the Internet. Most common used parameters for capture include:

-# A packet number is printed on every line.
-c Exit the dump after the specified number of packets.
-D Print all available interfaces for capture.
-e Print also the link-layer header of a packet (e.g., to see the vlan tag).
-i Interface to dump from (e.g., eth0 or in the example enp0s31f6).
-n Do not resolve the addresses to names (e.g., IP reverse lookup).
-q Shorter output (for small terminals).
-v Be a little bit verbose to see more packet information.
-w Write the captured traffic to a file.

The following example shows an ICMP echo request and echo reply combination as common for *ping*:

```
$ tcpdump -i enp0s31f6 -n -e icmp
tcpdump: verbose output suppressed, use -v or -vv for full protocol decode
listening on enp0s31f6, link-type EN10MB (Ethernet), capture size 262144 bytes
06:44:34.409654 4c:de:ad:ff:be:ef > 28:de:ad:0b:ee:fc, ethertype 802.1Q (0
    x8100), length 102: vlan 717, p 0, ethertype IPv4, 172.31.56.3 >
    172.31.56.95: ICMP echo request, id 31996, seq 2, length 64
06:44:34.411150 28:de:ad:0b:ee:fc > 4c:de:ad:ff:be:ef, ethertype 802.1Q (0
    x8100), length 102: vlan 717, p 0, ethertype IPv4, 172.31.56.95 >
    172.31.56.3: ICMP echo reply, id 31996, seq 2, length 64
```

Directly visible from the example are:

- MAC-Address of sender and receiver
- Ethertype field (here vlan-tagged packet)
- Vlan 717 with no priority tags (p 0)
- Subethertype (here ipv4)
- Source and destination IPv4 address
- Type of packet (here ICMP echo request/reply)
- ID of packet
- Length of packet

27.6.2 Wireshark

Wireshark currently is the most widely used open-source network scanner and protocol analyzer. It is published under the GNU General Public License and is available for several operating systems, for example, Windows, Linux, and macOS. *Wireshark* is able to read, record, and analyze data traffic on various interfaces, such as Ethernet, WiFi, Bluetooth, or USB. By analyzing network protocols *Wireshark* can be beneficial for solving network errors and monitoring network traffic and security. Since the payload of the packets can be evaluated, for example, VoIP traffic, this tool should be handled with reasonable responsibility. As a packet-oriented analysis tool, *Wireshark* can recognize large numbers of different protocols and subsequently present the most important information of their headers in a comprehensible way. The actual task of capturing network frame and packet information is performed by programs like *dumpcap*, *usbpcap* (Unix), or *winpcap* (Windows), which typically are low-level. *Dumpcap* needs some attention in terms of security as we will see in one of the next subsections. Preconditions for a successful monitoring process at the desired interface are a suitable preinstalled capture routine and sufficient access authorization for the user. To limit the elaborate data to a reasonable amount, special filters can be used. Those capture filters can be created, for instance, for specific IP or MAC addresses, protocols, protocol messages, or other parameters. Other powerful parts of *Wireshark* are the possibility to trace and analyze different TCP or UDP datastreams and possibility to perform statistical analysis with respect to nearly all captured parameters.

27.6.2.1 Main features

The most important features of *Wireshark* can be summarized as follows:

- Aid in the capturing live packet data from many different network media.
- Display the content of captured network packets with fine-grained protocol information.
- Filter packets and search for packets based on many criteria.
- Perform various statistical analysis.

Besides, packet data captured with *Wireshark* can be saved and exported to be analyzed by other tools.

27.6.2.2 Installation

After a standard installation of *Wireshark* on UNIX-OSes, nonroot users have no permission to capture packets. Either *Wireshark* has to be started with root privilege, for example, `sudo wireshark`, or the Wireshark package needs to be reconfigured by running `sudo dpkg-reconfigure wireshark-common` in a terminal. This provides the option to allow nonroot users to capture packets, for which a group *wireshark* is added to the system. As all members of that group are allowed to capture, users have to be added carefully, because they all can *sniff* the network. For other possible solutions of this privilege problem, we refer to the Wireshark user guide. Installation on Windows can simply be processed by downloading and running the installer program. Default settings should work for most purposes. The corresponding capture software *winpcap* will be automatically installed alongside without any privilege issues.

27.6.2.3 User interface

In this short overview, we focus on four topics:

Usage of the user interface: The usage of the user interface can be best seen from the provided screenshot in Fig. 27.3. The packets captured in Fig. 27.3 repre-

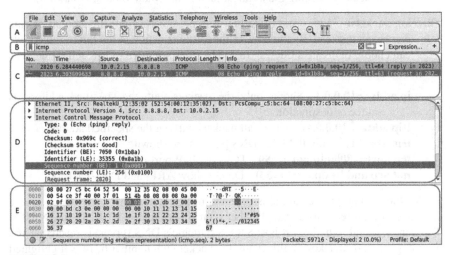

FIGURE 27.3

A typical *Wireshark* window.

sent a simple ping request from a host with IP address 10.0.2.15 to the host with address 8.8.8.8 (Google's DNS server), which responds with a ping reply. The user interface basically consists of a well-structured menu and toolbars to create actions and three main panes to display the results:

- The *menu* in the top can be used to start actions.
- The *main toolbar* (Part A) enables quick access to frequently used menu items.

- The *filter toolbar* (Part B) can be used to set display filters to control, which kind of packets should be displayed.
- The *packet list pane* (Part C) presents an overview of all packets captured. By clicking on specific packets the content for the following two panes is selected.
- The *packet details pane* (Part D) provides a more detailed insight into the packet selected in packet list pane.
- The *packet bytes pane* (Part E) displays data from the current packet in hexdump style and, if possible, the corresponding ASCII code.
- The *status bar* at the bottom presents further information about program state and captured data.

Packet capture: Capturing packets is possible by three different methods, depending on which one the user prefers:

1. Double-clicking on an interface in the main window,
2. Using the capture interface dialog box, or
3. Immediately starting a capture process by choosing Capture > Start from the menu or clicking the leftmost icon in the Main Toolbar (shark fin).

Filtering packets: With packet filtering, it is possible to limit the amount of data displayed or stored. By clicking on the leftmost button in the Filter Toolbar we can choose or create a filtering rule. The rules are more or less self-evident, as is seen from the following examples:

ip.addr==10.0.0.1   Shows only the packets with the specified address as source or destination.

!(ip.addr==10.0.0.1)   Shows all packets without the specified address.

icmp or dns   Only ICMP or DNS packets are shown.

tcp.port==80||udp.port==80   TCP- or UDP-port is 80.

not arp and !(udp.port==53)   Do not show ARP- and DNS-packets.

Another simple ping example is depicted in Fig. 27.4, where we can observe a sequence of ping requests and replies between IP-Addresses 10.0.2.15 and 8.8.8.8. This can be useful to find out if host 8.8.8.8 (Google's DNS server) is reachable and how long a roundtrip takes. In this figure the packet with number 6214 is highlighted in the packet list pane, which can be analyzed as ICMP-Message type 0 (ping reply) in the packet details pane. With the display filter preset to *icmp or arp*, the output in the packet list pane is limited to these protocols.

27.7 Rapid Python prototyping – Jupyter

The Jupyter project was originally intended to enable rapid prototyping and presentation in the field of data science using powerful scripting languages such as Python and R. Unlike the other tools described previously in this chapter, Jupyter is not ex-

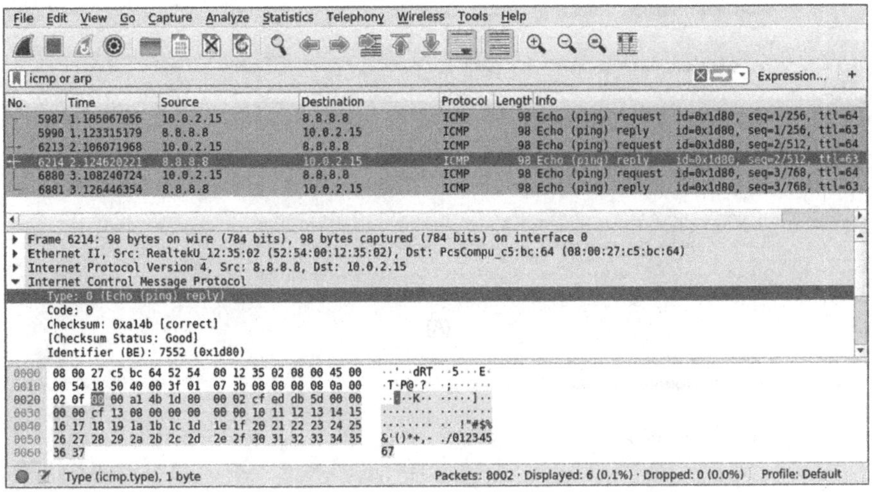

FIGURE 27.4

Another *ICMP (ping)* example.

actly a networking tool. Jupyter notebook is used in the previous chapters to explain and demonstrate network technologies, such as Network Coding using Python. This section intents to give an overall understanding of the `jupyter notebook` and its basic usage. Note that Jupyter notebook is simply one of the building blocks of the project Jupyter [368], which comprises a suite of tools that provide an extensible environment for interactive and reproducible computing. JupyterLab serves as the next-generation user interface, which can be extended using standard npm packages. Although this section exclusively covers the usage of only Jupyter notebook, we urge the readers to familiarize themselves with the entire Jupyter suite of tools.

Jupyter notebooks are documents that can be displayed in any standard browser, containing both text elements and executable code. Each notebook consists of multiple discrete elements called cells. Each cell may contain either a code snippet or markdown formatted text. Although Jupyter supports different programming languages, any given notebook can only contain one programming language, which must be chosen during the creation of the notebook through the field *kernel type*.

The *ComNetsEmu* has a built-in Jupyter server preinstalled, with only one kernel option, *python3*. The Jupyter notebook within the ComNetsEmu is deployed as a Docker instance. By default the port 8888 is exposed by the ComNetsEmu, which is mapped to the Jupyter container running within the VM. The Jupyter notebook can be accessed by simply navigating to `https://localhost:8888` from any standard browser.

Once the Jupyter environment is open, new notebooks can be created using *file->New Notebook->Python 3*. Note that it is also possible to navigate through the file system, create, modify, rename, and delete from this page. A typical use of Jupyter notebooks is for researchers or educators to create a notebook (which has a `.ipynb`

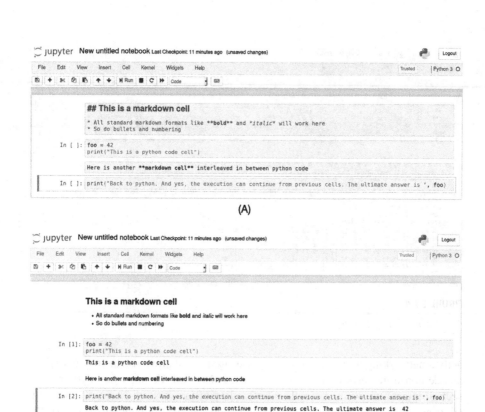

FIGURE 27.5

Example Jupyter notebook. (A) Raw; (B) Executed.

extension) and share it with a class of students or other peers to read, modify, test, and execute the program. Changes made to the notebook can be made persistent by either saving it from the file menu or even downloading the ipynb file directly.

A typical Jupyter python notebook looks similar to that depicted by the screenshot in Fig. 27.5. As mentioned earlier, a notebook is comprised of multiple cells, each of which may contain either code or Markdown-formatted text. As shown in Fig. 27.5, the Python cells and the Markdown cells may be interleaved between each other to provide a detailed explanation within the code. It is also possible to insert images within the Markdown block and draw plots and other data visualization charts from within the code block using libraries, such as plotly or matplotlib.

Each cell may be executed individually using the `Run Cell` option found on the tool bar. Alternatively, we can also press `Ctrl+Enter` to achieve the same result. Figs. 27.5A and 27.5B show the Jupyter cells before and after their respective execution in the example. Note that the Markdown cells are executed and their results are shown in place, whereas the Python cells print their output (stdout and stderr) appended to the source cell. To execute the cell and advance to the next cell, Shift+Enter can be used. One of the key advantages of the Jupyter notebooks is the autocomplete and autosuggest options, which can be used through the Tab key.

27.8 **Hands-on example to tie all tools together**

In the previous sections, we discussed different tools and their respective usage. Now we will discuss an example in which we use most of the tools discussed earlier in an effort to exercise all the tools in a practical use case.

The end objective of this example is to study the effect of latency and losses in a TCP stream. We will achieve this by emulating a TCP connection *(generated using iperf)* between two virtual hosts *(created using ip commands)* while recording the traffic flow *(using tcpdump)* to visually analyze it at a later time *(with Wireshark)* and manipulating the link between the virtual hosts *(using tc)* to study the reactions in the TCP flow. All the following snippets from this example were executed and excerpted from the *ComNetsEmu*. Initially, we consider the interfaces present in the ComNetsEmu.

```
vagrant@comnetsemu:~$ ip link
1: lo: <LOOPBACK,UP,LOWER_UP> mtu 65536 qdisc noqueue state UNKNOWN mode
    DEFAULT group default qlen 1000
    link/loopback 00:00:00:00:00:00 brd 00:00:00:00:00:00
2: eth0: <BROADCAST,MULTICAST,UP,LOWER_UP> mtu 1500 qdisc fq_codel state UP
    mode DEFAULT group default qlen 1000
    link/ether 08:00:27:c5:bc:64 brd ff:ff:ff:ff:ff:ff
3: docker0: <NO-CARRIER,BROADCAST,MULTICAST,UP> mtu 1500 qdisc noqueue state
    DOWN mode DEFAULT group default
    link/ether 02:42:ef:db:98:cb brd ff:ff:ff:ff:ff:ff
```

Listing 27.17: Available interfaces in ComNetsEmu.

Next, we create a virtual Ethernet pair and assign IP addresses to them. Note that once the veth pair is created, it is not possible to assign IP addresses in the same subnet. This is because the interfaces are created in the default network namespace, and within the same network namespace, Linux does not permit interfaces to have IP addresses with the same subnet. Therefore we must also create a new namespace and move one of the newly created veth interface to it.

```
vagrant@comnetsemu:~$ sudo ip link add H1 type veth peer name H2
vagrant@comnetsemu:~$ ip link show type veth
4: H2@H1: <BROADCAST,MULTICAST,M-DOWN> mtu 1500 qdisc noop state DOWN mode
    DEFAULT group default qlen 1000
    link/ether 42:d7:09:e3:71:b6 brd ff:ff:ff:ff:ff:ff
```

```
5: H1@H2: <BROADCAST,MULTICAST,M-DOWN> mtu 1500 qdisc noop state DOWN mode
    DEFAULT group default qlen 1000
    link/ether 2e:c5:09:2f:89:28 brd ff:ff:ff:ff:ff:ff

vagrant@comnetsemu:~$ sudo ip netns add ns-remote
vagrant@comnetsemu:~$ sudo ip link set H2 netns ns-remote
```

Listing 27.18: Create virtual ethernet pair.

Once an interface moved into a different namespace, commands cannot be directly executed on it. Instead, the intended commands must be prefixed with `ip netns exec <namespace>`.

```
vagrant@comnetsemu:~$ sudo ip addr add dev H1 10.42.0.1/24
vagrant@comnetsemu:~$ sudo ip l set H1 up
vagrant@comnetsemu:~$ sudo ip netns exec ns-remote ip addr add dev H2
    10.42.0.2/24
vagrant@comnetsemu:~$ sudo ip netns exec ns-remote ip l set H2 up
vagrant@comnetsemu:~$ ping -c 1 -I H1 10.42.0.2
PING 10.42.0.2 (10.42.0.2) from 10.42.0.1 H1: 56(84) bytes of data.
64 bytes from 10.42.0.2: icmp_seq=1 ttl=64 time=0.109 ms

--- 10.42.0.2 ping statistics ---
1 packets transmitted, 1 received, 0% packet loss, time 0ms
rtt min/avg/max/mdev = 0.109/0.109/0.109/0.000 ms
```

Listing 27.19: Configure virtual ethernet pair.

We now can use `tcpdump` to inspect the live traffic during the ping message:

```
vagrant@comnetsemu:~$ sudo tcpdump -i H1 -n
tcpdump: verbose output suppressed, use -v or -vv for full protocol decode
listening on H1, link-type EN10MB (Ethernet), capture size 262144 bytes
11:21:09.782949 IP 10.42.0.1 > 10.42.0.2: ICMP echo request, id 16237, seq 1,
    length 64
11:21:09.783002 IP 10.42.0.2 > 10.42.0.1: ICMP echo reply, id 16237, seq 1,
    length 64
```

Listing 27.20: Flow monitoring using TCPDump.

Next, we measure the throughput between the two interfaces. The easiest way to achieve this is using `iperf` as in the following example:

```
vagrant@comnetsemu:~$ sudo ip netns exec ns-remote iperf -s -D
Running Iperf Server as a daemon
vagrant@comnetsemu:~$ iperf -c 10.42.0.2
```

```
------------------------------------------------------------
Client connecting to 10.42.0.2, TCP port 5001
TCP window size:  765 KByte (default)
------------------------------------------------------------
[  3] local 10.42.0.1 port 59738 connected with 10.42.0.2 port 5001
[ ID] Interval        Transfer     Bandwidth
[  3]  0.0-10.0 sec  47.1 GBytes  40.5 Gbits/sec
```

Listing 27.21: Quantifying link capacity using iperf.

The throughput is extremely high, which can be expected since both interfaces are on the same physical host and the available bandwidth is only limited by the processing speed of the host computer. Also, there are virtually no losses. This, however, is not a realistic representation of any real-world network. Hence this setup cannot be used directly for any network emulation experiments in its current state. A typical network that one would want to emulate should have latencies in the order of a few milliseconds, and the throughput would be limited to a few hundred Mbits/s in case of wired links. Wireless links may also require the emulation losses. To emulate such realistic links, we use tc to add corresponding Qdiscs to the virtual interfaces.

For our example, we limit the throughput to 1 Mbps. Following the addition of the qdisc-tbf, the drop in the throughput can be verified using iperf.

```
vagrant@comnetsemu:~$ sudo tc qdisc add dev H1 root tbf rate 1Mbit buffer 10K
    latency 10ms
vagrant@comnetsemu:~$ iperf -c 10.42.0.2
------------------------------------------------------------
Client connecting to 10.42.0.2, TCP port 5001
TCP window size: 85.0 KByte (default)
------------------------------------------------------------
[  3] local 10.42.0.1 port 47810 connected with 10.42.0.2 port 5001
[ ID] Interval        Transfer     Bandwidth
[  3]  0.0-10.5 sec  1.25 MBytes  1.00 Mbits/sec
```

Listing 27.22: Limit throughput using tc.

Next, we examine the *Wireshark* output for the ping example between the two namespaces illustrated in Fig. 27.6. We can observe a ping request sent by the host with IP address 10.42.0.1 toward the host with address 10.42.0.2, which responds with a ping reply.

As the ARP cache of the originating host has been emptied before, the first two packets in the corresponding Wireshark window are created by ARP. The ARP request is sent to a broadcast address. When the request arrives at the target, the host generates the ARP reply containing the corresponding MAC address. In the packet details pane, the sender IP address, as part of the ARP protocol, has been highlighted, and the corresponding hexadecimal representation can be seen in the packet bytes pane.

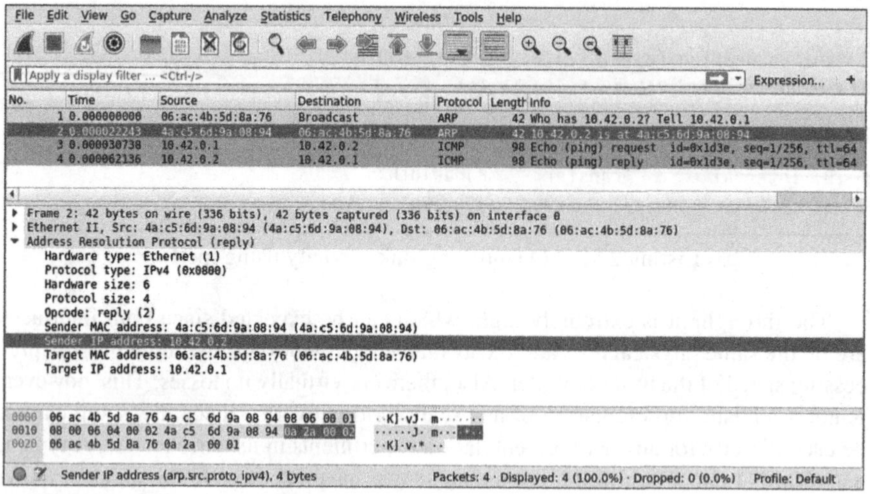

FIGURE 27.6

The *Wireshark* window for the ping example.

Wireshark is not merely a packet visualization tool. It can also perform statistical analysis on the overall traffic. Let us consider Figs. 27.7 and 27.8, referring to our example with limited bandwidth. Whereas in Fig. 27.7 the limited throughput of 1 Mbps and the current TCP segment length can be seen, Fig. 27.8 shows the transmitted packets (black line) and the receiver window (green line; light gray in print version), which opens at the beginning and remains constant during the transmission. These two graphs can be obtained by simply clicking on *Statistics* in the *Wireshark* menu bar and choosing *TCP Stream Graphs*.

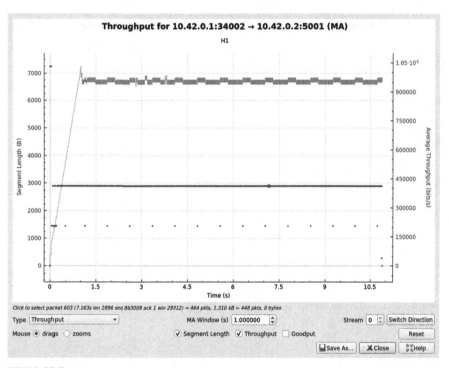

FIGURE 27.7

Throughput between 10.42.0.1:34002 and 10.42.0.2:5001.

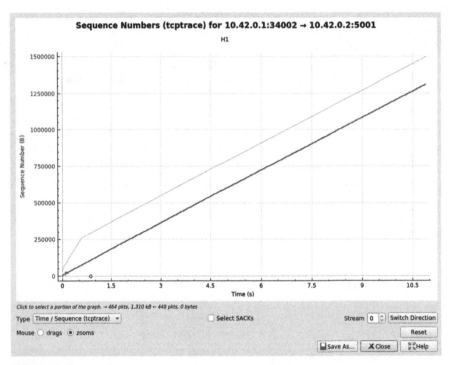

FIGURE 27.8

tcptrace between 10.42.0.1:34002 and 10.42.0.2:5001.

Bibliography

[1] Baran P. On Distributed Communications: I. Introduction to Distributed Communications Networks. No. RM-3420-PR. RAND Corporation; 1964.

[2] Boehm SP, Baran P. On Distributed Communications: II. Digital Simulation of Hot-Potato Routing in a Broadband Distributed Communications Network. No. RM-3103-PR. RAND Corporation; 1964.

[3] Smith JW. On Distributed Communications: III. Determination of Path-Lengths in a Distributed Network. No. RM-3578-PR. RAND Corporation; 1964.

[4] Baran P. On Distributed Communications: IV. Priority, Precedence, and Overload. No. RM-3638-PR. RAND Corporation; 1964.

[5] Baran P. On Distributed Communications: V. History, Alternative Approaches, and Comparisons. No. RM-3097-PR. RAND Corporation; 1964.

[6] Baran P. On Distributed Communications: VI. Mini-Cost Microwave. No. RM-3762-PR. RAND Corporation; 1964.

[7] Baran P. On Distributed Communications: VII. Tentative Engineering Specifications and Preliminary Design for a High-Data-Rate Distributed Network Switching Node. No. RM-3763-PR. RAND Corporation; 1964.

[8] Baran P. On Distributed Communications: VIII. The Multiplexing Station. No. RM-3764-PR. RAND Corporation; 1964.

[9] Baran P. On Distributed Communications: IX. Security, Secrecy, and Tamper-Free Considerations. No. RM-3765-PR. RAND Corporation; 1964.

[10] Baran P. On Distributed Communications: X. Cost Estimate. No. RM-3766-PR. RAND Corporation; 1964.

[11] Baran P. On Distributed Communications: XI. Summary Overview. No. RM-3767-PR. RAND Corporation; 1964.

[12] Kleinrock L. Models for computer networks. In: Proceedings of the IEEE International Conference on Communications (ICC); 1969.

[13] Kleinrock L, Tobagi F. Packet switching in radio channels: part I – carrier sense multiple-access modes and their throughput-delay characteristics. IEEE Transactions on Communications 1975;23(12):1400–16.

[14] Tobagi F, Kleinrock L. Packet switching in radio channels: part II – the hidden terminal problem in carrier sense multiple-access and the busy-tone solution. IEEE Transactions on Communications 1975;23(12):1417–33.

[15] Tobagi F, Kleinrock L. Packet switching in radio channels: part III – polling and (dynamic) split-channel reservation multiple access. IEEE Transactions on Communications 1976;24(8):832–45.

[16] Tobagi F, Kleinrock L. Packet switching in radio channels: part IV – stability considerations and dynamic control in carrier sense multiple access. IEEE Transactions on Communications 1977;25(10):1103–19.

[17] Cerf V, Kahn R. A protocol for packet network intercommunication. IEEE Transactions on Communications 1974;22(5):637–48.

[18] Xia Z, Chen Z, Ming Z, Liu J. A multipath TCP based on network coding in wireless mesh networks. In: Proceedings of the International Conference on Information Science and Engineering (ICISE); 2009.

[19] Ford A, Raiciu C, Handley MJ, Bonaventure O. TCP extensions for multipath operation with multiple addresses. Tech. Rep. RFC 6824. IETF; 2013.

[20] Sundararajan JK, Shah D, Médard M, Jakubczak S, Mitzenmacher M. Network coding meets TCP: theory and implementation. Proceedings of the IEEE 2011;99(3):490–512.

457

[21] Viterbi AJ. CDMA: Principles of Spread Spectrum Communication. Addison Wesley Longman; 1995.

[22] Soni J, Goodman R. A Mind at Play: How Claude Shannon Invented the Information Age. Simon & Schuster; 2017.

[23] Radiocommunication Sector of International Telecommunication Union (ITU), IMT vision – Framework and overall objectives of the future development of IMT for 2020 and beyond. Recommendation ITU-R M.2083-0. M Series – Mobile, radiodetermination, amateur and related satellite services; 2015.

[24] 5G Lab Germany. https://5glab.de.

[25] Cheshire S. It's the latency, stupid; 1996.

[26] Xiang Z, Gabriel F, Urbano Pérez E, Nguyen GT, Reisslein M, Fitzek FHP. Reducing latency in virtual machines: enabling Tactile Internet for human–machine co-working. IEEE Journal on Selected Areas in Communications 2019;37(5):1098–116.

[27] Sterbenz JPG, Hutchison D, Çetinkaya EK, Jabbar A, Rohrer JP, Schöller M, et al. Resilience and survivability in communication networks: strategies, principles, and survey of disciplines. Computer Networks 2010;54(8):1245–65.

[28] Kullback S. Information Theory and Statistics. Courier Corporation; 1997.

[29] Cisco. Cisco visual networking index: global mobile data traffic forecast update, 2017–2022. White paper. 2019.

[30] Ericsson mobility report. https://www.ericsson.com/assets/local/mobility-report/documents/2018/ericsson-mobility-report-june-2018.pdf, 2018.

[31] Raza U, Kulkarni P, Sooriyabandara M. Low power wide area networks: an overview. IEEE Communications Surveys & Tutorials 2017;19(2):855–73.

[32] Ikpehai A, Adebisi B, Rabie KM, Anoh K, Ande RE, Hammoudeh M, et al. Low-power wide area network technologies for Internet-of-Things: a comparative review. IEEE Internet of Things Journal 2019;6(2):2225–40.

[33] ETSI. 5G; Service requirements for next generation new services and markets (3GPP TS 22.261, version 15.7.0, Release 15). Tech. Rep. ETSI TS 122 261, v15.7.0. 2019.

[34] BATMAN. Better approach to mobile ad-hoc networking (B.A.T.M.A.N.). https://www.open-mesh.org/projects/open-mesh/wiki. [Accessed December 2019].

[35] Han H, Shakkottai S, Hollot CV, Srikant R, Towsley D. Multi-path TCP: a joint congestion control and routing scheme to exploit path diversity in the Internet. IEEE/ACM Transactions on Networking 2006;14(6):1260–71.

[36] Gabriel F, Rischke J, Fitzek FHP, Mühleisen M. No plan survives contact with the enemy: on gains of coded multipath over MPTCP in dynamic settings. In: Proceedings of the IEEE Wireless Communications and Networking Conference (WCNC); 2019.

[37] Braun PJ, Pandi S, Schmoll RS, Fitzek FHP. On the study and deployment of mobile edge cloud for Tactile Internet using a 5G gaming application. In: Proceedings of the IEEE Consumer Communications & Networking Conference (CCNC); 2017.

[38] Pandi S, Schmoll RS, Braun PJ, Fitzek FHP. Demonstration of mobile edge cloud for Tactile Internet using a 5G gaming application. In: Proceedings of the IEEE Consumer Communications & Networking Conference (CCNC); 2017.

[39] Schmoll RS, Pandi S, Braun PJ, Fitzek FHP. Demonstration of VR/AR offloading to mobile edge cloud for low latency 5G gaming application. In: Proceedings of the IEEE Consumer Communications & Networking Conference (CCNC); 2018.

[40] Tsokalo IA, Wu H, Nguyen GT, Salah H, Fitzek FHP. Mobile edge cloud for robot control services in industry automation. In: Proceedings of the IEEE Consumer Communications & Networking Conference (CCNC); 2019.

[41] Kropp A, Schmoll RS, Nguyen GT, Fitzek FHP. Demonstration of a 5G multi-access edge cloud enabled smart sorting machine for Industry 4.0. In: Proceedings of the IEEE Consumer Communications & Networking Conference (CCNC); 2019.

[42] Zhdanenko O, Liu J, Torre R, Mudriievskiy S, Salah H, Nguyen GT, et al. Demonstration of mobile edge cloud for 5G connected cars. In: Proceedings of the IEEE Consumer Communications & Networking Conference (CCNC); 2019.

[43] Revsbech K, Heide Møller J, Højgaard-Hansen KR, Perrucci GP, Fitzek FHP. Energy saving potential using active networking on Linux mobile phones. In: Proceedings of the European Wireless Conference (EW); 2009.

[44] Taghouti M, Waurick T, Tömösközi M, Chorppath AK, Fitzek FHP. On the joint design of compressed sensing and network coding for wireless communications. Transactions on Emerging Telecommunications Technologies 2019:e3645, pp. 1–17.

[45] Taghouti M, Chorppath AK, Waurick T, Fitzek FHP. Practical compressed sensing and network coding for intelligent distributed communication networks. In: Proceedings of the International Wireless Communications and Mobile Computing Conference (IWCMC); 2018.

[46] Guttman E. 5G standardization in 3GPP. https://www.itu.int/en/ITU-T/Workshops-and-Seminars/201807/Documents/3_Erik_Guttman.pdf, 2018.

[47] 3GPP. 3rd Generation Partnership Project; Technical Specification Group Services and System Aspects; System architecture for the 5G system (5GS); Stage 2 (Release 15). Tech. Rep. 3GPP TS 23.501 v15.8.0. 2019.

[48] Kekki S, Featherstone W, Fang Y, Kuure P, Li A, Ranjan A, et al. MEC in 5G networks. Tech. Rep. White paper no. 28. ETSI; 2018.

[49] 3GPP. 3rd Generation Partnership Project; Technical Specification Group Services and System Aspects; Telecommunication management; Study on management and orchestration of network slicing for next generation network (Release 15). Tech. Rep. 3GPP TR 28.801 v15.1.0. 2018.

[50] ETSI. Network Functions Virtualisation (NFV), Release 3; Evolution and Ecosystem; Report on Network Slicing Support with ETSI NFV Architecture Framework. Tech. Rep. ETSI GR NFV-EVE 012, v3.1.1. 2017.

[51] ETSI. Network Functions Virtualization (NFV); Architectural Framework. Tech. Rep. ETSI GS NFV 002, v1.2.1. 2014.

[52] ETSI. Network Functions Virtualization (NFV); Management and Orchestration. Tech. Rep. ETSI GS NFV-MAN 001, v1.1.1. 2014.

[53] Boucadair M, Jacquenet C. Software-defined networking: a perspective from within a service provider environment. Tech. Rep. RFC 7149. IETF; 2014.

[54] Haleplidis E, Pentikousis K, Denazis S, Salim JH, Meyer D, Koufopavlou O. Software-defined networking (SDN): layers and architecture terminology. Tech. Rep. RFC 7426. IETF; 2015.

[55] Bhuvaneswaran V, Basil A, Tassinari M, Manral V, Banks S. Benchmarking methodology for software-defined networking (SDN) controller performance. Tech. Rep. RFC 8456. IETF; 2018.

[56] Corson S, Macker J. Mobile ad hoc networking (MANET): routing protocol performance issues and evaluation considerations. Tech. Rep. RFC 2501. IETF; 1999.

[57] Welzl M, Mühlhäuser M. Scalability and quality of service: a trade-off? IEEE Communications Magazine 2003;41(6):32–6.

[58] Blokdyk G. VLAN – A Complete Guide. 5STARCooks; 2018.

[59] Buresh B, Jansen DED, Gmitter J, Ostermiller J, Moreno J, Lei K, et al. A Modern, Open and Scalable Fabric – VXLAN EVPN. Cisco Press; 2016.

[60] PlanetLab. https://www.planet-lab.org/. [Accessed November 2019].

[61] Sherwood R, Gibb G, Yap KK, Appenzeller G, Casado M, McKeown N, et al. FlowVisor: a network virtualization layer. Tech. Rep. OPENFLOW-TR-2009-1. Open Networking Foundation; 2009.

[62] Open Networking Foundation (ONC). OpenVirteX. https://openvirtex.com/. [Accessed November 2019].

[63] Peng S, Chen R, Mirsky G. Packet network slicing using segment routing. Tech. Rep. Draft-peng-lsr-network-slicing-00. IETF; 2019.

[64] NEC. Making 5G a reality. https://www.nec.com/en/global/solutions/nsp/5g_vision/doc/wp2018ar.pdf, 2018.

[65] Son HJ, Yoo C. E2E network slicing – Key 5G technology: What is it? Why do we need it? How do we implement it? https://www.netmanias.com/en/post/blog/8325/5g-iot-network-slicing-sdn-nfv/e2e-network-slicing-key-5g-technology-what-is-it-why-do-we-need-it-how-do-we-implement-it, 2015.

[66] Wu J, Zhang Z, Hong Y, Wen Y. Cloud radio access network (C-RAN): a primer. IEEE Network 2015;29(1):35–41.

[67] Hoang DT, Lee C, Niyato D, Wang P. A survey of mobile cloud computing: architecture, applications, and approaches. Wireless Communications and Mobile Computing 2013;13(18):1587–611.

[68] Patel M, Hu Y, Hédé P, Joubert J, Thornton C, Naughton B, et al. Mobile-edge computing. Tech. Rep. Introductory technical White paper. ETSI; 2014.

[69] ETSI. Multi-access Edge Computing (MEC); Phase 2: Use Cases and Requirements. Tech. Rep. ETSI GS MEC 002, v2.1.1. 2018.

[70] Hu YC, Patel M, Sabella D, Sprecher N, Young V. Mobile Edge Computing – A key technology towards 5G. Tech. Rep. White paper no. 11. ETSI; 2015.

[71] Satyanarayanan M, Bahl P, Caceres R, Davies N. The case for VM-based cloudlets in mobile computing. IEEE Pervasive Computing 2009;8(4):14–23.

[72] Clark C, Fraser K, Hand S, Hansen JG, Jul E, Limpach C, et al. Live migration of virtual machines. In: Proceedings of the USENIX Symposium on Networked Systems Design and Implementation (NSDI); 2005.

[73] Mirkin A, Kuznetsov A, Kolyshkin K. Containers checkpointing and live migration. In: Proceedings of the Ottawa Linux Symposium (OLS); 2008.

[74] Nobach L, Rimac I, Hilt V, Hausheer D. Statelet-based efficient and seamless NFV state transfer. IEEE Transactions on Network and Service Management 2017;14(4):964–77.

[75] Kumar K, Lu YH. Cloud computing for mobile users: can offloading computation save energy? Computer 2010;43(4):51–6.

[76] Wang Y, Sheng M, Wang X, Wang L, Li J. Mobile-edge computing: partial computation offloading using dynamic voltage scaling. IEEE Transactions on Communications 2016;64(10):4268–82.

[77] Huang D, Wang P, Niyato D. A dynamic offloading algorithm for mobile computing. IEEE Transactions on Wireless Communications 2012;11(6):1919–95.

[78] Abbas N, Zhang Y, Taherkordi A, Skeie T. Mobile edge computing: a survey. IEEE Internet of Things Journal 2018;5(1):450–65.

[79] ETSI. Multi-access Edge Computing (MEC); Framework and Reference Architecture. Tech. Rep. ETSI GS MEC 003, v2.1.1. 2019.

[80] Doan TV, Fan Z, Nguyen GT, You D, Kropp A, Salah H, et al. Seamless service migration framework for autonomous driving in mobile edge cloud. In: Proceedings of the IEEE Consumer Communications & Networking Conference (CCNC); 2020.

[81] Docker. https://www.docker.com/. [Accessed October 2019].

[82] etcd. https://etcd.io/. [Accessed October 2019].

[83] Udacity. A self-driving car simulator built with Unity. https://github.com/udacity/self-driving-car-sim. [Accessed October 2019].

[84] NGINX. https://www.nginx.com/. [Accessed October 2019].

[85] confd. https://github.com/kelseyhightower/confd/. [Accessed October 2019].

[86] Salah H. Measuring, understanding, and improving content distribution technologies. Ph.D. thesis. Technische Universität Darmstadt; 2016.

[87] Cisco. Cisco visual networking index: forecast and trends, 2017–2022. White paper. 2019.

[88] Seeling P, Reisslein M. Video traffic characteristics of modern encoding standards: H.264/AVC with SVC and MVC extensions and H.265/HEVC. The Scientific World Journal 2014;2014(189481):1–16.

[89] Brunnström K, Beker S, De Moor K, Dooms A, Egger S, Garcia MN, et al. Qualinet White paper on definitions of quality of experience. v1.2, hal-00977812; 2013.

[90] Fiedler M, Hoßfeld T. A generic quantitative relationship between quality of experience and quality of service. IEEE Network 2010;24(2):36–41.

[91] Reichl P, Tuffin B, Schatz R. Logarithmic laws in service quality perception: where microeconomics meets psychophysics and quality of experience. Telecommunication Systems 2013;52(2):587–600.

[92] Bentaleb A, Taani B, Begen AC, Timmerer C, Zimmermann R. A survey on bitrate adaptation schemes for streaming media over HTTP. IEEE Communications Surveys & Tutorials 2019;21(1):562–85.

[93] Jacobson V, Smetters DK, Thornton JD, Plass MF, Briggs NH, Braynard RL. Networking named content. In: Proceedings of the ACM International Conference on Emerging Networking Experiments and Technologies (CoNEXT); 2009.

[94] Passarella A. A survey on content-centric technologies for the current Internet: CDN and P2P solutions. Computer Communications 2012;35(1):1–32.

[95] Xylomenos G, Ververidis CN, Siris VA, Fotiou N, Tsilopoulos C, Vasilakos X, et al. A survey of information-centric networking research. IEEE Communications Surveys & Tutorials 2014;16(2):1024–49.

[96] Akamai Technologies. https://www.akamai.com/. [Accessed July 2019].

[97] MaxCDN. https://www.stackpath.com/maxcdn/. [Accessed July 2019].

[98] CoralCDN. https://www.coralcdn.org/. [Accessed July 2019].

[99] Liu W. Research on DoS attack and detection programming. In: Proceedings of the International Symposium on Intelligent Information Technology Application (IITA); 2009.

[100] Zanjirani Farahani R, Hekmatfar M. Facility Location: Concepts, Models, Algorithms and Case Studies. Physica-Verlag; 2009.

[101] Karlsson M, Karamanolis C. Choosing replica placement heuristics for wide-area systems. In: Proceedings of the IEEE International Conference on Distributed Computing Systems (ICDCS); 2004.

[102] Sidiropoulos A, Pallis G, Katsaros D, Stamos K, Vakali A, Manolopoulos Y. Prefetching in content distribution networks via Web communities identification and outsourcing. World Wide Web 2008;11(1):39–70.

[103] Fujita N, Ishikawa Y, Iwata A, Izmailov R. Coarse-grain replica management strategies for dynamic replication of Web contents. Computer Networks 2004;45(1):19–34.

[104] Chen Y, Qiu L, Chen W, Nguyen L, Katz RH. Efficient and adaptive Web replication using content clustering. IEEE Journal on Selected Areas in Communications 2003;21(6):979–94.

[105] Freedman MJ, Freudenthal E, Mazières D. Democratizing content publication with Coral. In: Proceedings of the USENIX Symposium on Networked Systems Design and Implementation (NSDI); 2004.

[106] Dilley J, Maggs BM, Parikh J, Prokop H, Sitaraman RK, Weihl WE. Globally distributed content delivery. IEEE Internet Computing 2002;6(5):50–8.

[107] Pan J, Hou YT, Li B. An overview of DNS-based server selections in content distribution networks. Computer Networks 2003;43(6):695–711.

[108] Partridge C, Mendez T, Milliken W. Host anycasting service. Tech. Rep. RFC 1546. IETF; 1993.

[109] Freedman MJ, Lakshminarayanan K, Mazières D. OASIS: anycast for any service. In: Proceedings of the USENIX Symposium on Networked Systems Design and Implementation (NSDI); 2006.

[110] Vakali A, Pallis G. Content delivery networks: status and trends. IEEE Internet Computing 2003;7(6):68–74.

[111] Hofmann M, Beaumont LR. Content Networking: Architecture, Protocols, and Practice. Elsevier; 2005.

[112] Named Data Networking project. http://named-data.net/. [Accessed July 2019].

[113] Koponen T, Chawla M, Chun BG, Ermolinskiy A, Kim KH, Shenker S, et al. A data-oriented (and beyond) network architecture. In: Proceedings of the ACM Conference on the Applications, Technologies, Architectures, and Protocols for Computer Communications (SIGCOMM); 2007.

[114] Lagutin D, Visala K, Tarkoma S. Publish/subscribe for Internet: PSIRP perspective. In: Tselentis G, Galis A, Gavras A, Krčo S, Lotz V, Simperl E, et al., editors. Towards the Future Internet – Emerging Trends from European Research. IOS Press; 2010. p. 75–85.

[115] Dannewitz C. NetInf: an information-centric design for the future Internet. In: Proceedings of the GI/ITG KuVS Workshop on the Future Internet; 2009.

[116] NSF Future Internet Architecture project. http://www.nets-fia.net/. [Accessed July 2019].

[117] Eugster PT, Felber PA, Guerraoui R, Kermarrec AM. The many faces of publish/subscribe. ACM Computing Surveys 2003;35(2):114–31.

[118] Salah H, Strufe T. Comon: an architecture for coordinated caching and cache-aware routing in CCN. In: Proceedings of the IEEE Consumer Communications & Networking Conference (CCNC); 2015.

[119] Perino D, Varvello M. A reality check for content centric networking. In: Proceedings of the ACM SIGCOMM Workshop on Information-Centric Networking (ICN); 2011.

[120] Zhang M, Luo H, Zhang H. A survey of caching mechanisms in information-centric networking. IEEE Communications Surveys & Tutorials 2015;17(3):1473–99.

[121] Fettweis GP. A 5G wireless communications vision. Microwave Journal 2012;55(12):24–31.

[122] Caswell E, Teare D, Tiso J. Designing Cisco Network Service Architectures (ARCH): Foundation Learning Guide. Foundation Learning Guides. Cisco Press; 2011.

[123] Hedrick CL. Routing information protocol. Tech. Rep. RFC 1058. IETF; 1988.

[124] Moy J. OSPF specification. Tech. Rep. RFC 1131. IETF; 1989.

[125] Lougheed K, Rekhter Y. Border gateway protocol (BGP). Tech. Rep. RFC 1105. IETF; 1989.

[126] Schlinker B, Kim H, Cui T, Katz-Bassett E, Madhyastha HV, Cunha I, et al. Engineering egress with edge fabric: steering oceans of content to the world. In: Proceedings of the ACM Conference on the Applications, Technologies, Architectures, and Protocols for Computer Communications (SIGCOMM); 2017.

[127] Strickx T. How Verizon and a BGP optimizer knocked large parts of the Internet offline today. https://blog.cloudflare.com/how-verizon-and-a-bgp-optimizer-knocked-large-parts-of-the-internet-offline-today/. [Accessed November 2019].

[128] Casado M, Freedman MJ, Pettit J, Luo J, McKeown N, Shenker S. Ethane: taking control of the enterprise. ACM SIGCOMM Computer Communication Review 2007;37(4):1–12.

[129] McKeown N, Anderson T, Balakrishnan H, Parulkar G, Peterson L, Rexford J, et al. OpenFlow: enabling innovation in campus networks. ACM SIGCOMM Computer Communication Review 2008;38(2):69–74.

[130] Open Networking Foundation (ONC). https://www.opennetworking.org/. [Accessed November 2019].

[131] Enns R, Björklund M, Schönwälder J, Bierman A. NETCONF configuration protocol. Tech. Rep. RFC 6241. IETF; 2011.

[132] Bosshart P, Daly D, Gibb G, Izzard M, McKeown N, Rexford J, et al. P4: programming protocol-independent packet processors. ACM SIGCOMM Computer Communication Review 2014;44(3):87–95.

[133] Floodlight is a Java-based OpenFlow controller. http://www.projectfloodlight.org/. [Accessed August 2019].

[134] OpenDaylight. http://www.opendaylight.org/. [Accessed November 2019].

[135] Bailey J, Stuart S. FAUCET: deploying SDN in the enterprise. Communications of the ACM 2017;60(1):45–9.

[136] Nippon Telegraph and Telephone Corporation. Ryu SDN framework. https://osrg.github.io/ryu/. [Accessed November 2019].

[137] Erickson D. The Beacon OpenFlow controller. In: Proceedings of the ACM SIGCOMM Workshop on Hot Topics in Software Defined Networking (HotSDN); 2013.

[138] POX network software platform. https://github.com/noxrepo/pox. [Accessed November 2019].

[139] Gude N, Koponen T, Pettit J, Pfaff B, Casado M, McKeown N, et al. NOX: towards an operating system for networks. ACM SIGCOMM Computer Communication Review 2008;38(3):105–10.

[140] Kreutz D, Ramos FMV, Veríssimo PE, Rothenberg CE, Azodolmolky S, Uhlig S. Software-defined networking: a comprehensive survey. Proceedings of the IEEE 2015;103(1):14–76.

[141] Pfaff B, Pettit J, Koponen T, Amidon K, Casado M, Shenker S. Extending networking into the virtualization layer. In: Proceedings of the ACM SIGCOMM Workshop on Hot Topics in Networks (HotNets); 2009.

[142] Christensen M, Kimball K, Solensky F. Considerations for Internet Group Management Protocol (IGMP) and Multicast Listener Discovery (MLD) snooping switches. Tech. Rep. RFC 4541. IETF; 2006.

[143] P4language. P4-16 language specification. https://github.com/p4lang/p4-spec/tree/master/p4-16/spec. [Accessed December 2019].

[144] Björklund M. YANG – a data modeling language for the network configuration protocol (NETCONF). Tech. Rep. RFC 6020. IETF; 2010.

[145] Nayak NG, Dürr F. Time-sensitive software-defined network (TSSDN) for real-time applications. In: Proceedings of the International Conference on Real-Time Networks and Systems (RTNS); 2016.

[146] Mijumbi R, Serrat J, Gorricho JL, Bouten N, De Turck F, Boutaba R. Network function virtualization: state-of-the-art and research challenges. IEEE Communications Surveys Tutorials 2016;18(1):236–62.

[147] Dötsch U, Doll M, Mayer H, Schaich F, Segel J, Sehier P. Quantitative analysis of split base station processing and determination of advantageous architectures for LTE. Bell Labs Technical Journal 2013;18(1):105–28.

[148] Aditya P, Akkuş IE, Beck A, Chen R, Hilt V, Rimac I, et al. Will serverless computing revolutionize NFV? Proceedings of the IEEE 2019;107(4):667–78.

[149] Matias J, Garay J, Toledo N, Unzilla J, Jacob E. Toward an SDN-enabled NFV architecture. IEEE Communications Magazine 2015;53(4):187–93.

[150] Duan Q, Ansari N, Toy M. Software-defined network virtualization: an architectural framework for integrating SDN and NFV for service provisioning in future networks. IEEE Network 2016;30(5):10–6.

[151] Granelli F, Bassoli R. Autonomic mobile virtual network operators for future generation networks. IEEE Network 2018;32(5):76–84.

[152] Huard JF, Lazar AA. A programmable transport architecture with QoS guarantees. IEEE Communications Magazine 1998;36(10):54–62.

[153] Akyildiz IF, Altunbaşak Y, Fekri F, Sivakumar R. AdaptNet: an adaptive protocol suite for the next-generation wireless Internet. IEEE Communications Magazine 2004;42(3):128–36.

[154] Zhou C, Chen H, Xiong N, Huang X, Vasilakos AV. Model-driven development of reconfigurable protocol stack for networked control systems. IEEE Transactions on Systems, Man, and Cybernetics, Part C (Applications and Reviews) 2012;42(6):1439–53.

[155] Guan Z, Bertizzolo L, Demirörs E, Melodia T. WNOS: an optimization-based wireless network operating system. CoRR. arXiv:1712.08667 [abs], 2017.

[156] Wen R, Feng G, Tan W, Ni R, Qin S, Wang G. Protocol function block mapping of software defined protocol for 5G mobile networks. IEEE Transactions on Mobile Computing 2018;17(7):1651–65.

[157] Mäder A, Lalam M, De Domenico A, Pateromichelakis E, Wübben D, Bartelt J, et al. Towards a flexible functional split for cloud-RAN networks. In: Proceedings of the European Conference on Networks and Communications (EuCNC); 2014.

[158] Turing AM. Computing machinery and intelligence. Mind 1950;LIX(236):433–60.

[159] Rosasco L, De Vito E, Caponnetto A, Piana M, Verri A. Are loss functions all the same? Neural Computation 2004;16(5):1063–76.

[160] Hennig C, Kutlukaya M. Some thoughts about the design of loss functions. REVSTAT Statistical Journal 2007;5(1):19–39.

[161] Hastie T, Tibshirani R, Friedman JH. The Elements of Statistical Learning. Springer; 2017.

[162] Boyd S, Vandenberghe L. Convex Optimization. Cambridge University Press; 2004.

[163] Liu J, Cosman PC, Rao BD. Robust linear regression via ℓ_0 regularization. IEEE Transactions on Signal Processing 2018;66(3):698–713.

[164] Cormen TH, Leiserson CE, Rivest RL, Stein C. Introduction to Algorithms: Data Mining, Inference, and Prediction. MIT Press; 2009.

[165] Powell WB. A unified framework for stochastic optimization. European Journal of Operational Research 2018;275(3):795–821.

[166] Bertsekas DP, Shreve SE. Stochastic Optimal Control: The Discrete-Time Case. Athena Scientific; 2007.

[167] Sutton RS, Barto AG. Reinforcement Learning: An Introduction. MIT Press; 2018.

[168] Watkins CJCH, Dayan P. Q-learning. Machine Learning 1992;8(3):279–92.

[169] Mnih V, Kavukcuoglu K, Silver D, Rusu AA, Veness J, Bellemare MG, et al. Human-level control through deep reinforcement learning. Nature 2015;518(7540):529–33.

[170] Ahlswede R, Cai N, Li SYR, Yeung RW. Network information flow. IEEE Transactions on Information Theory 2000;46(4):1204–16.

[171] Li SYR, Yeung RW, Cai N. Linear network coding. IEEE Transactions on Information Theory 2003;49(2):371–81.

[172] Kötter R, Médard M. An algebraic approach to network coding. IEEE/ACM Transactions on Networking 2003;11(5):782–95.

[173] Ho T, Médard M, Kötter R, Karger DR, Effros M, Shi J, et al. A random linear network coding approach to multicast. IEEE Transactions on Information Theory 2006;52(10):4413–30.

[174] Médard M, Fitzek FHP, Montpetit MJ, Rosenberg C. Network coding mythbusting: why it is not about butterflies anymore. IEEE Communications Magazine 2014;52(7):177–83.

[175] Kibler M. Galois Fields and Galois Rings Made Easy. Elsevier; 2017.

[176] Jetzek U. Galois Fields, Linear Feedback Shift Registers and Their Applications. Carl Hanser Verlag; 2018.

[177] Heide Møller J, Pedersen MV, Fitzek FHP, Larsen T. Network coding for mobile devices – systematic binary random rateless codes. In: Proceedings of the IEEE International Conference on Communications (ICC), Workshop on Cooperative Mobile Networks; 2009.

[178] Lucani Rötter DE, Médard M, Stojanović M. Random linear network coding for time-division duplexing: field size considerations. In: Proceedings of the IEEE Global Telecommunications Conference (GLOBECOM); 2009.

[179] Heide Møller J, Pedersen MV, Fitzek FHP, Médard M. On code parameters and coding vector representation for practical RLNC. In: Proceedings of the IEEE International Conference on Communications (ICC); 2011.

[180] Paramanathan A, Pedersen MV, Lucani Rötter DE, Fitzek FHP, Katz MD. Lean and mean: network coding for commercial devices. IEEE Wireless Communications 2013;20(5):54–61.

[181] Trullols-Cruces O, Barceló Ordinas JM, Fiore M. Exact decoding probability under random linear network coding. IEEE Communications Letters 2011;15(1):67–9.

[182] Zhao X. Notes on 'Exact decoding probability under random linear network coding'. IEEE Communications Letters 2012;16(5):720–1.

[183] Sørensen CW, Paramanathan A, Cabrera Guerrero JA, Pedersen MV, Lucani Rötter DE, Fitzek FHP. Leaner and meaner: network coding in SIMD enabled commercial devices. In: Proceedings of the IEEE Wireless Communications and Networking Conference (WCNC); 2016.

[184] Wunderlich S, Cabrera Guerrero JA, Fitzek FHP, Reisslein M. Network coding in heterogeneous multicore IoT nodes with DAG scheduling of parallel matrix block operations. IEEE Internet of Things Journal 2017;4(4):917–33.

[185] Wunderlich S, Cabrera Guerrero JA, Fitzek FHP, Pedersen MV. Network coding parallelization based on matrix operations for multicore architectures. In: Proceedings of the IEEE International Conference on Ubiquitous Wireless Broadband (ICUWB); 2015.

[186] Xiao M, Aulin T, Médard M. Systematic binary deterministic rateless codes. In: Proceedings of the IEEE International Symposium on Information Theory (ISIT); 2008.

[187] Heide Møller J, Pedersen MV, Fitzek FHP, Médard M. A perpetual code for network coding. In: Proceedings of the IEEE Vehicular Technology Conference. VTC Spring; 2014.

[188] Pahlevani P, Crisóstomo S, Lucani Rötter DE. An analytical model for perpetual network codes in packet erasure channels. In: Madsen TK, Nielsen JJ, Pratas NK, editors. Multiple Access Communications. Lecture Notes in Computer Science, vol. 10121. Springer; 2016. p. 126–35.

[189] Garrido P, Lucani Rötter DE, Agüero R. Markov chain model for the decoding probability of sparse network coding. IEEE Transactions on Communications 2017;65(4):1675–85.

[190] Sørensen CW, Badr AS, Cabrera Guerrero JA, Lucani Rötter DE, Heide Møller J, Fitzek FHP. A practical view on tunable sparse network coding. In: Proceedings of the European Wireless Conference (EW); 2015.

[191] Feizi S, Lucani Rötter DE, Médard M. Tunable sparse network coding. In: Proceedings of the International Zurich Seminar on Communications (IZS); 2012.

[192] Feizi S, Lucani Rötter DE, Sørensen CW, Makhdoumi A, Médard M. Tunable sparse network coding for multicast networks. In: Proceedings of the International Symposium on Network Coding (NetCod); 2014.

[193] Garrido P, Lucani Rötter DE, Agüero R. How to tune sparse network coding over wireless links. In: Proceedings of the IEEE Wireless Communications and Networking Conference (WCNC); 2017.

[194] Sundararajan JK, Shah D, Médard M, Mitzenmacher M, Barros J. Network coding meets TCP. In: Proceedings of the IEEE International Conference on Computer Communications (INFOCOM); 2009.

[195] Wunderlich S, Gabriel F, Pandi S, Fitzek FHP. We don't need no generation – a practical approach to sliding window RLNC. In: Proceedings of the Wireless Days Conference (WD); 2017.

[196] Szabó D, Csoma A, Megyesi P, Gulyás A, Fitzek FHP. Network coding as a service. CoRR. arXiv: 1601.03201 [abs], 2016.

[197] Candès EJ, Tao T. Decoding by linear programming. CoRR. arXiv:abs/math/0502327, 2005.

[198] Donoho DL. Compressed sensing. IEEE Transactions on Information Theory 2006;52(4):1289–306.

[199] Donoho DL, Elad M. Optimally sparse representation in general (nonorthogonal) dictionaries via ℓ^1 minimization. Proceedings of the National Academy of Sciences 2003;100(5):2197–202.

[200] Strohmer T, Heath Jr RW. Grassmannian frames with applications to coding and communication. Applied and Computational Harmonic Analysis 2003;14(3):257–75.

[201] Cohen A, Dahmen W, DeVore RA. Compressed sensing and best k-term approximation. Journal of the American Mathematical Society 2009;22(1):211–31.

[202] Candès EJ. The restricted isometry property and its implications for compressed sensing. Comptes Rendus Mathematique 2008;346(9–10):589–92.

[203] Koiran P, Zouzias A. Hidden cliques and the certification of the restricted isometry property. IEEE Transactions on Information Theory 2014;60(8):4999–5006.

[204] Tillmann AM, Pfetsch ME. The computational complexity of the restricted isometry property, the nullspace property, and related concepts in compressed sensing. IEEE Transactions on Information Theory 2013;60(2):1248–59.

[205] DeVore RA. Deterministic constructions of compressed sensing matrices. Journal of Complexity 2007;23(4–6):918–25.

[206] Chen SS, Donoho DL, Saunders MA. Atomic decomposition by basis pursuit. SIAM Review 2001;43(1):129–59.

[207] Tibshirani R. Regression shrinkage and selection via the lasso. Journal of the Royal Statistical Society: Series B (Methodological) 1996;58(1):267–88.

[208] Efron B, Hastie T, Johnstone I, Tibshirani R. Least angle regression. The Annals of Statistics 2004;32(2):407–99.

[209] Donoho DL, Maleki A, Montanari A. Message-passing algorithms for compressed sensing. Proceedings of the National Academy of Sciences 2009;106(45):18914–9.

[210] Mallat SG, Zhang Z. Matching pursuits with time-frequency dictionaries. IEEE Transactions on Signal Processing 1993;41(12):3397–415.

[211] DeVore RA, Temlyakov VN. Some remarks on greedy algorithms. Advances in Computational Mathematics 1996;5(1):173–87.

[212] Cai TT, Wang L. Orthogonal matching pursuit for sparse signal recovery with noise. IEEE Transactions on Information Theory 2011;57(7):4680–8.

[213] Carrillo RE, Barner KE. Lorentzian based iterative hard thresholding for compressed sensing. In: Proceedings of the IEEE International Conference on Acoustics, Speech, and Signal Processing (ICASSP); 2011.

[214] Beck A, Teboulle M. A fast iterative shrinkage-thresholding algorithm for linear inverse problems. SIAM Journal on Imaging Sciences 2009;2(1):183–202.

[215] Blanchard JD, Tanner J. Performance comparisons of greedy algorithms in compressed sensing. Numerical Linear Algebra with Applications 2015;22(2):254–82.

[216] Indyk P, Ružić M. Near-optimal sparse recovery in the ℓ_1 norm. In: Proceedings of the IEEE Symposium on Foundations of Computer Science (FOCS); 2008.

[217] Kutyniok G, Lim WQ. Compactly supported shearlets are optimally sparse. Journal of Approximation Theory 2011;163(11):1564–89.

[218] Qin Z, Fan J, Liu Y, Gao Y, Li GY. Sparse representation for wireless communications: a compressive sensing approach. IEEE Signal Processing Magazine 2018;35(3):40–58.

[219] Marcellin MW, Gormish MJ, Bilgin A, Boliek MP. An overview of JPEG-2000. In: Proceedings of the Data Compression Conference (DCC); 2000.

[220] Engan K, Aase SO, Husøy JH. Method of optimal directions for frame design. In: Proceedings of the IEEE International Conference on Acoustics, Speech, and Signal Processing (ICASSP); 1999.

[221] Gribonval R, Schnass K. Dictionary identification – sparse matrix-factorization via ℓ_1-minimization. IEEE Transactions on Information Theory 2010;56(7):3523–39.

[222] Geng Q, Wright J. On the local correctness of ℓ^1-minimization for dictionary learning. In: Proceedings of the IEEE International Symposium on Information Theory (ISIT); 2014.

[223] Aharon M, Elad M, Bruckstein A. K-SVD: an algorithm for designing overcomplete dictionaries for sparse representation. IEEE Transactions on Signal Processing 2006;54(11):4311–22.

[224] Duarte MF, Sarvotham S, Baron D, Wakin MB, Baraniuk RG. Distributed compressed sensing of jointly sparse signals. In: Proceedings of the Asilomar Conference on Signals, Systems, and Computers (Asilomar); 2005.

[225] Qiao W, Liu B, Chen CW. JSM-2 based joint ECG compressed sensing with partially known support establishment. In: Proceedings of the IEEE International Conference on e-Health Networking, Applications and Services (Healthcom); 2012.

[226] Wang B, Ge Y, He C, Wu Y, Zhu Z. Study on communication channel estimation by improved SOMP based on distributed compressed sensing. EURASIP Journal on Wireless Communications and Networking 2019;2019(121):1–8.

[227] Tropp JA, Gilbert AC, Strauss MJ. Simultaneous sparse approximation via greedy pursuit. In: Proceedings of the IEEE International Conference on Acoustics, Speech, and Signal Processing (ICASSP); 2005.

[228] Rao X, Lau VKN. Compressive sensing with prior support quality information and application to massive MIMO channel estimation with temporal correlation. IEEE Transactions on Signal Processing 2015;63(18):4914–24.

[229] Duarte MF, Baraniuk RG. Kronecker compressive sensing. IEEE Transactions on Image Processing 2012;21(2):494–504.

[230] Leinonen M, Codreanu M, Juntti M. Sequential compressed sensing with progressive signal reconstruction in wireless sensor networks. IEEE Transactions on Wireless Communications 2014;14(3):1622–35.

[231] Shakeri Z, Sarwate AD, Bajwa WU. Identifiability of Kronecker-structured dictionaries for tensor data. IEEE Journal of Selected Topics in Signal Processing 2018;12(5):1047–62.

[232] Dantas CF, Da Costa MN, da Rocha Lopes R. Learning dictionaries as a sum of Kronecker products. IEEE Signal Processing Letters 2017;24(5):559–63.

[233] Duarte MF, Baraniuk RG. Kronecker product matrices for compressive sensing. In: Proceedings of the IEEE International Conference on Acoustics, Speech, and Signal Processing (ICASSP); 2010.

[234] Rivenson Y, Stern A. An efficient method for multi-dimensional compressive imaging. In: Proceedings of the Computational Optical Sensing and Imaging (COSI); 2009.

[235] Association for Computing Machinery (ACM). Artifact review and badging. https://www.acm.org/publications/policies/artifact-review-badging. [Accessed October 2019].

[236] Saucez D, Iannone L, Bonaventure O. Evaluating the artifacts of SIGCOMM papers. ACM SIGCOMM Computer Communication Review 2019;49(2):44–7.

[237] Heller B. Reproducible network research with high-fidelity emulation. Ph.D. thesis. Stanford University; 2013.

[238] Lantz B, Heller B, McKeown N. A network in a laptop: rapid prototyping for software-defined networks. In: Proceedings of the ACM SIGCOMM Workshop on Hot Topics in Networks (HotNets); 2010.

[239] Yan L, McKeown N. Learning networking by reproducing research results. ACM SIGCOMM Computer Communication Review 2017;47(2):19–26.

[240] Linux Foundation. Open vSwitch: an open virtual switch. https://www.openvswitch.org/. [Accessed August 2019].

[241] veth – Virtual Ethernet device. http://man7.org/linux/man-pages/man4/veth.4.html. [Accessed August 2019]

[242] Mininet Wiki: Introduction to Mininet. https://github.com/mininet/mininet/wiki/Introduction-to-Mininet. [Accessed August 2019].

[243] Mininet Python API reference manual. http://mininet.org/api/annotated.html. [Accessed August 2019].

[244] netem – Network emulator. http://man7.org/linux/man-pages/man8/tc-netem.8.html. [Accessed August 2019].

[245] Overview of Linux namespaces. http://man7.org/linux/man-pages/man7/namespaces.7.html. [Accessed August 2019].

[246] Containers unplugged: Linux namespaces. http://man7.org/conf/meetup/Linux-namespaces--jambit-Kerrisk-2019-05-20.pdf. [Accessed September 2019].

[247] Containers unplugged: an introduction to control groups (cgroups) v1. http://man7.org/conf/meetup/conf_cgroups_v1--jambit-Kerrisk-2019-05-29.pdf. [Accessed September 2019].

[248] Poettering L, Sievers K, Leemhuis T. Control centre: the systemd Linux init system. The H Open 2012.

[249] perf – Linux profiling with performance counters. https://perf.wiki.kernel.org/index.php/Main_Page. [Accessed August 2019].

[250] Brown MA. Traffic control HOWTO. http://linux-ip.net/articles/Traffic-Control-HOWTO/index.html. [Accessed August 2019].

[251] Hubert B. Linux man-pages: tc. https://linux.die.net/man/8/tc. [Accessed November 2019].

[252] tc-hfcs – hierarchical fair service curve. http://man7.org/linux/man-pages/man7/tc-hfsc.7.html. [Accessed August 2019].

[253] Pfaff B, Pettit J, Koponen T, Jackson E, Zhou A, Rajahalme J, et al. The design and implementation of Open vSwitch. In: Proceedings of the USENIX Symposium on Networked Systems Design and Implementation (NSDI); 2015.

[254] Pfaff B, Davie B. The Open vSwitch database management protocol. Tech. Rep. RFC 7047. IETF; 2013.

[255] ComNetsEmu. A holistic testbed/emulator for the book Computing in Communication Networks: From Theory to Practice. https://git.comnets.net/public-repo/comnetsemu. [Accessed November 2019].

[256] runc – a lightweight universal container runtime. https://github.com/opencontainers/runc. [Accessed August 2019].

[257] VMware. https://www.vmware.com/. [Accessed August 2019].

[258] VirtualBox. https://www.virtualbox.org/. [Accessed August 2019].

[259] Linux KVM. https://www.linux-kvm.org/page/Main_Page. [Accessed August 2019].

[260] Linux containers. https://linuxcontainers.org/. [Accessed August 2019].

[261] Kubernetes. https://kubernetes.io/. [Accessed August 2019].

[262] OpenStack. https://www.openstack.org/. [Accessed August 2019].

[263] Containernet. https://containernet.github.io/. [Accessed December 2019].

[264] Peuster M, Karl H, van Rossem S. MeDICINE: rapid prototyping of production-ready network services in multi-PoP environments. In: Proceedings of the IEEE Conference on Network Function Virtualization and Software Defined Networks (NFV-SDN); 2016.

[265] ACM SIGCOMM. Mininet Hackathon Wiki 2018. https://github.com/acmsigcomm18hackathon/hackathonprojects/wiki/Mininet. [Accessed August 2019].

[266] Vagrant. https://www.vagrantup.com/. [Accessed August 2019].

[267] Ubuntu Server. https://ubuntu.com/server/. [Accessed August 2019].

[268] Ryu SDN framework community. https://osrg.github.io/ryu/. [Accessed August 2019].

[269] Docker SDK for Python. https://docker-py.readthedocs.io/en/stable/. [Accessed August 2019].

[270] Ben-Yehuda M, Day MD, Dubitzky Z, Factor M, Har'El N, Gordon A, et al. The Turtles project: design and implementation of nested virtualization. In: Proceedings of the USENIX Symposium on Operating Systems Design and Implementation (OSDI); 2010.

[271] Kubernetes. POD overview. https://kubernetes.io/docs/concepts/workloads/pods/pod-overview/. [Accessed August 2019].

[272] Docker: run reference. https://docs.docker.com/engine/reference/run/. [Accessed August 2019].

[273] Docker in Docker. https://github.com/jpetazzo/dind. [Accessed August 2019].

[274] Wu H, Nguyen GT, Chorppath AK, Fitzek FHP. Network slicing for conditional monitoring in the industrial Internet of Things. IEEE Softwarization 2018.

[275] Wu H, Tsokalo IA, Kuß D, Salah H, Pingel L, Fitzek FHP. Demonstration of network slicing for flexible conditional monitoring in industrial IoT networks. In: Proceedings of the IEEE Consumer Communications & Networking Conference (CCNC); 2019.

[276] ComNetsEmu. MEC application example. https://git.comnets.net/public-repo/comnetsemu/tree/master/app/realizing_mobile_edge_clouds. [Accessed August 2019].

[277] Leiserson CE. Fat-trees: universal networks for hardware-efficient supercomputing. IEEE Transactions on Computers 1985;c-34(10):892–901.

[278] Mogul JC. Broadcasting Internet datagrams in the presence of subnets. Tech. Rep. RFC 922. IETF; 1984.

[279] Open Networking Foundation (ONC). OpenFlow switch specification, version 1.5.1 (wire protocol 0x01); 2015.

[280] Phemius K, Bouet M. Monitoring latency with OpenFlow. In: Proceedings of the International Conference on Network and Service Management (CNSM); 2013.

[281] Seeling P, Reisslein M, Fitzek FHP, Hendrata S. Video quality evaluation for wireless transmission with robust header compression. In: Proceedings of the Joint Conference of the International Conference on Information, Communications and Signal Processing (ICICS) and the Pacific Rim Conference on Multimedia (PCM); 2003.

[282] Thubert P, Hui J. Compression format for IPv6 datagrams over IEEE 802.15.4-based networks. Tech. Rep. RFC 6282. IETF; 2011.

[283] Minaburo A, Toutain L, Gomez C, Barthel D, Zúñiga JC. Static context header compression (SCHC) and fragmentation for LPWAN, application to UDP/IPv6. Tech. Rep. Draft-ietf-lpwan-ipv6-static-context-hc-24. IETF; 2019.

[284] Tömösközi M, Seeling P, Ekler P, Fitzek FHP. Robust header compression version 2 power consumption on Android devices via tunnelling. In: Proceedings of the IEEE International Conference on Communications (ICC), Workshop on Green Networking and Transport; 2017.

[285] Iyengar J, Thomson M. QUIC: a UDP-based multiplexed and secure transport. Tech. Rep. Draft-ietf-quic-transport-25. IETF; 2020.

[286] Jacobson V. Compressing TCP/IP headers for low-speed serial links. Tech. Rep. RFC 1144. IETF; 1990.

[287] Martensson A, Wiebke T, Burmeister C, Hakenberg R, Fukushima H, Yoshimura T, et al. RObust Header Compression (ROHC): framework and four profiles: RTP, UDP, ESP, and uncompressed. Tech. Rep. RFC 3095. IETF; 2001.

[288] Sandlund K, Pelletier G. RObust Header Compression version 2 (ROHCv2): profiles for RTP, UDP, IP, ESP and UDP-Lite. Tech. Rep. RFC 5225. IETF; 2008.

[289] Tömösközi M, Seeling P, Ekler P, Fitzek FHP. Performance evaluation of network header compression schemes for UDP, RTP and TCP. Periodica Polytechnica Electrical Engineering and Computer Science 2016;60(3):151–62.

[290] Tömösközi M, Luo M, Fitzek FHP, Ekler P. Initial concept of an Oracle-Structured Stream Compression protocol for arbitrary network flows. In: European Wireless 2019 (EW 2019); 2019.

[291] Nagle J. Congestion control in IP/TCP internetworks. Tech. Rep. RFC 896. IETF; 1984.

[292] Stevens WR. TCP/IP Illustrated (Vol. 1): The Protocols. Addison Wesley Longman; 1993.

[293] Allman M, Paxson V, Blanton E. TCP congestion control. Tech. Rep. RFC 5681. IETF; 2009.

[294] Kleinrock L. Internet congestion control using the power metric: keep the pipe just full, but no fuller. Ad Hoc Networks 2018;80:142–57.

[295] Cardwell N, Cheng Y, Gunn CS, Yeganeh SH, Jacobson V. BBR: congestion-based congestion control. Communications of the ACM 2017;60(2):58–66.

[296] Jay N, Rotman NH, Godfrey PB, Schapira M, Tamar A. Internet congestion control via deep reinforcement learning. CoRR. arXiv:1810.03259 [abs], 2019.

[297] Plappert M. keras-rl. https://github.com/keras-rl/keras-rl. [Accessed November 2019].

[298] Cisco. VNI global fixed and mobile Internet traffic forecasts. https://www.cisco.com/c/en/us/solutions/service-provider/visual-networking-index-vni/index.html. [Accessed May 2019].

[299] Pedestrian direction recognition dataset. http://www.rovit.ua.es/dataset/pedirecog/index.html. [Accessed August 2019].

[300] Hui J. Real-time object detection with YOLO, YOLOv2 and now YOLOv3. https://medium.com/@jonathan_hui/real-time-object-detection-with-yolo-yolov2-28b1b93e2088. [Accessed May 2019].

[301] Girshick RB, Donahue J, Darrell T, Malik J. Rich feature hierarchies for accurate object detection and semantic segmentation. CoRR. arXiv:1311.2524 [abs], 2014.

[302] Ren S, He K, Girshick RB, Sun J. Faster R-CNN: towards real-time object detection with region proposal networks. CoRR. arXiv:1506.01497 [abs], 2016.

[303] Redmon J, Divvala SK, Girshick RB, Farhadi A. You only look once: unified, real-time object detection. CoRR. arXiv:1506.02640 [abs], 2016.

[304] Halpern J, Pignataro C. Service function chaining (SFC) architecture. Tech. Rep. RFC 7665. IETF; 2015.

[305] TensorFlow. https://www.tensorflow.org. [Accessed February 2019].

[306] Karpathy A. CS231n: convolutional neural networks for visual recognition. http://cs231n.github.io/convolutional-networks/. [Accessed May 2019].

[307] Redmon J. Darknet: open source neural networks in C. http://pjreddie.com/darknet/, 2013–2016. [Accessed November 2019].

[308] Krizhevsky A, Sutskever I, Hinton GE. ImageNet classification with deep convolutional neural networks. In: Proceedings of the Conference on Neural Information Processing Systems (NIPS); 2012.

[309] Lin TY, Maire M, Belongie SJ, Bourdev LD, Girshick RB, Hays J, et al. Microsoft COCO: common objects in context. CoRR. arXiv:1405.0312 [abs], 2015.

[310] ComNetsEmu. Machine learning for object detection application example. https://git.comnets.net/public-repo/comnetsemu/tree/master/app/machine_learning_for_object_detection. [Accessed August 2019].

[311] Haßlinger G, Hohlfeld O. The Gilbert-Elliott model for packet loss in real time services on the Internet. In: Proceedings of the GI/ITG Conference on Measurement, Modelling and Evaluation of Computer and Communication Systems (MMB); 2008.

[312] Park K, Wang W. AFEC: an adaptive forward error correction protocol for end-to-end transport of real-time traffic. In: Proceedings of the IEEE International Conference on Computer Communications and Networks (ICCCN); 1998.

[313] Ho T, Kötter R, Médard M, Karger DR, Effros M. The benefits of coding over routing in a randomized setting. In: Proceedings of the IEEE International Symposium on Information Theory (ISIT); 2003.

[314] Wunderlich S, Gabriel F, Pandi S, Fitzek FHP, Reisslein M. Caterpillar RLNC (CRLNC): a practical finite sliding window RLNC approach. IEEE Access 2017;5:20183–97.

[315] Pandi S, Gabriel F, Cabrera Guerrero JA, Wunderlich S, Reisslein M, Fitzek FHP. PACE: redundancy engineering in RLNC for low-latency communication. IEEE Access 2017;5:20477–93.

[316] Krigslund J, Hansen J, Lucani Rötter DE, Fitzek FHP, Médard M. Network coded software defined networking: design and implementation. In: Proceedings of the European Wireless Conference (EW); 2015.

[317] Hansen J, Lucani Rötter DE, Krigslund J, Médard M, Fitzek FHP. Network coded software defined networking: enabling 5G transmission and storage networks. IEEE Communications Magazine 2015;53(9):100–7.

[318] Michel F, De Coninck Q, Bonaventure O. QUIC-FEC: bringing the benefits of forward erasure correction to QUIC. In: Proceedings of the IFIP Networking Conference (NETWORKING); 2019.

[319] Rischke J, Gabriel F, Pandi S, Nguyen GT, Salah H, Fitzek FHP. Improving communication reliability efficiently: adaptive redundancy for RLNC in SDN. In: Proceedings of the IEEE Conference on Network Softwarization (NetSoft); 2019.

[320] Buringh E, van Zanden JL. Charting the 'Rise of the West': manuscripts and printed books in Europe, A long-term perspective from the sixth through eighteenth centuries. The Journal of Economic History 2009;69(2):409–45.

[321] Patterson DA, Gibson G, Katz RH. A case for redundant arrays of inexpensive disks (RAID). In: Proceedings of the ACM SIGMOD International Conference on Management of Data (SIGMOD); 1988.

[322] Kijima M, Morimura H, Suzuki Y. Periodical replacement problem without assuming minimal repair. European Journal of Operational Research 1988;37(2):194–203.

[323] Mohajer S, Tandon R. Exact repair for distributed storage systems: partial characterization via new bounds. In: Proceedings of the Information Theory and Applications Workshop (ITA); 2015.

[324] Dimakis AG, Godfrey PB, Wu Y, Wainwright MJ, Ramchandran K. Network coding for distributed storage systems. IEEE Transactions on Information Theory 2010;56(9):4539–51.

[325] Wicker SB, Bhargava VK. Reed-Solomon Codes and Their Applications. Wiley; 1999.

[326] Fitzek FHP, Toth T, Szabados A, Pedersen MV, Lucani Rötter DE, Sipos M, et al. Implementation and performance evaluation of distributed cloud storage solutions using random linear network coding. In: Proceedings of the IEEE International Conference on Communications Workshops (ICC); 2014.

[327] Pedregosa F, Varoquaux G, Gramfort A, Michel V, Thirion B, Grisel O, et al. scikit-learn: machine learning in Python. Journal of Machine Learning Research 2011;12:2825–30.

[328] Gebel R. KL1p – a portable C++ library for compressed sensing, v0.4.2. http://kl1p.sourceforge.net/home.html, 2012.

[329] Madden S. Intel lab data. http://db.csail.mit.edu/labdata/labdata.html, 2004.

[330] nftables. http://netfilter.org/projects/nftables/. [Accessed September 2019].

[331] Donenfeld JA. WireGuard: next generation kernel network tunnel. In: Proceedings of the Network and Distributed System Security Symposium (NDSS); 2017.

[332] Bernstein DJ. ChaCha, a variant of Salsa20. Tech. Rep. University of Illinois at Chicago, Department of Mathematics, Statistics, and Computer Science; 2008.

[333] Bernstein DJ. The Poly1305-AES message-authentication code. In: Gilbert H, Handschuh H, editors. Fast Software Encryption. Lecture Notes in Computer Science, vol. 3557. Springer; 2005. p. 32–49.

[334] Dowling B, Paterson KG. A cryptographic analysis of the WireGuard protocol. In: Preneel B, Vercauteren F, editors. Applied Cryptography and Network Security. Lecture Notes in Computer Science, vol. 10892. Springer; 2018. p. 3–21.

[335] IEEE Instrumentation and Measurement Society. IEEE standard for a precision clock synchronization protocol for networked measurement and control systems. IEEE Std 1588-2008 (Revision of IEEE Std 1588-2002) 2008.

[336] IEEE Computer Society. IEEE standard for local and metropolitan area networks – timing and synchronization for time-sensitive applications in bridged local area networks. IEEE Std 8021AS-2011 2011.

[337] IEEE Computer Society. IEEE standard for local and metropolitan area networks – virtual bridged local area networks, Amendment 12: forwarding and queuing enhancements for time-sensitive streams. IEEE Std 8021Qav-2009 2010.

[338] IEEE Computer Society. IEEE standard for local and metropolitan area networks – bridges and bridged networks, Amendment 25: enhancements for scheduled traffic. IEEE Std 8021Qbv-2015 2016.

[339] IEEE Computer Society. IEEE standard for local and metropolitan area networks – bridges and bridged networks, Amendment 28: per-stream filtering and policing. IEEE Std 8021Qci-2017 2017.

[340] IEEE Computer Society. IEEE standard for Ethernet, Amendment 5: specification and management parameters for interspersing express traffic. IEEE Std 8023br-2016 2016.

[341] Gomes VC. net/sched: introduce the taprio scheduler. https://lwn.net/Articles/767383/. [Accessed December 2019].

[342] InnoRoute GmbH. TrustNode – an expandable, robust network evaluation and development platform. https://innoroute.com/?p=2143. [Accessed October 2019].

[343] Dürr F. Software TSN-switch with Linux. https://www.frank-durr.de/?p=376. [Accessed October 2019].

[344] Lackey RI, Upmal DW. Speakeasy: the military software radio. IEEE Communications Magazine 1995;33(5):56–61.

[345] Sullivan GJ, Ohm J, Han W, Wiegand T. Overview of the high efficiency video coding (HEVC) standard. IEEE Transactions on Circuits and Systems for Video Technology 2012;22(12):1649–68.

[346] Compta PT, Lucani Rötter DE, Fitzek FHP. Network coding is the 5G key enabling technology: effects and strategies to manage heterogeneous packet lengths. Transactions on Emerging Telecommunications Technologies 2015;26(1):46–55.

[347] Lucani Rötter DE, Pedersen MV, Ruano D, Sørensen CW, Fitzek FHP, Heide Møller J, et al. Fulcrum: flexible network coding for heterogeneous devices. IEEE Access 2018;6:77890–910.

[348] Berrou C, Glavieux A, Thitimajshima P. Near Shannon limit error-correcting coding and decoding: turbo-codes (1). In: Proceedings of the IEEE International Conference on Communications (ICC); 1993.

[349] Gallager R. Low-density parity-check codes. IRE Transactions on Information Theory 1962;8(1):21–8.

[350] Arikan E. Channel polarization: a method for constructing capacity-achieving codes for symmetric binary-input memoryless channels. IEEE Transactions on Information Theory 2009;55(7):3051–73.

[351] Proakis JG, Salehi M. Digital Communications. McGraw-Hill; 2008.

[352] Collins TF, Getz R, Pu D, Wyglinski AM. Software-Defined Radios for Engineers. Artech House; 2018.

[353] Rice M. Digital Communications: A Discrete-Time Approach. Prentice Hall; 2008.

[354] Rathinakumar SM, Radunović B, Marina MK. CPRecycle: recycling cyclic prefix for versatile interference mitigation in OFDM based wireless systems. In: Proceedings of the ACM International Conference on Emerging Networking Experiments and Technologies (CoNEXT); 2016.

[355] Fettweis GP, Krondorf M, Bittner S. GFDM – generalized frequency division multiplexing. In: Proceedings of the IEEE Vehicular Technology Conference (VTC Spring); 2009.

[356] Myung HG, Lim J, Goodman DJ. Single carrier FDMA for uplink wireless transmission. IEEE Vehicular Technology Magazine 2006;1(3):30–8.

[357] Paredes MCP, García MJFG. The problem of peak-to-average power ratio in OFDM systems. CoRR. arXiv:1503.08271 [abs], 2015.

[358] Nisar MD, Nottensteiner H, Hindelang T. On performance limits of DFT spread OFDM systems. In: Proceedings of the IST Mobile and Wireless Communications Summit; 2007.

[359] GNU radio project. https://github.com/gnuradio/gnuradio. [Accessed October 2019].

[360] Truong NB. Latency analysis in GNU radio/USRP-based software radio platforms. In: Proceedings of the IEEE Military Communications Conference (MILCOM); 2013.

[361] Kerrisk M. The Linux Programming Interface: A Linux and UNIX System Programming Handbook. No Starch Press; 2010.

[362] Kerrisk M. Linux man-pages: ping. https://linux.die.net/man/8/ping. [Accessed November 2019].

[363] Litvak M. Linux man-pages: ip. https://linux.die.net/man/8/ip. [Accessed November 2019].

[364] Durganm J. Linux man-pages: iperf. https://linux.die.net/man/1/iperf. [Accessed November 2019].

[365] Fenski B, Muhammad H, Launchbury V. Linux man-pages: htop. https://linux.die.net/man/1/htop. [Accessed November 2019].

[366] Jacobson V, Leres C, McCanne S. Linux man-pages: tcpdump. https://linux.die.net/man/8/tcpdump. [Accessed November 2019].

[367] Wireshark – Official documentation. https://www.wireshark.org/docs/. [Accessed November 2019].

[368] Jupyter – Official documentation. https://jupyter.org/documentation. [Accessed November 2019].

[369] Postel J. Internet control message protocol. Tech. Rep. RFC 792. IETF; 1981.

Index